London Mathematical Society Lecture Note Series. 288

Surveys in Combinatorics, 2001

Edited by

J. W. P. Hirschfeld
University of Sussex

CAMBRIDGE
UNIVERSITY PRESS

CAMBRIDGE UNIVERSITY PRESS
Cambridge, New York, Melbourne, Madrid, Cape Town,
Singapore, São Paulo, Delhi, Mexico City

Cambridge University Press
The Edinburgh Building, Cambridge CB2 8RU, UK

Published in the United States of America by Cambridge University Press, New York

www.cambridge.org
Information on this title: www.cambridge.org/9780521002707

First published 2001

A catalogue record for this publication is available from the British Library

ISBN 978-0-521-00270-7 Paperback

LONDON MATHEMATICAL SOCIETY LECTURE NOTE SERIES

Managing Editor: Professor N.J. Hitchin, Mathematical Institute,
University of Oxford, 24–29 St Giles, Oxford OX1 3LB, United Kingdom

The titles below are available from booksellers, or, in case of difficulty, from Cambridge University Press.

This book is dedicated to Crispin Nash-Williams

19 December 1932 – 20 January 2001

Contents

Preface

On the occasion of the 18th British Combinatorial Conference at the University of Sussex, 1 to 6 July, 2001, this book comprises the survey papers by the nine invited speakers and a memoire of Crispin Nash-Williams, past chairman of the British Combinatorial Committee.

The survey papers range across many parts of modern combinatorics.

Martin Aigner discusses the ideas of Penrose on the 4-colour problem, as well as the application of Penrose polynomials to other combinatorial structures.

Ian Anderson surveys some of the key ideas in the study of cyclic designs, including some of the classical results of the past 150 years as well as some very recent developments.

Robert Calderbank and Ayman Naguib show the connection between the practice of wireless communication with the mathematics of quadratic forms developed by Radon and Hurwitz about a hundred years ago. This occurs through orthogonal designs, known as space-time block codes in the communications literature.

Leslie Goldberg surveys the computational problems of randomly sampling unlabelled combinatorial structures, and of counting and approximately counting unlabelled structures.

Bojan Mohar considers the interplay between graph minors and graphs embedded in surfaces.

Michael Molloy surveys the progress on two fundamental problems in random graphs and random boolean formulae. The first is the question of how many edges must be added to a random graph until it is not almost surely k-colourable.

James Oxley considers aspects of the interplay between graphs and matroids, and shows the fruitfulness for both fields of applying results from the other.

Joseph Thas considers the geometrical structures fundamental to finite simple groups, namely finite classical polar spaces, and the properties of their substructures.

Douglas Woodall discusses two problems of graph theory associated to colourings of a graph in which each vertex receives a colour from a prescribed list of colours.

The conference is grateful for the support of the London Mathematical Society, the Institute of Combinatorics and its Applications, Hewlett Packard, and AT&T.

James Hirschfeld
University of Sussex
4 March 2001

Crispin Nash-Williams

John Sheehan

Abstract

This tribute consists of an appreciation from 1996, some further thoughts, and a list of Nash-Williams' publications.

1 An appreciation written on his retirement in 1996

I arrived in Aberdeen in 1965 to start my academic career as an assistant lecturer. I had become interested in graph theory and in particular in a series of papers with such resonant titles as

On well-quasi-ordering infinite trees
By C. ST. J.A. NASH-WILLIAMS
King's College, Aberdeen

So it is with pleasure that I am writing this appreciation of Professor Nash-Williams in the Quincentennial Year of the University of Aberdeen.

In 1967, after some ten years at Aberdeen, Nash-Williams moved to the University of Waterloo, returning to Aberdeen as Professor of Mathematics in 1972. In 1975 he took up a Professorship of Mathematics at the University of Reading where he joined a flourishing group of combinatorialists which included Richard Rado (then recently retired), David Daykin and Anthony Hilton. He has remained at Reading ever since, apart from a year in West Virginia and frequent visits to Waterloo.

It is not my intention to give a full appreciation of Nash-Williams' contribution to graph theory. How could I? This will, I hope, be done elsewhere with a complete edition of his papers. I shall content myself with a few random remarks on his work.

Nash-Williams is a graph theorist. Amongst combinatorialists this is an unusually positive statement. Though to call oneself simply a combinatorialist is too unspecific; after all, what mathematician is not to some extent a combinatorialist? Not only is he a graph theorist but even more unusual for a combinatorialist he claims [3] it as his first interest, and I quote:

> ... This cannot but provoke one's own early recollections, a favourite topic of conversation among graph-theorists being what first drew one's attention to graph theory. My own answer is 'Nothing: I just invented it'. In other words, when starting work as a research student, also at Cambridge, I felt that there ought to be a branch of mathematics dealing with this kind of thing, and, if there was not, I would create it. (References to binary relations in algebra courses might have helped to foster this idea.) it is a measure

1

of the little known state of graph theory at that time that it took
me some weeks to discover that I was not its first inventor and to
hear of the one existing textbook on the subject Koenig's *Theorie
der endlichen und unendlichen Graphen*, published eighteen years
earlier in 1936.

The areas of graph theory in which Nash-Williams has made central contributions include: well-quasi-ordering of graphs – both finite and infinite –
relative to subdivisions; Menger's Theorem; the theory of transversals – both
finite and infinite; orientations of graphs and the theory of infinite graphs in
general. In recent years he has published a series of papers on detachments
of graphs, the theory of which in a certain sense generalizes Eulerian graph
theory.

Nash-Williams' work has applications outside graph theory, for example
in the theory of well-quasi-ordering transfinite sets, and most recently he,
together with David White, have become interested in the interaction of graph
theory with the rearrangements of conditionally convergent real series.

One senses, however, that throughout his career, Nash-Williams has been
infected more than most by two of the three graphical diseases [2]: Hamiltonian
circuits and the Reconstruction Problem. These are interests to which he
continually returns. He appears, at least up to now, to have developed an
immunity to the third disease - the Four Colour Theorem.

The really surprising thing is that he has any time at all for research. During the term he has always devoted the majority of his time to teaching. This
is surely one of his most valuable contributions; there are countless students
who have given testimony to the clarity of his teaching and the generosity with
which he makes time for them. A particular characteristic is his willingness to
help the weakest as well as the more able students. All of us appreciate the
excellence of his lectures at conferences; this excellence is repeated day in and
day out during the term.

Nash-Williams is also much in demand to serve on committees and seems
not to be able to say no. He has served on the British Combinatorial Committee for many years and was its Chairman from 1987 to 1992.

As an example of the variety of his activities he has always been an active
member of the Association of University Teachers and was indeed the President
of the Reading Branch of the Association in 1987-89. This was a somewhat
troubled time involving threats of industrial action. I am sure that few would
believe that he was described by someone in high office at this time as an 'out
and out militant'.

Finally, no tribute would be complete without mentioning the time and
energy Nash-Williams spends in refereeing: many of us have been grateful
recipients of his constructive critiques which are sometimes as lengthy and
certainly more meticulous than the original articles. It is probably no secret
that many of a long series of papers concerning well-quasi-ordering of graphs by

relationship of one graph being a minor of another, which have been appearing in the Journal of Combinatorial Theory in the last decade, have been refereed by Nash-Williams. The amount of selfless work this has involved is probably not universally appreciated; graph theorists owe him an inestimable debt for this alone.

Professor Tutte, when also nearing retirement, remarked [4]:

> What is Mathematics? You seem to have three choices. Mathematics is the Humanity that hymns eternal logic. It is the Science that studies the phenomenon called logic. It is the Art that fashions structures of ethereal beauty out of the raw material called logic. It is all of these and more. Much more, I can assure you, for Mathematics is Fun.

Crispin St John Alvah, and I leave it until now to use his full name, has contributed in great measure to this fun and I finish with an anecdote from Blanche Descartes [1]:

> I think of the occasion at a crowded conference when Crispin was standing, quite happily it seemed, in a stream of hot air from a grill on the floor. He was urged to come away on the grounds that he was 'crispin' and singein'.

2   Further thoughts

Professor C St-J A Nash-Williams died on the 20th of January 2001. The stereotypical 'English Gentleman' is oft maligned. Crispin was an English Gentleman in the good old sense of this phrase. He was a gentle, unassuming and kind man with very strong principles which he defended courageously. As a teacher he is remembered and loved for his consideration for the weak students as well as the strong. I remember one such student in particular – necessarily nameless – who even by today's standard would be considered at risk. He slavishly struggled to get him to pass and eventually, after several resits, indeed he did. I have met very few teachers, if any, who could have risen to this particular challenge. It was all achieved with unflappable patience and I was never once aware of him mentioning it. A truly modest man.

Crispin retired in 1996 and since then he seemed to be at his happiest. He had been out of sympathy with many of the recent developments in University Education. With no administrative and teaching chores he was able to pursue his research interests which he enjoyed so much. He continued to collaborate with his colleagues in the department, in particular including David White and Anthony Hilton, until almost the very end. Professor Crispin St-J Alvah Nash-Williams was a good man: incapable of pettiness, generous and noble of spirit.

Our Combinatorial Community will always remember him with affection and pride that he took such a delight in our subject and our company.

3 The publications of C.St.J.A. Nash-Williams

1. Random walk and electric currents in networks, *Proc. Cambridge Philos. Soc.* **55** (1959), 181–194.

2. Abelian groups, graphs and generalised knights, *Proc. Cambridge Philos. Soc.* **55** (1959), 232–238.

3. Decomposition of graphs into closed and endless chains, *Proc. London Math. Soc.* **10** (1960), 221–238.

4. On orientations, connectivity and odd-vertex-pairings in finite graphs, *Canad. J. Math.* **12** (1960), 555–567.

5. Decomposition of finite graphs into open chains, *Canad. J. Math.* **13** (1961), 157–166.

6. Decomposition of the n-dimensional lattice graph into Hamiltonian lines, *Proc. Edinburgh Math. Soc.* **12** (1961), 123–131.

7. Edge disjoint spanning trees, *J. London Math. Soc.* **36** (1961), 445–450.

8. Decomposition of graphs into two-way infinite paths, *Canad. J. Math.* **15** (1963), 479–485.

9. On well-quasi-ordering finite trees, *Proc. Cambridge Philos. Soc.* **59** (1963), 833–835.

10. Decomposition of finite graphs into forests, *J. London Math. Soc.* **39** (1964), 12.

11. On well-quasi-ordering lower sets of finite trees, *Proc. Cambridge Philos. Soc.* **60** (1964), 369–384.

12. On well-quasi-ordering trees, *Theory of Graphs and its Applications*, Proc. Sympos. Smolenice, 1963, Czechoslovak Academy of Sciences, Prague (1964), 83–84.

13. Hamiltonian lines in products of infinite trees, *J. London Math. Soc.* **40** (1965), 37–40.

14. On well-quasi-ordering transfinite sequences, *Proc. Cambridge Philos. Soc.* **61** (1965), 33–39.

15. On well-quasi-ordering infinite trees, *Proc. Cambridge Philos. Soc.* **61** (1965), 697–720.

16. On Eulerian and Hamiltonian graphs and line graphs (with F. Harary), *Canad. Math. Bull.* **8** (1965), 701–709.

17. On Hamiltonian circuits in finite graphs, *Proc. Amer. Math. Soc.* **17** (1966), 466–467.

18. Euler lines in infinite directed graphs, *Canad. J. Math.* **18** (1966), 692–714.

19. Infinite graphs – a survey, *J. Combin. Theory* **3** (1967), 286–301.

20. An application of matroids to graph theory, *Theory of Graphs*, International Symposium, Rome, July 1966, Gordon and Breach (1967), 263–265.

21. On well-quasi-ordering trees, *A Seminar on Graph Theory*, Holt, Rinehart and Winston, New York (1967), 79–82.

22. On better-quasi-ordering transfinite sequences, *Proc. Cambridge Philos. Soc.* **64** (1968), 273–290.

23. Euler lines in infinite directed graphs, *Theory of Graphs*, Proc. Coll., Tihany, Hungary, September 1966, Hungarian Academy of Sciences (1968), 243–249.

24. Hamiltonian circuits in graphs and digraphs, *The Many Facets of Graph Theory*, Proc. Conf., Western Mich. Univ., Kalamazoo, Mich., 1968, Springer, Berlin (1969), 237–243.

25. Well-balanced orientations of finite graphs and unobtrusive odd-vertex-pairings, *Recent Progress in Combinatorics*, Proc. Third Waterloo Conf. on Combinatorics, 1968, Academic Press, New York (1969), 133–149.

26. Counterexamples in the theory of well-quasi-ordered sets (with T.A. Jenkyns), *Proof Techniques in Graph Theory*, Proc. Second Ann Arbor Graph Theory Conf., Ann Arbor, Mich., 1968, Academic Press, New York (1969), 87–91.

27. A survey of graph theory, *Combinatorics, Graph Theory and Computing*, Proc. Conf., Louisiana State Univ., Baton Rouge, 1970, Louisiana State Univ., Baton Rouge (1970), 383–444.

28. A survey of the theory of well-quasi-ordered sets, *Combinatorial structures and Their Applications*, Proc. Calgary International Conference, Gordon and Breach (1970), 293–299.

29. Hamiltonian lines in graphs whose vertices have sufficiently large valencies, *Combinatorial theory and its applications, III*, Proc. Colloq., Balatonfüred, 1969, North-Holland, Amsterdam (1970), 813–819.

30. Hamiltonian arcs and circuits, *Recent Trends in Graph Theory*, Proc. Conf., New York, 1970, Lecture Notes in Math. **186**, Springer, Berlin (1971), 197–210.

31. Possible directions in graph theory, *Combinatorial Mathematics and its Applications*, Proc. Conf., Oxford, 1969, Academic Press, London (1971), 191–200.

32. Edge-disjoint Hamiltonian circuits in graphs with vertices of large valency, *Studies in Pure Mathematics (Presented to Richard Rado)*, Academic Press, London (1971), 157–183.

33. Hamiltonian lines in infinite graphs with few vertices of small valency, *Aequationes Math.* **7** (1971), 59–81.

34. Simple constructions for balanced incomplete block designs with block size three, *J. Combin. Theory Ser. A* **13** (1972), 1–6.

35. Plane curves with many inscribed rectangles, *J. London Math. Soc.* **5** (1972), 417–418.

36. Which infinite set-systems have transversals? - a possible approach, *Combinatorics*, Proc. Conf. Combinatorial Math., Math. Inst., Oxford, 1972, Inst. Math. Appl., Southend (1972), 237–253.

37. Unexplored and semi-explored territories in graph theory, *New Directions in the Theory of Graphs*, Proc. Third Ann Arbor Conf., Univ. Michigan, Ann Arbor, Mich., 1971, Academic Press, New York (1973), 149–186.

38. The square of a block is Hamiltonian connected (with G. Chartrand, A.M. Hobbs, H.A. Jung and S.F. Kapoor) *J. Combin. Theory Ser. B* **16** (1974), 290–292.

39. Marriage in denumerable societies, *Recent Advances in Graph Theory*, Proc. Second Czechoslovak Sympos., Prague, 1974, Academia, Prague (1975), 393–397.

40. Marriage in denumerable societies, *J. Combin. Theory Ser. A* **19** (1975), 335–366.

41. Hamiltonian circuits, *Studies in Graph Theory*, Math. Assoc. Amer. (1975), 301–360.

42. Editor (with J. Sheehan), *Proceedings of the Fifth British Combinatorial Conference*, Congr. Numer. **15**, Utilitas Math., Winnipeg (1976).

43. Hamiltonian circuits, *Studies in Graph Theory, Part II*, Studies in Math. **12**, Math. Assoc. Amer., Washington, D.C. (1975), 301–360.

44. More proofs of Menger's theorem (with W.T. Tutte), J. *Graph Theory* **1** (1977), 13–17.

45. Should axiomatic set theory be translated into graph theory?, *Combinatorics Vol. II*, Proc. Fifth Hungarian Colloq., Keszthely, 1976, Colloq. Math. Soc. János Bolyai, 18, North-Holland, Amsterdam (1978), 743–757.

46. Another criterion for marriage in denumerable societies, *Ann. Discrete Math.* **3** (1978), 165–179.

47. Acyclic detachments of graphs, *Graph Theory and Combinatorics*, Proc. Conf., Open Univ., Milton Keynes, 1978, 87–97.

48. Marriage in infinite societies in which each woman knows countably many men, *Proceedings of the Tenth Southeastern Conference on Combinatorics, Graph Theory and Computing*, Florida Atlantic Univ., Boca Raton, Fla., 1979, *Congr. Numer.* **23/24**, Utilitas Math., Winnipeg (1979), 103–115.

49. A note on some of Professor Tutte's mathematical work, *Graph Theory and Related Topics*, Proc. Conf., Univ. Waterloo, Waterloo, Ont., 1977, Academic Press, New York, 1979, xxv–xxviii.

50. A glance at graph theory – Part I, *Bull. London Math. Soc.* **14** (1982), 177–212.

51. A glance at graph theory – Part II, *Bull. London Math. Soc.* **14** (1982), 294–328.

52. A general criterion for the existence of transversals (with R. Aharoni and S. Shelah), *Proc. London Math. Soc.* **47** (1983), 43–68.

53. Another form of a criterion for the existence of transversals (with R. Aharoni, and S. Shelah), *J. London Math. Soc.* **29** (1984), 193–203.

54. Marriage in infinite societies (with R. Aharoni, and S. Shelah), *Progress in Graph Theory*, Waterloo, 1982, Academic Press, Toronto (1984), 71–79.

55. Forward to: W. T. Tutte, *Graph Theory*, Addison-Wesley, Reading, Mass., 1984.

56. Connected detachments of graphs and generalised Euler trails, *J. London Math. Soc.* **31** (1985), 17–29.

57. Detachments of graphs and generalised Euler trails, *Surveys in Combinatorics 1985*, Tenth British Combin. Conf., Glasgow, 1985, Cambridge Univ. Press, Cambridge (1985), 137–151.

58. Reconstruction of locally finite connected graphs with at least three infinite wings, *J. Graph Theory* **11** (1987), 497–505.

59. Amalgamations of almost regular edge-colourings of simple graphs, *J. Combin. Theory Ser. B* **43** (1987), 322–342.

60. Another proof of a theorem concerning detachments of graphs, *European J. Combin.* **12** (1991), 245–247.

61. Reconstruction of infinite graphs, *Discrete Math.* **95** (1991), 221–229.

62. Reconstruction of locally finite connected graphs with two infinite wings, *Discrete Math.* **92** (1991), 227–249.

63. Introduction to *Directions in Infinite Graph Theory and Combinatorics*, Proc. Conf. Cambridge, England, 1989, Topics in Discrete Math. **3**, North-Holland, Amsterdam (1992), 3–4.

64. Reconstructing the number of copies of a valency-labeled finite graph in an infinite graph (with A.J.H. King), *J. Graph Theory* **18** (1994), 109–117.

65. Infinite digraphs with nonreconstructable outvalency sequences, *J. Graph Theory* **18** (1994), 535–537.

66. A direct proof of a theorem on detachments of finite graphs, *J. Combin. Math. Combin. Comput.* **19** (1995), 314–318.

67. Strongly connected mixed graphs and connected detachments of graphs, *J. Combin. Math. Combin. Comput.* **19** (1995), 33–47.

68. An application of network flows to rearrangement of series (with D.J. White), *J. London Math. Soc.* **59** (1999), 637–646.

69. Obituary: Eric Charles Milner, *Bull. London Math. Soc.* **32** (2000), 91–104.

70. Rearrangement of vector series. I (with D.J. White), *Math. Proc. Cambridge Philos. Soc.* **130** (2001), 89–109.

71. Rearrangement of vector series. II (with D.J. White), *Math. Proc. Cambridge Philos. Soc.* **130** (2001), 111–134.

72. Hamiltonian double latin squares (with A.J.W. Hilton, M. Mays, and C.A. Rodger), in preparation.

References

[1] B. Descartes, Private communication.

[2] F. Harary, The Four Colour Conjecture and other graphical diseases, *Proof Techniques in Graph Theory*, (ed. F. Harary), Academic Press, New York (1969), 1–9.

[3] C.St.J.A. Nash-Williams, Foreword to *Graph Theory* by W.T. Tutte, Addison-Wesley, Reading, Mass. (1984), xv–xviii.

[4] W.T. Tutte, What is mathematics? A University Lecture given at the University of Waterloo.

Department of Mathematical Sciences
University of Aberdeen
Old Aberdeen AB24 3UE
United Kingdom
js@maths.abdn.ac.uk

The Penrose polynomial of graphs and matroids

Martin Aigner

1 Introduction

In 1971 Roger Penrose published in a Conference Proceedings a paper entitled "Applications of negative-dimensional tensors". Perhaps it was this somewhat cryptic title that prevented the paper from becoming widely known at first. Still, as is to be expected from Penrose, it contained a wealth of original ideas on the theory of plane graphs. If one studies the paper more closely one finds that the main object is the enumeration of certain colourings of plane graphs. And pretty soon one surmises that Penrose wanted, in fact, to solve the 4-colour conjecture. Remember that in 1971 it was still a problem; it was only 1976 that it became the 4-colour theorem.

Penrose implicitly defines in his paper a polynomial for 3-regular plane graphs, now called the *Penrose polynomial*, and deduces four equivalent formulations of the 4-colour theorem. His ideas were taken up in the eighties by several people, foremost by the late François Jaeger, and generalized in various ways. In this article we survey at a leisurely pace how the Penrose polynomial is connected to some famous conjectures in graph theory, to binary spaces, Hopf algebras and polynomial invariants in knot theory.

2 The Penrose polynomial

Let $G = (V, E, F)$ be a plane connected graph with vertex-set V, edge-set E, and face-set F. To avoid trivialities we always assume that G contains at least one edge.

Look at the plane graph G in Figure 1:

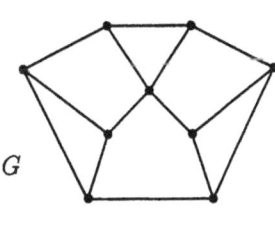

G

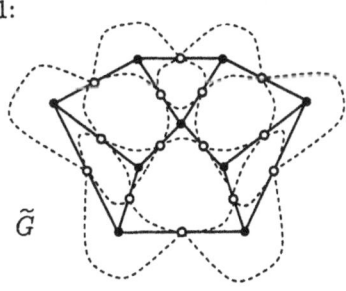

$\tilde{G}$

Figure 1   Figure 2

In every face (including the outer face) we draw a closed curve near the boundary which touches every boundary edge, where we identify the two touching points in an edge. Figure 2 shows the construction. The resulting graph $\widetilde{G}$ with the touching points as vertices and the dashed lines L as edges is called the *medial graph* $\widetilde{G} = (E, L)$, where we identify the vertex-set of $\widetilde{G}$ with the edge-set E of G. Clearly, $\widetilde{G}$ is again plane connected, and it is a 4-regular graph, since every vertex of $\widetilde{G}$; that is, every edge of G, has 4 neighbouring edges (possibly not all distinct).

We come to the definition of the Penrose polynomial of G. Consider $\widetilde{G} = (E, L)$ and an arbitrary subset $A \subseteq E$. Let us call A the *crossing vertices* and $E \setminus A$ the *non-crossing vertices*. We run through L in the following manner: We start with an arbitrary edge $\ell \in L$ and give it an orientation. If the end-vertex of ℓ is in A, then we cross into the other face containing ℓ, and if ℓ is in $E \setminus A$, then we continue inside the same face (which contains ℓ). Continuing in this way, we eventually come back to the starting edge ℓ, and the edge-trail is closed. If there is an edge of L left (that is, not in the trail) then we start a new trail, until all edges of $\widetilde{G}$ have been traversed. It is easy to see that it does not matter with which edge to begin or in what direction - we always get the same decomposition of L into closed trails. In particular, the *number* $c(A)$ of trails is uniquely defined.

As an illustration look at Figure 3 and take A to be the shaded dots. Starting with the edge labelled 1 we obtain the trail in the figure containing 10 edges. Altogether we get $c(A) = 3$.

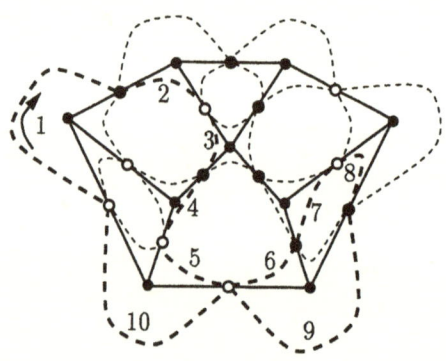

Figure 3

Definition The Penrose polynomial $P_G(\lambda)$ of a plane connected graph $G = (V, E, F)$ is

$$P_G(\lambda) = \sum_{A \subseteq E} (-1)^{|A|} \lambda^{c(A)}.$$

The following figure depicts the two smallest graphs with $|E| = 1$: the bridge and the loop.

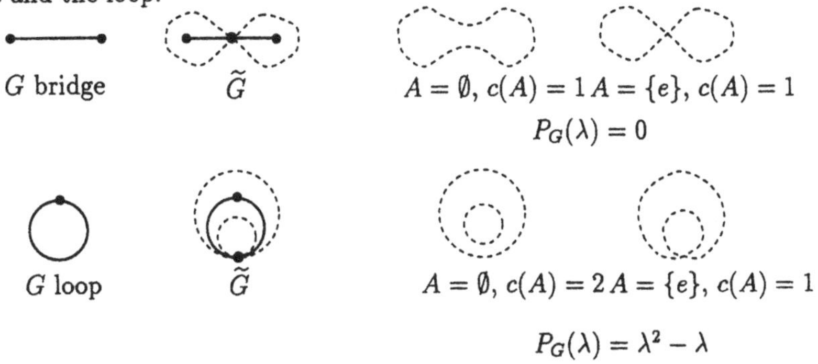

G bridge $\qquad \widetilde{G}$ $\qquad A = \emptyset, c(A) = 1 \; A = \{e\}, c(A) = 1$

$$P_G(\lambda) = 0$$

G loop $\qquad \widetilde{G}$ $\qquad A = \emptyset, c(A) = 2 \; A = \{e\}, c(A) = 1$

$$P_G(\lambda) = \lambda^2 - \lambda$$

Figure 4

Let us note two easy facts:

(2.1) $P_G(\lambda) \equiv 0 \iff G$ *contains a bridge* ($= $ *cut-edge*);

(2.2) *if* $G = (V, E, F)$ *has no bridge, then the Penrose polynomial* $P_G(\lambda)$ *has degree* $|F|$.

In what follows, we will study $P_G(\lambda)$ from two different perspectives, a geometric and an algebraic viewpoint. In the last section we will return to the 4-colour theorem and some great open problems.

3   Geometry of the Penrose polynomial

Without further mention $G = (V, E, F)$ will always be a plane connected graph, and $\widetilde{G} = (E, L)$ its medial graph. As $\widetilde{G}$ is 4-regular, we can colour the faces of $\widetilde{G}$ with two colours, white and black, such that faces with a common boundary receive different colours. We always colour the outer face white, the rest of the colouring is then fixed. Figure 5 shows our example:

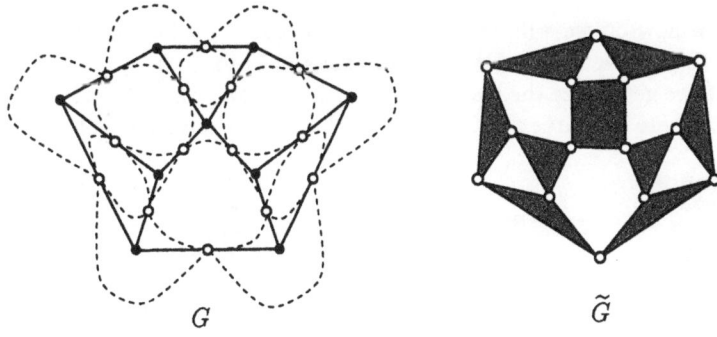

$G \qquad\qquad\qquad\qquad \widetilde{G}$

Figure 5

We note the following:

(3.1) the *black* faces of $\widetilde{G}$ correspond to the *vertices* of G, and the number of edges in a black face equals the degree of the corresponding vertex in G;

(3.2) the *white* faces of $\widetilde{G}$ correspond to the *faces* of G with the same number of face edges;

(3.3) every *vertex* of $\widetilde{G}$ is incident to two white and two black faces, which may not be distinct;

(3.4) two vertices of G are adjacent if and only if the corresponding black faces in $\widetilde{G}$ have a vertex in common; dually, two faces of G have a common edge precisely if the corresponding white faces in $\widetilde{G}$ have a common vertex.

The 2-*coloured* medial graph $\widetilde{G}$ corresponds therefore uniquely to the underlying graph G. Hence we may recover the trails that we constructed in the definition of $P_G(\lambda)$ within the medial graph $\widetilde{G}$. Consider $A \subseteq E$:

- if we come to a vertex $e \in A$, then we cross;

- if we arrive at $e \notin A$, then we continue along the white face;

- we never go along a black face.

Let us look at one more example to make things clear. The crossing vertices are drawn black.

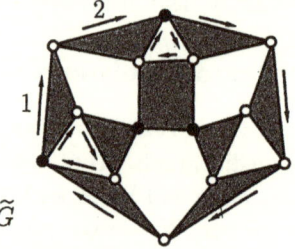

$\widetilde{G}$

Figure 6

From now on we will work with the 2-coloured medial graph $\widetilde{G}$ and will then interpret the results for G via the facts (3.1) to (3.4).

Now we generalize the trail decompositions in two ways. Consider a vertex e in $\widetilde{G}$. There are three possibilities to split e into two vertices of degree 2:

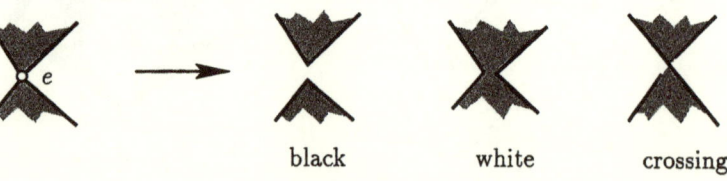

black   white   crossing

Figure 7

We speak of a *transition* $p(e)$ at e of *black*, *white*, or *crossing type*. Choosing a transition $p(e)$ at every vertex $e \in E$, we call $p = \{p(e) : e \in E\}$ a *transition system*. Thus there are $3^{|E|}$ different transition systems. A transition system plainly decomposes the edge-set L of $\widetilde{G}$ into disjoint cycles. Let us denote by $c(p)$ the number of these cycles.

For the second generalization we assign *weights* to every type (in a field of characteristic 0):

$$W(p(e)) = \begin{cases} \alpha & & \text{black} \\ \beta & \text{if } p(e) \text{ is} & \text{white} \\ \gamma & & \text{crossing,} \end{cases}$$

and set

$$W(p) = \prod_{e \in E} W(p(e)).$$

Definition The *transition polynomial* $Q(\widetilde{G}, W, \lambda)$ of $\widetilde{G}$ with respect to the weighting W is

$$Q(\widetilde{G}, W, \lambda) = \sum_p W(p) \lambda^{c(p)}.$$

A moment's thought should convince the reader that the weighting $\alpha = 0$ (no black transitions), $\beta = 1$ and $\gamma = -1$ yields precisely the Penrose polynomial; that is,

$$P_G(\lambda) = Q(\widetilde{G}, W_{\alpha=0, \beta=1, \gamma=-1}, \lambda). \tag{3.5}$$

The idea of the transition polynomial has proved very useful in many diverse problems, see [21] for a nice survey. We now look at $P_G(\lambda)$ in more detail.

3.1 Evaluation of $P_G(\lambda)$ at positive integers

As a first result we are going to show that $P_G(k)$ *counts* something for $\lambda = k$, implying $P_G(k) \geq 0$ for all non-negative integers k.

Let $G = (V, E, F)$ be plane and $\widetilde{G} = (E, L)$ its 2-coloured medial graph. A *k-valuation* of $\widetilde{G}$ is a map $f : L \longrightarrow \{1, 2, \ldots, k\}$ with the property that at every vertex e exactly two different numbers $i \neq j$ appear, and that one of the following cases holds:

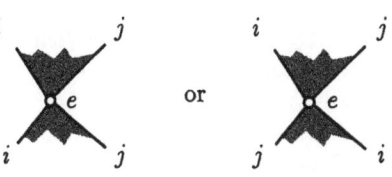

Figure 8

In the first case we say e is a *white* vertex with respect to f, and in the second case e is a *crossing* vertex.

Theorem 3.1 *For all $k \geq 0$,*

$$P_G(k) = \#k\text{-valuations of } \widetilde{G}.$$

In particular, $P_G(k) \geq 0$ for all $k \geq 0$.

Proof We relax the definition of a k-valuation to allow also vertices $e \in E$ where all four edges incident with e carry the same integer i. We say in this situation that e is *total* with respect to f. Let us denote by Ω the enlarged set of mappings. For $e \in E$, let W_e, C_e and T_e be the subsets of mappings f in Ω such that e is white, crossing respectively total with respect to f. Next we show that

$$\#k\text{-valuations} = \sum_{A \subseteq E} (-1)^{|A|} \left| \bigcap_{e \in A} \overline{W}_e \cap \bigcap_{e \notin A} \overline{C}_e \right|, \qquad (3.6)$$

where the complements are taken with respect to Ω. Take any map $f \in \Omega$, and let W, C, T be the sets of white, crossing and total vertices. The map f is contained in $\bigcap_{e \in A} \overline{W}_e \cap \bigcap_{e \notin A} \overline{C}_e$ precisely when $C \subseteq A \subseteq C \cup T$. Hence if $T \neq \varnothing$, that is, if f is not a k-valuation, then the count on the right-hand side of (3.6) is $\sum_{C \subseteq A \subseteq C \cup T} (-1)^{|A|} = 0$, so both sides of (3.6) contribute 0.

On the other hand, if $T = \varnothing$, then f is a k-valuation, and the left-hand count is 1. On the right-hand side we obtain $(-1)^{|C|}$. But since $\widetilde{G}$ is a plane graph, the decomposition of $\widetilde{G}$ into cycles (carrying the same symbol) has an even number of crossings, and so we find $(-1)^{|C|} = 1$.

It remains to show that

$$k^{c(A)} = \left| \bigcap_{e \in A} \overline{W}_e \cap \bigcap_{e \notin A} \overline{C}_e \right|.$$

Now $k^{c(A)}$ equals the number of k-colourings of the $c(A)$ trails in the decomposition associated with A. This equals the number of $f \in \Omega$, where $e \in A$ implies $f \in C_e \cup T_e = \overline{W}_e$ and where $e \notin A$ implies $f \in W_e \cup T_e = \overline{C}_e$. Thus $f \in \bigcap_{e \in A} \overline{W}_e \cap \bigcap_{e \notin A} \overline{C}_e$, and the proof is complete. $\square$

Theorem 3.1 says that $P_G(\lambda)$ is the *counting function* of λ-valuations. As we will see, $P_G(\lambda)$ is closely connected to another counting function, namely the chromatic function $\chi_G(\lambda) = \#\lambda$-colourings of G. A classical result states that $\chi_G(\lambda)$ is a polynomial of degree $|V|$. Hence if G^* denotes the dual graph of G, then $\chi_{G^*}(\lambda)$ is a polynomial of degree $|F|$. The 4-colour theorem is thus equivalent to $\chi_{G^*}(4) > 0$ for all bridgeless plane graphs G.

Note, however, one important difference. While $\chi_{G^\bullet}(\lambda)$ is defined as a *counting* function which is then proved to be a polynomial, with $P_G(\lambda)$ the situation is the other way around: it is defined as a *polynomial*, and we proved that it counts something. We will return to this connection as we go along.

From the definition of a valuation we can draw two immediate conclusions.

$$0 \leq P_G(1) \leq P_G(2) \leq P_G(3) \leq \cdots \tag{3.7}$$

Clearly, every k-valuation is also a $k + 1$-valuation.

$$P_G(k) \geq \chi_{G^\bullet}(k). \tag{3.8}$$

The k-valuations which contain only white vertices are precisely the k-colourings of the faces of G.

Let us look more closely at the first values $\lambda = k$. By the definition of a valuation we certainly have

$$P_G(0) = P_G(1) = 0. \tag{3.9}$$

A graph is *Eulerian* if all vertices have even degree.

$$P_G(2) = \begin{cases} 2^{|V|} & \text{if } G \text{ is Eulerian} \\ 0 & \text{otherwise.} \end{cases} \tag{3.10}$$

For the proof we note from Figure 8 that a 2-valuation assumes the values 1 and 2 alternately around a black face of $\widetilde{G}$. This is only possible when all black faces have an even number of edges, or what is the same by (3.1), when G is Eulerian. In this case we have two possibilities for each black face, and $P_G(2) = 2^{|V|}$ results.

The next value $\lambda = 3$ yields Penrose's first result. Let us call $g : V \cup F \longrightarrow \{0,1,2,3\}$ a *special* 4-colouring of G if adjacent faces are coloured differently, and also incident vertices and faces; adjacent vertices may receive the same colour.

Theorem 3.2 *We have*

$$P_G(3) = \# \text{ special 4-colourings of } G, \text{ with the outer}$$
$$\text{face coloured 0.}$$

Proof We take as colour-set the Klein 4-group $\{0, a, b, c\}$ and as valuation-set $\{a, b, c\}$. Let $g : V \cup F \longrightarrow \{0, a, b, c\}$ be a special 4-colouring. For the medial graph $\widetilde{G}$ this means that we have the situation in the left figure:

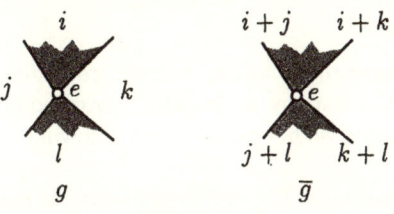

Figure 9

Now we define $\bar{g} : L \longrightarrow \{0, a, b, c\}$ as in the right figure. Note that, because of $i \neq j$, $i \neq k$, $j \neq \ell$, $k \neq \ell$, all elements are $\neq 0$. If $i = \ell$, then $i + j = j + \ell \neq i + k = k + \ell$, hence e is a white vertex with respect to $\bar{g}$. On the other hand, if $i \neq \ell$, then $i + j + k + \ell = 0$, and we obtain $i + j = k + \ell \neq i + k = j + \ell$. Thus e is a crossing vertex, and we find that $\bar{g}$ is a 3-valuation of $\tilde{G}$. Now it is easy to see that $g \longrightarrow \bar{g}$ is a bijection, and the result follows.   □

A special 4-colouring g induces, of course, a 4-face colouring on F, when we restrict g to F. If G is 3-regular, then a 4-face colouring (always with the outer face coloured 0) can be uniquely extended to a special 4-colouring of G by assigning to a vertex v the colour that is not used for the three incident faces. Now one of the earliest colouring theorems (Theorem of Tait) states that for 3-regular plane graphs the number of 4-face colourings (outer face coloured 0) equals the number of 3-edge colourings. This is precisely the first formula of Penrose [30].

Theorem 3.3 *If G is a 3-regular plane graph, then*

$$P_G(3) = \#3\text{-edge colourings of } G.$$

What about arbitrary plane graphs? Since we noted that every special 4-colouring is also a 4-face colouring, we certainly have

$$P_G(3) > 0 \Longrightarrow G \text{ is 4-face colourable,}$$

or in other words: $P_G(3) > 0$ for *all* bridgeless plane graphs implies the 4-colour theorem. Assume, conversely, the 4-colour theorem. We associate to G a 3-regular graph $\hat{G}$ by replacing every vertex by a face as in the following figure:

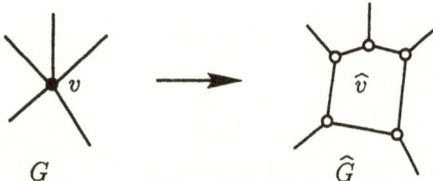

Figure 10

The graph $\hat{G}$ is clearly plane again, and we immediately note that a 4-face colouring of $\hat{G}$ (which exists by assumption) corresponds precisely to a special 4-colouring of G. Hence we have shown the following equivalence.

Theorem 3.4 *The 4-colour theorem is equivalent to the statement: $P_G(3) > 0$ for all bridgeless plane graphs.*

The second result of Penrose [30] that

$$P_G(4) > 0 \iff G \text{ is 4-face colourable} \qquad (3.11)$$

will be discussed more generally for binary spaces in Theorem 4.2. In summary, we have the following equivalent statements:

$$\text{4-colour theorem} \iff P_G(3) > 0 \iff P_G(4) > 0, \qquad (3.12)$$

for all bridgeless plane graphs.

Examples Let us look at a few small examples. For $G = K_4$ the Penrose polynomial $P_{K_4}(\lambda)$ has degree $4 = |F|$. We know from (3.9) and (3.10) that $P_{K_4}(0) = P_{K_4}(1) = P_{K_4}(2) = 0$, and further from Theorem 3 that $P_{K_4}(3) = 6$. Furthermore, the highest coefficient is 1, which gives

$$P_{K_4}(\lambda) = \lambda^4 - 5\lambda^3 + 8\lambda^2 - 4\lambda = \lambda(\lambda - 1)(\lambda - 2)^2.$$

In the same way one computes the following polynomials:

$$
\begin{aligned}
P_{4-\text{wheel}}(\lambda) &= \lambda^5 - 8\lambda^4 + 27\lambda^3 - 40\lambda^2 + 20\lambda \\
&= \lambda(\lambda - 1)(\lambda - 2)(\lambda^2 - 5\lambda + 10); \\
P_{\text{prism}}(\lambda) &= \lambda^5 - 7\lambda^4 + 18\lambda^3 - 20\lambda^2 + 8\lambda = \lambda(\lambda - 1)(\lambda - 2)^3; \\
P_{\text{cube}}(\lambda) &= \lambda^6 - 12\lambda^5 + 83\lambda^4 - 284\lambda^3 + 420\lambda^2 - 208\lambda \\
&= \lambda(\lambda - 1)(\lambda - 2)(\lambda^3 - 9\lambda^2 + 54\lambda - 104).
\end{aligned}
$$

We will make some remarks about the coefficients in the last section.

3.2 Connection to the Tutte polynomial and $\lambda = -2$

We come to the most remarkable of Penrose's formulae, the case $\lambda = -2$. He noted another equivalence to the 4-colour theorem:

$$P_G(-2) \neq 0 \iff G \text{ is 4-face colourable}.$$

Jaeger showed in [18] that this equivalence can be elegantly shown by means of the Tutte polynomial.

Let us fix some notation. First we need the usual concepts of *reduction* and *contraction* [45,50]. Let $G = (V, E)$ be an arbitrary graph, and e an edge. We denote the reduction by $G \setminus e$ and the contraction through e by G/e. A *minor* of G is a graph H that is obtained from G by a sequence of reductions and contractions. The *Tutte polynomial* of $G = (V, E)$ is the polynomial $T_G(x, y)$ in two variables x and y, defined recursively as follows:

(3.13) (i) if $E = \varnothing$, then $T_G(x, y) = 1$;

  (ii) if $e \in E$ is a bridge or a loop, then $T_G(x, y) = xT_{G/e}(x, y)$ or $T_G(x, y) = yT_{G/e}(x, y)$;

(iii) in the case that e is neither a bridge nor a loop, we have

$$T_G(x,y) = T_{G\backslash e}(x,y) + T_{G/e}(x,y).$$

Tutte showed in [46] that the polynomial is uniquely defined by (3.13). As an example, for the triangle $G = K_3$ we obtain $T_G(K_3) = x^2 + x + y$.

The importance of the Tutte polynomial derives (among other things) from the following universal property. Let f be a function which associates with every graph an element in a field K. The function is called a (generalized) *Tutte-Grothendieck invariant*, T–G invariant for short, if there exist elements $a \neq 0$, $b \neq 0$ in K such that the following holds. Let $A = f(\text{bridge})$ and $B = f(\text{loop})$:

(3.14)   (i) if $e \in E$ is a bridge or a loop, then $f(G) = Af(G/e)$ or $f(G) = Bf(G/e)$;

(ii) if $e \in E$ is neither a bridge nor a loop, then

$$f(G) = af(G \backslash e) + bf(G/e).$$

The following result now holds (see [11]).

(3.15)   *If f is a T–G invariant with $f(bridge) = A$ and $f(loop) = B$, and $a \neq 0$, $b \neq 0$ as in (3.14), then we have for any connected graph $G = (V, E)$*

$$f(G) = a^{|E|-|V|+1} b^{|V|-1} T_G\left(\frac{A}{b}, \frac{B}{a}\right).$$

The classical example of a T–G invariant is again the chromatic polynomial $\chi_G(\lambda)$. It is easy to see that $\frac{\chi_G(\lambda)}{\lambda}$ is a T–G invariant with $A = \lambda - 1$, $B = 0$, $a = 1$, $b = -1$. Hence we obtain from (3.15) that

$$\chi_G(\lambda) = (-1)^{|V|-1} \lambda T_G(1 - \lambda, 0). \qquad (3.16)$$

For a plane graph $G = (V, E, F)$ we thus find for the face-colouring polynomial $\chi_{G^*}(\lambda)$:

$$\chi_{G^*}(\lambda) = (-1)^{|F|-1} \lambda T_G(0, 1 - \lambda). \qquad (3.17)$$

Taking the factor λ into account we infer in particular that

(3.18)   *#4-face colourings of $G = (V, E, F)$ with outer face coloured 0*
$$= (-1)^{|F|-1} T_G(0, -3).$$

Now what is the connection of the Tutte polynomial to the Penrose polynomial? As always let $G = (V, E, F)$ be a plane connected graph and $\widetilde{G} = (E, L)$ its 2-coloured medial graph. We consider the transition polynomial $Q(\widetilde{G}, W, \lambda)$ with respect to the weighting

$$W_{\text{black}} = \alpha \neq 0, \quad W_{\text{white}} = \beta \neq 0, \quad W_{\text{crossing}} = 0.$$

For this weighting we find

$$Q(\widetilde{\text{bridge}}) = \alpha\lambda^2 + \beta\lambda, \quad Q(\widetilde{\text{loop}}) = \alpha\lambda + \beta\lambda^2.$$

Let $e \in E$ be neither a bridge nor a loop in G. The following figure shows the operations $G \setminus e$ and G/e, and their effect on the transition polynomials:

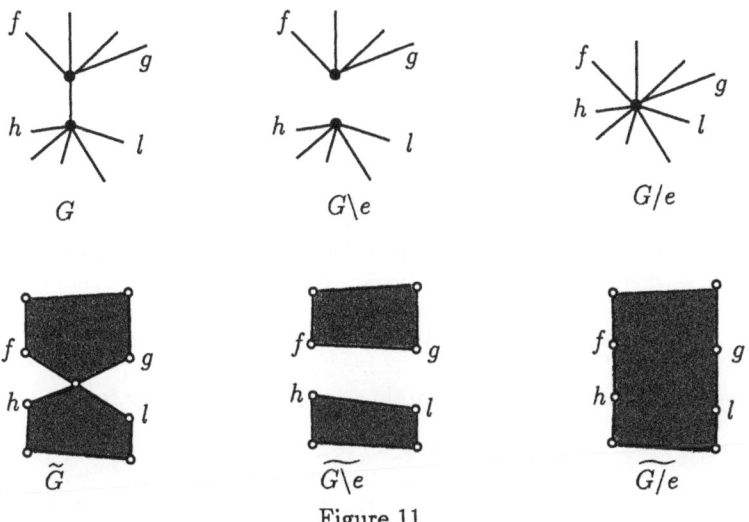

Figure 11

We see that the reduction corresponds precisely to a *black* transition at e, while the contraction corresponds to a *white* transition. Hence

$$Q(\widetilde{G}, W, \lambda) = \alpha Q(\widetilde{G \setminus e}, W, \lambda) + \beta Q(\widetilde{G/e}, W, \lambda).$$

Using (3.14) it is readily seen that $\frac{1}{\lambda}Q(\widetilde{G}, W, \lambda)$ is a T–G invariant. So with (3.15) and the Euler formula $|E| = |V| + |F| - 2$ we obtain the following important result [18].

Theorem 3.5 *Let $G = (V, E, F)$ be a plane graph and $\widetilde{G}$ its medial graph, $\alpha \neq 0$, $\beta \neq 0$, then*

$$Q(\widetilde{G}, W_{\alpha,\beta,0}, \lambda) = \alpha^{|F|-1}\beta^{|V|-1}\lambda T_G(1 + \frac{\alpha}{\beta}\lambda, 1 + \frac{\beta}{\alpha}\lambda).$$

We still do not see the connection to the Penrose polynomial, but here it comes. The transition polynomials of different weightings W and W' are, of course, different in general. Suppose that W and W' differ only by an additive constant m; that is,

$$\alpha' = \alpha + m, \quad \beta' = \beta + m, \quad \gamma' = \gamma + m.$$

Again we will have $Q(\widetilde{G}, W, \lambda) \neq Q(\widetilde{G}, W', \lambda)$, but as Penrose showed, they always agree at the point $\lambda = -2$.

(3.19)   *If two weightings W, W' satisfy $W' = W + m$, then*

$$Q(\widetilde{G}, W, -2) = Q(\widetilde{G}, W', -2).$$

The proof of (3.19) is an easy induction on the number of vertices. Consider now the weightings

$$W: \quad \alpha = 0, \quad \beta = 1, \quad \gamma = -1$$
$$W': \quad \alpha' = 1, \quad \beta' = 2, \quad \gamma' = 0,$$

thus $W' = W + 1$. The transition polynomial with respect to W is by (3.5) precisely the Penrose polynomial. For W' we use Theorem 3.5 and obtain with (3.19)

$$P_G(-2) = Q(\widetilde{G}, W_{1,2,0}, -2) = -2^{|V|} T_G(0, -3).$$

Taking (3.18) into account we have thus proved the following result.

Theorem 3.6 *For a plane graph $G = (V, E, F)$,*

$$P_G(-2) = (-1)^{|F|} 2^{|V|} \cdot (\#4\text{-face colourings of } G \text{ with outer face coloured } 0).$$

For 3-regular plane graphs we have $2|E| = 3|V|$, whence we obtain by Euler's formula the third result of Penrose [30]:

(3.20)   *For a 3-regular plane graph $G = (V, E, F)$,*
$$P_G(-2) = (-4)^{\frac{|V|}{2}} \cdot (\# 3\text{-edge colourings of } G).$$

Theorem 3.6 implies now the equivalence announced earlier.

Theorem 3.7 *The 4-colour theorem is equivalent to $P_G(-2) \neq 0$ for all bridgeless plane graphs G.*

Perhaps a word on the standing of the 4-colour theorem is in order. The opinions of mathematicians on the importance of this theorem are traditionally divided. There are those who think that the original problem was attractive and that the efforts towards a solution produced a whole new field – graph theory – and also that the proofs [6,31] to date with their massive use of computers point far into the future. On the other hand, there are also those who think the theorem a topological curiosity and the proofs monstrous, if the computer calculations are accepted as proofs at all. The work of Penrose reveals a third aspect: it appears that we have not really understood the true meaning of the 4-colour theorem, but that at any rate it seems to be a focus point of plane topology. Reviewing once more the line of reasoning which led up to Theorem 3.7 we note three things:

(i) only for weightings with $\gamma = 0$ can the transition polynomial be expressed via the Tutte polynomial;

(ii) the result of Penrose regarding two weightings with $W' = W + m$ holds only for $\lambda = -2$;

(iii) the connection between T_G and P_G works only for $T_G(0, -3)$, and this value counts precisely the 4-face colourings.

That the 4-colour theorem lies precisely in the intersection of these three facts should owe to some deeper topological property yet to be discovered.

Penrose proved in his paper still another equivalence. If we restrict the sets $A \subseteq E$ to the cocycle-set $\mathcal{K}$ (see Section 4), we obtain the polynomial

$$R_G(\lambda) = \sum_{A \in \mathcal{K}} (-1)^{|A|} \lambda^{c(A)}.$$

Penrose [30] now showed for 3-regular plane graphs (but it holds in general):

(3.21) *4-colour theorem $\Longleftrightarrow R_G(2) > 0$ for all bridgeless plane graphs.*

We will prove this last equivalence for arbitrary binary matroids in Section 4.1.

There is another very interesting case where Theorem 3.5 applies. Consider the weightings $W : \alpha = -1, \beta = -1, \gamma = 0$ and $W' : \alpha' = 0, \beta' = 0, \gamma' = 1$. For W we obtain as transition polynomial with Euler's formula

$$Q(\widetilde{G}, W, \lambda) = (-1)^{|E|} \lambda T_G(1 + \lambda, 1 + \lambda).$$

The second weighting W' says that we consider the *unique* transition system where all transitions are of crossing type. Hence

$$Q(\widetilde{G}, W', \lambda) = \lambda^c \qquad (c = \text{number of cycles}).$$

With (3.19) we thus find

$$(-2)^{c-1} = (-1)^{|E|} T_G(-1, -1), \qquad (3.22)$$

and in particular,

$$c = 1 \iff |T_G(-1,-1)| = 1. \tag{3.23}$$

As an example, we obtain for $G = K_4$ three cycles in the "crossing" decomposition of $\widetilde{K}_4$, in agreement with $T_{K_4}(x,y) = x^3 + 3x^2 + 2x + 4xy + 2y + 3y^2 + y^3$, and $T_{K_4}(-1,-1) = 4 = (-2)^2$.

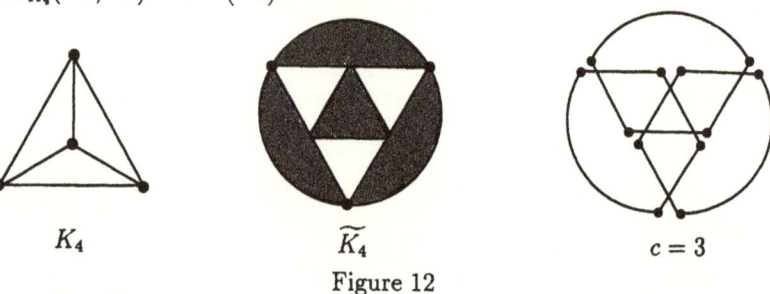

$$K_4 \qquad\qquad \widetilde{K}_4 \qquad\qquad c = 3$$

Figure 12

3.3   Knot invariants

Consider again $G = (V, E, F)$ and its 2-coloured medial graph $\widetilde{G} = (E, L)$. The graph $\widetilde{G}$ is 4-regular and can thus be interpreted as a diagram of a knot (or, more generally, of a link), by observing the convention "left over right" seen from the black country.

$$\widetilde{G} \qquad\qquad\qquad \text{link } L_G$$

Figure 13

As an example, we get from K_3 the trefoil:

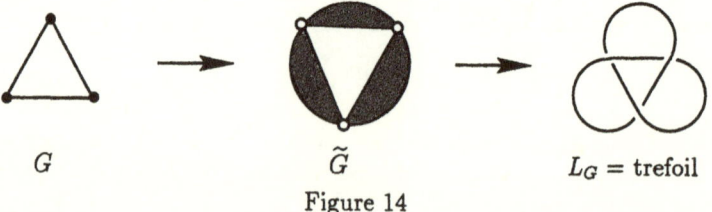

$$G \qquad\qquad \widetilde{G} \qquad\qquad L_G = \text{trefoil}$$

Figure 14

In this manner we obtain all *alternating* knots or links where the crossing points are alternately "over" and "under".

In the past years, a number of polynomial invariants have been studied, that is, polynomials which are the same for equivalent links. A particularly nice example is the *bracket polynomial* $\langle L \rangle$ due to Kauffman [23]. It is a Laurent polynomial in the variable A defined recursively as follows:

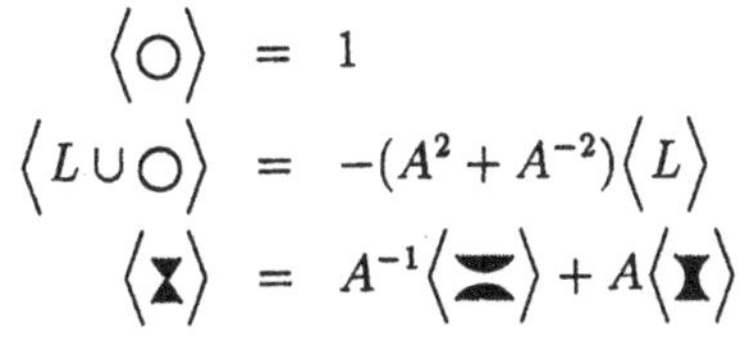

$$\langle \bigcirc \rangle = 1$$
$$\langle L \cup \bigcirc \rangle = -(A^2 + A^{-2})\langle L \rangle$$
$$\langle \mathbf{X} \rangle = A^{-1}\langle \rangle + A\langle \mathbf{I} \rangle$$

Figure 15

The resolution of a crossing into a black and white part suggests immediately that $\langle L \rangle$ is a T–G invariant. With (3.15) we easily compute, for $G = (V, E, F)$,

$$G \quad \longrightarrow \quad \tilde{G} \quad \longrightarrow \quad L_G$$

graph   medial   link

that

$$\langle L_G \rangle = A^{|V|-|F|} T_G(-A^{-4}, -A^4).$$

For the example $G = K_3$ we already know $T_{K_3}(x, y) = x^2 + x + y$. Hence we calculate, for the trefoil L of Figure 14,

$$\langle \text{trefoil} \rangle = -A^5 - A^{-3} + A^{-7}.$$

The bracket polynomial is invariant under the Reidemeister moves II and III. Introducing a winding factor $\omega(L)$ for oriented links K (see [17,25,49]), the polynomial

$$f_K(A) = (-A^3)^{-\omega(L)}\langle L \rangle$$

is also invariant under the Reidemeister move I and hence a knot invariant. Finally, the substitution $A = x^{-1/4}$ yields the Jones polynomial [22]:

$$V_K(x) = f_K(x^{-1/4}).$$

For arbitrary knots and links (not just alternating) there exists a nice theory involving *signed* 4-regular plane graphs, where $+1$ means "left over right" and -1 means "right over left". The details and the connections to the *signed* Tutte polynomial can be found in [40]. For other fascinating connections to combinatorial questions the reader is referred to [3,7].

3.4   The Gauss problem

Let us digress for a moment to consider a beautiful problem raised by Gauss which fits into our general framework, and its two equally beautiful solutions. Let C be a closed curve in the plane (without triple points) as in the figure.

Denote the crossing points by letters $A, B, C, \ldots$ and write down the word W as we run through the curve.

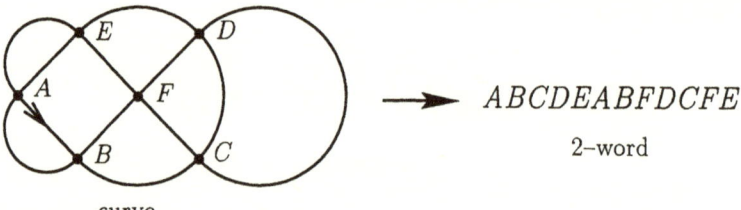

$$ABCDEABFDCFE$$

2–word

curve

Figure 16

Thus we obtain a (cyclic) 2-word W where every letter appears exactly twice. Gauss asked for a characterization of those 2-words which are *realizable* as the crossing points of a plane closed curve.

To any 2-word W on letters V we associate its *interlace graph* H. The vertices of H are V, and we join letters A and B if they interlace, meaning that they appear in the order $\ldots A \ldots B \ldots A \ldots B \ldots$ in W. In other words, the interlace graphs are just what are usually called the *circle graphs* (see, for example, [15]): the word is arranged around a circle, equal letters are joined by a chord, and two distinct letters are adjacent if and only if their chords intersect. The next figure shows the interlace graph of our example, and its description around a circle.

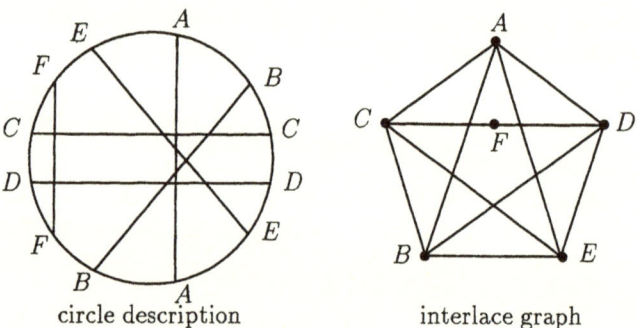

circle description   interlace graph

Figure 17

Call an interlace graph *realizable* if its 2-word is realizable. Gauss himself suggested (without proof) that realizable interlace graphs are Eulerian, that is, all letters interlace with an even number of other letters. The proof of this is easy. Let $W = A\alpha A\beta$ where α, β are the two sub-words between A. Consider the sub-word $A\alpha A$ and regard it as a plane graph G with the crossing points as vertices. The graph G is 4-regular (or a free loop if there are no crossing points), with the letters interlaced with A marked on the curve. The faces of G can thus be 2-coloured. Now we run through $A\beta A$. Since we start and end in the same face of G and change face colours every time we cross an interlaced symbol, there must be an even number of them.

The first solution to the Gauss problem is due to Lovász and Marx [26] and is reminiscent of Kuratowski's characterization of planar graphs. They introduced the following operations which preserve realizability. Suppose $W = A\alpha A\beta$ is realizable. By β^{-1} we denote the word β turned around. Now remove the crossing point A by choosing the other two possible transitions, as suggested in the figure, and denote the resulting words by W' and W'', where in W'' we also remove all letters interlaced with A. Clearly, W' and W'' are again realizable.

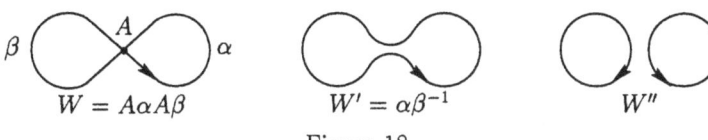

$$W = A\alpha A\beta \qquad W' = \alpha\beta^{-1} \qquad W''$$

Figure 18

For the interlace graph H of an arbitrary 2-word W these operations are easily seen to correspond to the following two graph operations:

Switch at A: Delete A and switch the adjacencies in the neighbourhood $N(A)$ of A, meaning that we delete the edges in $N(A)$ and insert them if they are not there. The rest of the graph remains unchanged.

Removal at A: Delete A and its neighbourhood and keep the rest unchanged.

By abuse of language let us call K a *minor* of H, denoted $K < H$, if K arises from H by a series of switches and removals. Clearly, the interlace graphs are closed with respect to $<$, and our observation above tells us that the realizable interlace graphs also form a closed class. So it remains to find the obstructions, that is the set of minor-minimal *non-realizable* interlace graphs. One set of obstructions is immediately found, the complete graphs K_n with n even, as these graphs are not Eulerian. The theorem of Lovász and Marx says that there are no others.

Theorem 3.8 *An interlace graph H is realizable if and only if $K_n \not< H$, with n even.*

Incidentally, there is another pleasing result along the same lines. Consider only the switch operation, and write $K <_s H$ if K arises from H by a series of switches.

Result *A graph G is Eulerian if and only if $K_n \not<_s G$, n even.*

We come to the second characterization provided by Rosenstiehl [33,34] in the same year. He gave a structural description of realizable interlace graphs. The condition "Eulerian" is certainly not enough to characterize realizable interlace graphs. The smallest counterexample (already known to Gauss) is the 5-cycle C_5. It is an interlace graph, but the removal at any vertex yields the

forbidden graph K_2. Generalizing this example we immediately find another
necessary condition. Suppose H is realizable, and A, B are non-neighbours.
Since the degree $d(B) = |N(A) \cap N(B)| + |N(B) \setminus N(A)|$ is even, as is the
degree $d''(B) = |N(B) \setminus N(A)|$ in the graph H'' arising from removal at A, we
infer that $|N(A) \cap N(B)|$ is even as well. In other words, any two non-adjacent
vertices have an even number of common neighbours. The octahedron graph
H shows that these two conditions still do not suffice. Also, H is an interlace
graph, and has again a K_2-minor as shown in the figure.

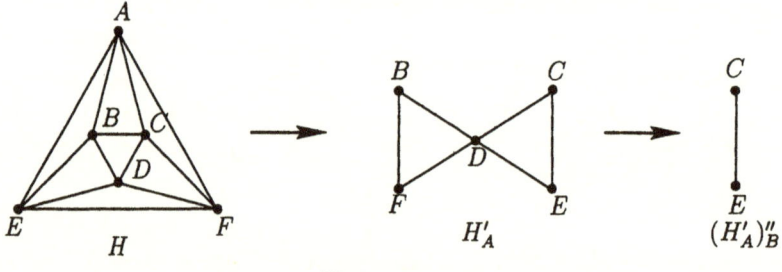

Figure 19

Here is Rosenstiehl's theorem.

Theorem 3.9 *An interlace graph H is realizable if and only if the following
conditions hold:*

(i) *H is Eulerian;*

(ii) *two non-adjacent vertices have an even number of common neighbours;*

(iii) *with an edge AB called* even *if A and B have an even number of common
neighbours, the set of even edges forms a cocycle.*

The reader will have noticed that a plane closed curve C can be interpreted
as the medial graph $\widetilde{G}$ of the underlying G such that all transitions are of
crossing type. Furthermore, the operations switch and removal correspond to
choosing the black or white transition at A instead.

For some interesting connections to directed Eulerian graphs, knots and
a new polynomial, called the *interlace polynomial*, the reader is referred to a
recent paper by Arratia, Bollobás and Sorkin [7].

4   Algebra of the Penrose polynomial

We have introduced the Penrose polynomial in Section 2 in a geometric
setting. But there is also a purely algebraic formulation which opens the way
to a description of the Penrose polynomial for arbitrary graphs and, more
generally, binary matroids.

Consider an arbitrary connected graph $G = (V, E)$, and let 2^E be the vector space over GF(2) of all subsets $A \subseteq E$, where we identify A with its characteristic vector. Hence, for $A, B \subseteq E$,

$$A + B = \text{symmetric difference of } A \text{ and } B;$$
$$A \cdot B = \begin{cases} 1 & \text{if } |A \cap B| \text{ is odd,} \\ 0 & \text{if } |A \cap B| \text{ is even.} \end{cases}$$

Next we define cycles and cocycles of G. $C \subseteq E$ is called a *cycle* if the subgraph $G = (V, C)$ is *Eulerian*; $K \subseteq E$ is a *cocycle* if (V, K) is a *bipartition*, that is, if there exists a decomposition $V = V_1 \cup V_2$ such that K consists precisely of the edges between V_1 and V_2. The cycles form a subspace C of 2^E, the *cycle space*, and the cocycles form the *cocycle space* $\mathcal{K}$.

A classical result of graph theory, dating back to Veblen [48], says that C and $\mathcal{K}$ are, in fact, orthogonal complements:

$$C^\perp = \mathcal{K}, \quad \mathcal{K}^\perp = C,$$

with

$$\dim C = |E| - |V| + 1, \quad \dim \mathcal{K} = |V| - 1. \tag{4.1}$$

Now let $G = (V, E, F)$ be a plane connected graph. Since cycles and cocycles are dual concepts, it follows, by identifying the edges of G and the dual graph G^*, that

$$C(G^*) = \mathcal{K}(G), \quad \mathcal{K}(G^*) = C(G).$$

In particular, we have $\dim C = |F| - 1$ which implies with (4.1) Euler's Formula $|V| - |E| + |F| = 2$.

It was noted by several authors (see, for example, [4,19]) that the theory of "left-right paths" in plane graphs as developed in [32,35,42] can be used to prove the following result.

Let $G = (V, E, F)$, and set, for $A \subseteq E$,

$$\mathcal{B}_G(A) = \{C \in C : C \cap A \in \mathcal{K}\} \subseteq 2^E.$$

The set $\mathcal{B}_G(A)$ is a subspace of 2^E, called the *bicycle space* of A. Let $c(A)$ be the number of trails with respect to A as in the definition of the Penrose polynomial. Then we have

$$c(A) = \dim \mathcal{B}_G(A) + 1, \tag{4.2}$$

and thus the following algebraic description of the Penrose polynomial.

Theorem 4.1 *For a plane graph $G = (V, E, F)$ we have*

$$P_G(\lambda) = \lambda \sum_{A \subseteq E} (-1)^{|A|} \lambda^{\dim B_G(A)}.$$

This result opens the way to define the Penrose polynomial for arbitrary binary matroids. We proceed as follows.

Let E be a finite set and $\mathcal{C} \subseteq 2^E$ a subspace, called the "*cycle space*"; the orthogonal complement $\mathcal{K} = \mathcal{C}^\perp$ is the "*cocycle space*". We call the pair $M = (E, \mathcal{C})$ a *binary matroid*, and $M^* = (E, \mathcal{K})$ the *dual matroid*.

Look for example at the *Fano matroid* F:

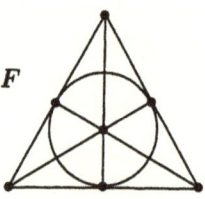

Figure 20

The ground-set E consists of the 7 points of the configuration, and the cycles are the empty set, the 7 three-point "lines" and the 8 complements. It is immediately checked that the cocycles of F are the empty set and the 7 complements of the 3-point lines. In particular, we have

$$\mathcal{K}(F) \subseteq \mathcal{C}(F) \qquad \text{and thus} \qquad \mathcal{C}(F^*) \subseteq \mathcal{K}(F^*). \tag{4.3}$$

Consider an arbitrary binary matroid $M = (E, \mathcal{C})$. As for graphs we define for $A \subseteq E$ the bicycle space

$$B_M(A) = \{C \in \mathcal{C} : C \cap A \in \mathcal{K}\}.$$

Definition The *Penrose polynomial* of a binary matroid $M = (E, \mathcal{C})$ is

$$P_M(\lambda) = \sum_{A \subseteq E} (-1)^{|A|} \lambda^{\dim B_M(A)}. \tag{4.4}$$

(It is convenient to delete the factor λ.)

Before going on let us fix the notation. Let $M = (E, \mathcal{C})$ be a binary matroid, $A \subseteq E$. The *reduction* $M|A = (A, \mathcal{C}|A)$ on A is given by

$$\mathcal{C}|A = \{C \subseteq A : C \in \mathcal{C}\}, \tag{4.5}$$

and the *contraction* $M \cdot A = (A, \mathcal{C} \cdot A)$ to A by

$$\mathcal{C} \cdot A = \{C \cap A : C \in \mathcal{C}\}. \tag{4.6}$$

For the cocycle space $\mathcal{K}$ we then have

$$\mathcal{K}|A = \{K \cap A : K \in \mathcal{K}\} \text{ and } \mathcal{K} \cdot A = \{K \subseteq A : K \in \mathcal{K}\}.$$

A *minor* of M is a matroid which can be obtained from M by a sequence of reductions and contractions. For graphs this corresponds to the former notions; see, for example, [29].

4.1   Evaluation of $P_M(\lambda)$

Let $P_M(\lambda)$ be the Penrose polynomial of the (non-empty) binary matroid $M = (E, \mathcal{C})$, and $\mathcal{K} = \mathcal{C}^\perp$. Now we consider the evaluation at $\lambda = 2^k$. Clearly, $P_M(1) = 0$, hence suppose $k \geq 1$. It follows from (4.4) that

$$P_M(2^k) = \sum_{A \subseteq E} (-1)^{|A|} 2^{k \dim \mathcal{B}_M(A)} = \sum_{A \subseteq E} (-1)^{|A|} |\mathcal{B}_M(A)|^k.$$

For $(C_1, \dots, C_k) \in \mathcal{C}^k$ set

$$\mathcal{K}(C_1, \dots, C_k) = \{A \subseteq E : A \cap C_1 \in \mathcal{K}, \dots, A \cap C_k \in \mathcal{K}\},$$

then we get

$$P_M(2^k) = \sum_{(C_1, \dots, C_k) \in \mathcal{C}^k} \sum_{A \in \mathcal{K}(C_1, \dots, C_k)} (-1)^{|A|}.$$

Since $\mathcal{K}(C_1, \dots, C_k)$ is a subspace, it either contains an equal number of odd and even sets (in which case the inner sum is 0), or all sets have even size. Hence setting

$$\widetilde{\mathcal{C}}_k = \{(C_1 \dots, C_k) \in \mathcal{C}^k : (A \in \mathcal{K}(C_1, \dots, C_k) \Longrightarrow |A| \text{ even})\}$$

we find

$$P_M(2^k) = \sum_{(C_1, \dots, C_k) \in \widetilde{\mathcal{C}}_k} |\mathcal{K}(C_1, \dots, C_k)|. \tag{4.7}$$

In other words, $P_M(2^k)$ is always nonnegative, and positive precisely when $\widetilde{\mathcal{C}}_k$ is not empty. It is clear that $P_M(2^k) \leq P_M(2^{k+1})$ holds for all k. From the definition of $\widetilde{\mathcal{C}}_k$ we immediately infer

$$(C_1, \dots, C_k) \in \widetilde{\mathcal{C}}_k \Longrightarrow C_1 \cup \dots \cup C_k = E. \tag{4.8}$$

To see this, suppose $e \in E \setminus (C_1 \cup \dots \cup C_k)$. Then $\{e\}$ lies in $\mathcal{K}(C_1, \dots, C_k)$, but has cardinality 1 which is odd.

Next we have

$$P_M(2) = \begin{cases} |\mathcal{K}| & \text{if } E \in \mathcal{C}, \\ 0 & \text{otherwise.} \end{cases} \tag{4.9}$$

Indeed, as we have just seen

$$P_M(2) > 0 \iff \tilde{C}_1 = \{E\},$$

and this latter fact is clearly equivalent to $E \in C$, in which case $\mathcal{K}(E) = \mathcal{K}$. In particular, we get back our graph result (3.10) by taking the factor $\lambda = 2$ into account.

We come to the most important case $\lambda = 4 = 2^2$.

Claim We have $(C_1, C_2) \in \tilde{C}_2 \iff C_1 \cup C_2 = E$.

We already know the direction $\Longrightarrow$. Suppose, conversely, $E = C_1 \cup C_2$ for $C_1, C_2 \in C$, and consider $A \in \mathcal{K}(C_1, C_2)$. Since $C_1 \cap A \in \mathcal{K}$ we find that $|(C_1 \cap A) \cap C_1| = |C_1 \cap A|$ is even. With $C_2 \cap A \in \mathcal{K}$ we have $(C_1 \cap A) + (C_2 \cap A) = (C_1 + C_2) \cap A \in \mathcal{K}$. This implies that $|(C_1 + C_2) \cap A \cap C_2| = |(C_2 \setminus C_1) \cap A|$ is even, and hence also $|A| = |C_1 \cap A| + |(C_2 \setminus C_1) \cap A|$, which gives $(C_1, C_2) \in \tilde{C}_2$.

In summary, the following has been proved.

Theorem 4.2 *Let $M = (E, C)$ be a binary matroid. Then we have*

$$P_M(4) > 0 \iff E = C_1 \cup C_2 \text{ with } C_1, C_2 \in C.$$

Now another classical graph theorem (see, for example, [2]) says that a bridgeless plane graph $G = (V, E, F)$ is 4-face colourable if and only if E can be covered by two Eulerian subgraphs, and this is precisely the statement $E = C_1 \cup C_2$. Hence we obtain the equivalence announced in (3.11).

Let us, finally, take a look at Penrose's fourth equivalence to the 4-colour theorem (3.21). Define for a binary matroid $M = (E, C)$ with $\mathcal{K} = C^\perp$ the polynomial

$$R_M(\lambda) = \sum_{K \in \mathcal{K}} (-1)^{|K|} \lambda^{\dim \mathcal{B}_M(K)}.$$

For $\lambda = 1$ we obtain $R_M(1) = \sum_{K \in \mathcal{K}} (-1)^{|K|}$. Hence either $R_M(1) = 0$ or $\mathcal{K}$ consists of even-sized subsets only in which case $R_M(1) = |\mathcal{K}|$. This latter possibility occurs if and only if $E \in C$, and we get an evaluation analogous to (4.9):

$$R_M(1) = \begin{cases} |\mathcal{K}| & \text{if } E \in C, \\ 0 & \text{otherwise.} \end{cases} \tag{4.10}$$

For arbitrary $\lambda = 2^k$ we derive, as in (4.7),

$$R_M(2^k) = \sum_{(C_1, \dots, C_k) \in \tilde{C}_k} |\mathcal{K}(C_1, \dots, C_k)|,$$

where $\mathcal{K}(C_1,\ldots,C_k)$ runs through the cocycles only, that is,

$$\mathcal{K}(C_1,\ldots,C_k) = \{K \in \mathcal{K} : K \cap C_1 \in \mathcal{K},\ldots,K \cap C_k \in \mathcal{K}\};$$
$$\tilde{C}_k = \{(C_1,\ldots,C_k) \in C^k : K \in \mathcal{K}(C_1,\ldots,C_k) \Longrightarrow |K| \text{ even}\}.$$

Thus we infer as before: $R_M(2^k) > 0 \Longleftrightarrow \tilde{C}_k \neq \varnothing$.

Lemma *The following conditions are equivalent for $C \in C$:*

(a) $C \in \tilde{C}_1$;

(b) $E \setminus C \in C \cdot (E \setminus C)$;

(c) $C = A + B$, $A, B \in C$ with $A \cup B = E$.

Proof (a) $\Longrightarrow$ (b). For $K \in \mathcal{K}$ with $K \cap C = \varnothing$ we plainly have $K \in \mathcal{K}(C)$, thus K is even-sized. This implies that $|(E \setminus C) \cap K| \equiv 0$ for all $K \subseteq E \setminus C$, which means $E \setminus C \in (\mathcal{K} \cdot (E \setminus C))^\perp = C \cdot (E \setminus C)$.

(b) $\Longrightarrow$ (c). Let $C = A + B$, $A, B \in C$ and choose A, B so that $|A \cup B|$ is maximal. Note that such representations exist, for example, $C = C + \varnothing$. We claim $A \cup B = E$. Suppose not, and take $e \in E \setminus A \cup B$. Let $D = E \setminus (A \cup B)$ as in the figure:

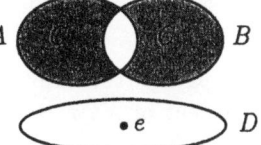

$$A \qquad\qquad B$$
$$\bullet\, e \qquad\qquad D$$

Figure 21

Since $A \cap B = A \cap (E \setminus C) \in C \cdot (E \setminus C)$ we have $D = (A \cap B) + (E \setminus C) \in C \cdot (E \setminus C)$. Hence $D = \hat{D} \cap (E \setminus C)$ with $\hat{D} \in C$, where $e \in \hat{D}$ and $\hat{D} \cap (A \cap B) = \varnothing$ (since $D \cap (A \cap B) = \varnothing$). Now set $\hat{A} = A + \hat{D}$, $\hat{B} = B + \hat{D}$. Then $\hat{A}$ and $\hat{B}$ are in C, and we have $\hat{A} + \hat{B} = A + B = C$. Since $\hat{A} \cap \hat{B} = (A \cap B) + \hat{D} = (A \cap B) \cup \hat{D} \supseteq A \cap B$ and $e \in \hat{A} \cup \hat{B}$ we obtain $\hat{A} \cup \hat{B} \supsetneq A \cup B$, a contradiction.

(c) $\Longrightarrow$ (a). Let $K \in \mathcal{K}$ with $K \cap C \in \mathcal{K}$. Then $|(A \setminus B) \cap K| = |A \cap (K \cap C)| \equiv 0$, and also $|B \cap K| \equiv 0$. Since $E = A \cup B$, we have $|K| = |(A \setminus B) \cap K| + |B \cap K| \equiv 0$, which means $C \in \tilde{C}_1$. $\qquad\square$

From the lemma, parts (a) and (c), we infer that $\tilde{C}_1 \neq \varnothing$ if and only if $E = C_1 \cup C_2$ with $C_1, C_2 \in C$, and thus the following result.

Theorem 4.3 *Let $M = (E, C)$ be a binary matroid. Then we have*

$$R_M(2) > 0 \Longleftrightarrow E = C_1 \cup C_2 \text{ with } C_1, C_2 \in C.$$

In the case of plane graphs we thus obtain the equivalence (3.21).

4.2   When is $P_M(\lambda) \equiv 0$?

We have observed for plane graphs G that

$$P_G(\lambda) \equiv 0 \iff G \text{ has a bridge.}$$

In a binary matroid $M = (E, \mathcal{C})$ a bridge is an element $e \in E$ that is contained in no cycle. If M has a bridge, then it is immediately seen that $P_M(\lambda) \equiv 0$: just note by (4.8) that all sets $\widetilde{C}_k$ must be empty. Suppose M is bridgeless, does it then follow that $P_M(\lambda) \not\equiv 0$? We will see that this is not so anymore, but that the matroids M with $P_M(\lambda) \equiv 0$ can be described in an elegant way.

From definition (4.4) we have $\deg P_M(\lambda) \leq \dim \mathcal{C}$, since all bicycle spaces are subspaces of $\mathcal{C}$. Maybe it comes as a surprise that the strict inequality already characterizes the matroids M with vanishing Penrose polynomial.

Theorem 4.4 *The following conditions are equivalent for a binary matroid* $M = (E, \mathcal{C})$:

(a) $P_M(\lambda) \equiv 0$;

(b) $\deg P_M(\lambda) < \dim \mathcal{C}$;

(c) *there exists* $A_0 \subseteq E$, $|A_0|$ *odd, with* $A_0 \cap C \in \mathcal{K}$ *for all* $C \in \mathcal{C}$.

Proof (a) $\Longrightarrow$ (b) is trivial.

(b) $\Longrightarrow$ (c) The bicycle space of the empty set is all of $\mathcal{C}$, hence $\mathcal{B}_M(\varnothing)$ contributes $+1$ to the coefficient of λ^r, $r = \dim \mathcal{C}$. So, there must be an odd-sized set $A_0 \subseteq E$ with also $\mathcal{B}_M(A_0) = \mathcal{C}$, and this is precisely (c).

(c) $\Longrightarrow$ (a) We show that (c) implies $\widetilde{C}_k = \varnothing$ for all k from which $P_M(\lambda) \equiv 0$ follows, since then $P_M(\lambda) = 0$ for infinitely many λ. Now A_0 clearly lies in all $\mathcal{K}(C_1, \ldots, C_k)$ as defined above from which $\widetilde{C}_k = \varnothing$ results. $\square$

Example . Suppose $M = (E, \mathcal{C})$ satisfies $\mathcal{C} \subseteq \mathcal{K} = \mathcal{C}^\perp$, and $|E|$ is odd. Clearly, in this case E is a set as specified in condition (c) of the previous theorem, whence $P_M(\lambda) \equiv 0$. The smallest example with this property (apart from a bridge) is the dual Fano matroid F^*. We have seen in (4.3) that $\mathcal{C}(F^*) \subseteq \mathcal{K}(F^*)$, and so $P_{F^*}(\lambda) \equiv 0$ since $|E| = 7$.

The following result gives a description of the matroids with $P_M(\lambda) \equiv 0$, for the details the reader is referred to [5].

Theorem 4.5 *Let* $M = (E, \mathcal{C})$ *be a bridgeless binary matroid with* $P_M(\lambda) \equiv 0$. *Then* M *contains the dual Fano matroid* F^* *as a minor.*

It is known (see, for example, [2]) that a graphic matroid $M_G = (E, \mathcal{C}_G)$ or a cographic matroid $M_G^* = (E, \mathcal{K}_G)$ never contains F^* as minor. So, the Penrose polynomial is not identically equal to zero for any bridgeless graph and dually for any bridgeless cographic matroid. We will return to Theorem 4.5 in the last section.

4.3   Isotropic systems and $\lambda = -2$

We discussed in Section 3.2 the most interesting formula of Penrose, the evaluation of $P_G(\lambda)$ at $\lambda = -2$. The argument there used the Tutte polynomial $T_G(x, y)$ and the transition polynomial $Q(\widetilde{G}, W, \lambda)$. The definition of the Tutte polynomial extends without difficulty to (arbitrary) matroids via the recursion

$$T_M(x, y) = T_{M \backslash e}(x, y) + T_{M/e}(x, y),$$

where $M \backslash e$ is shorthand for the reduction $M|(E \backslash e)$ and M/e for the contraction $M \cdot (E \backslash e)$; see [12]. But for $Q(\widetilde{G}, W, \lambda)$ we needed the geometric structure of G, and this we do not have at our disposal for binary matroids.

To proceed in this case we make use of the interesting theory of isotropic systems, as developed by Bouchet in [8,9]; for details see [28]. Let us collect the basics of the theory.

Let $K = \{0, x, y, z\}$ be the Klein four-group which we interpret as 2-dimensional vector space over GF(2), and set $K' = K \backslash \{0\}$. For $a, b \in K$ define the inner product $\langle a, b \rangle = 1$ if $0 \neq a \neq b \neq 0$, and 0 otherwise; the group K together with this symplectic form is the binary hyperbolic plane. For a finite set E consider K^E as a vector space of dimension $2|E|$ over GF(2). This means that, for $A = (A_e)$, $B = (B_e) \in K^E$,

$$A \perp B \iff |\{e \in E : 0 \neq A_e \neq B_e \neq 0\}| \text{ is even.}$$

In particular, every vector $A \in K^E$ is orthogonal to itself. A subspace $\mathcal{L} \subseteq K^E$ is called *totally isotropic* if $\mathcal{L} \subseteq \mathcal{L}^{\perp}$ holds.

Definition   An *isotropic system* is a pair $S = (E, \mathcal{L})$ where $\mathcal{L} \subseteq K^E$ is a totally isotropic subspace of dimension $|E|$ over GF(2).

Bouchet now showed that any pair of dual binary matroids gives rise to an isotropic system. This goes as follows. For $X \in K^E$ and $P \subseteq E$, let $X(P)$ be the *restriction* of X to P, that is,

$$(X(P))_e = \begin{cases} X_e & \text{if } e \in P \\ 0 & \text{otherwise.} \end{cases}$$

The set $\widehat{X} = \{X(P) : P \subseteq E\}$ is clearly a totally isotropic subspace of K^E. The vector $X \in K^E$ is *complete* if $X_e \neq 0$ for all $e \in E$, and two vectors $X, Y \in K^E$ are *supplementary* if $0 \neq X_e \neq Y_e \neq 0$ for all $e \in E$. We then write $X \# Y$.

It is immediately seen that for supplementary vectors $A = (A_e), B = (B_e) \in K^E$ and $P, Q \subseteq E$ we have

$$\langle A(P), B(Q) \rangle = P \cdot Q \quad (= |P \cap Q| \mod 2), \tag{4.11}$$

and

$$\hat{A} \cap \hat{B} = \{0\}. \tag{4.12}$$

For a complete vector $X \in K^E$ and any subspace $\mathcal{U} \subseteq 2^E$ we set

$$X(\mathcal{U}) = \{X(U) : U \in \mathcal{U}\}.$$

Here, $X(\mathcal{U})$ is clearly isomorphic to $\mathcal{U}$ via the canonical map $U \mapsto X(U)$. Now we have all the ingredients for Bouchet's theorem.

Theorem 4.6 *Let $M = (E, \mathcal{C})$ be a binary matroid, $\mathcal{K} = \mathcal{C}^\perp$, and $X \# Y \in K^E$. Then $S = (E, \mathcal{L})$ with*

$$\mathcal{L} = X(\mathcal{K}) + Y(\mathcal{C})$$

forms an isotropic system.

That $\mathcal{L}$ is totally isotropic follows directly from (4.11), and for the dimension we compute by (4.12):

$$\dim(X(\mathcal{K}) + Y(\mathcal{C})) = \dim X(\mathcal{K}) + \dim Y(\mathcal{C}) = \dim \mathcal{K} + \dim \mathcal{C} = |E|.$$

Bouchet gave a number of other examples and extended the theory to so-called multimatroids [10].

Now, fix the notation. In $K = \{0, x, y, z\}$ set $x = (1, 0)$, $y = (0, 1)$, $z = (1, 1)$, and $X = (x, \dots, x) \in K'^E$, $Y = (y, \dots, y) \in K'^E$, $Z = (z, \dots, z) \in K'^E$. Hence $Z = X + Y$. The elements x, y and z now play the roles of the three *transitions* in the geometric definition of the transition polynomial; that is, we choose to every pair (i, e), $i \in K'$, $e \in E$, a weight

$$W(i, e) = \begin{cases} \alpha & \text{if } i = x \\ \beta & \text{if } i = y \\ \gamma & \text{if } i = z, \end{cases}$$

and set, for a complete vector $A = (A_e) \in K'^E$,

$$W(A) = \prod_{e \in E} W(A_e, e).$$

Definition Let $M = (E, \mathcal{C})$ be a binary matroid and $S = (E, \mathcal{L} = X(\mathcal{K}) + Y(\mathcal{C}))$ the associated isotropic system. Then we call

$$Q(S, W, \lambda) = \sum_{A \in K'^E} W(A) \lambda^{\dim(\mathcal{L} \cap \hat{A})} \tag{4.13}$$

the *transition polynomial* of S with respect to the weighting W.

Special cases of transition polynomials are for example the *Martin polynomials*; see [13,24,27].

The following formula gives the connection to the bicycle spaces. Suppose $M = (E, \mathcal{C})$ is a binary matroid, $\mathcal{L} = X(\mathcal{K}) + Y(\mathcal{C})$, and $A \subseteq E$. Then

$$\dim \mathcal{B}_M(A) = \dim(\mathcal{L} \cap Z(\widehat{A}) + Y(E \setminus A)). \tag{4.14}$$

For the proof, check that the map $\varphi : \mathcal{B}_M(A) \longrightarrow \mathcal{L} \cap Z(\widehat{A}) + Y(E \setminus A)$ with $\varphi(C) = Z(C \cap A) + Y(C \cap (E \setminus A))$ is a bijection.

To establish the connection to the Tutte polynomial we need a recursion of isotropic systems with respect to reduction and contraction.

Let $S = (E, \mathcal{L})$ be an arbitrary isotropic system, $e \in E$. We set $\mathcal{L}|_i^e = \{A \in \mathcal{L} : A_e = i\}$ and $\mathcal{L}_i^e = p_e(\mathcal{L}|_0^e \cup \mathcal{L}|_i^e)$, where $p_e : K^E \longrightarrow K^{E \setminus e}$ is the canonical projection. The pair $S_i^e = (E \setminus e, \mathcal{L}_i^e)$ is again an isotropic system, called an *elementary minor*.

For a binary matroid $M = (E, \mathcal{C})$ and its associated isotropic system $S = (E, \mathcal{L} = X(\mathcal{K}) + Y(\mathcal{C}))$ one finds without difficulty that the reduction $M \setminus e$ has S_x^e as its associated system, and the contraction M/e the system S_y^e. Now it is an easy matter to transfer Theorem 3.5 of Section 3.2 to this general case.

Theorem 4.7 *Let $M = (E, \mathcal{C})$ be a binary matroid and $S = (E, X(\mathcal{K}) + Y(\mathcal{C}))$ its associated isotropic system. Then, for $\alpha \neq 0, \beta \neq 0$,*

$$Q(S, W_{\alpha, \beta, \gamma = 0}, \lambda) = \alpha^{\dim \mathcal{C}} \beta^{\dim \mathcal{K}} T_M(1 + \frac{\alpha}{\beta}\lambda, 1 + \frac{\beta}{\alpha}\lambda).$$

Also, Penrose's "binor" formula (3.19) holds in this case (see [28] for details):

(4.15) *If W, W' are two weightings with $W' = W + m$, then*

$$Q(S, W', -2) = Q(S, W, -2).$$

Finally, we look at the special weighting W with $\alpha = 0, \beta = 1, \gamma = -1$. By definition (4.13), the vector $A = (A_e) \in K'^E$ gives a non-zero contribution to $Q(S, W, \lambda)$ only when $A_e \in \{y, z\}$ for all $e \in E$. Setting $P = \{e \in E : A_e = z\}$, we have $A = Z(P) + Y(E \setminus P)$. By (4.14) this yields

$$\begin{aligned}
Q(S, W_{\alpha=0, \beta=1, \gamma=-1}, \lambda) &= \sum_{P \subseteq E} (-1)^{|P|} \lambda^{\dim(\mathcal{L} \cap Z(\widehat{P}) + Y(E \setminus P))} \\
&= \sum_{P \subseteq E} (-1)^{|P|} \lambda^{\dim \mathcal{B}_M(P)} = P_M(\lambda).
\end{aligned}$$

In summary, Theorem 4.7 (for the weighting $\alpha = 1, \beta = 2, \gamma = 0$) and (4.15) give

$$P_M(-2) = |\mathcal{K}| T_M(0, -3).$$

Now, it is known (see [11]) that

$$(-1)^{\dim \mathcal{C}} T_M(0, -3) = |\{(C_1, C_2) \in \mathcal{C}^2 : C_1 \cup C_2 = E\}|,$$

and so we obtain the precise generalization of Theorem 3.6.

Theorem 4.8 *Let $M = (E, \mathcal{C})$ be a binary matroid, $\mathcal{K} = \mathcal{C}^\perp$. Then we have*

$$P_M(-2) = (-1)^{\dim \mathcal{C}} |\mathcal{K}| \cdot \#\{(C_1, C_2) \in \mathcal{C}^2 : C_1 \cup C_2 = E\}.$$

In particular,

$$P_M(4) = 0 \iff P_M(-2) = 0. \tag{4.16}$$

Another way to arrive at Theorem 4.8 is the following; see [19], [44]. The *flow polynomial* of a binary matroid $M = (E, \mathcal{C})$ with $\mathcal{K} = \mathcal{C}^\perp$ is

$$F(\lambda) = (-1)^{\dim \mathcal{C}} T_M(0, 1 - \lambda);$$

see [11]. It counts for $\lambda = m$ the nowhere-zero m-flows of M. In particular, it is known that

$$F(2^k) = \#\{(C_1, \dots, C_k) \in \mathcal{C}^k : C_1 \cup \dots \cup C_k = E\}.$$

Using the definition of the Tutte polynomial via the rank function (see [11]) it can again be derived that

$$P_M(-2) = (-1)^{\dim \mathcal{C}} |\mathcal{K}| F_M(4),$$

which gives precisely Theorem 4.8.

Szegedy showed in [44] another nice connection between $P_M(\lambda)$ and $F_M(\lambda)$ where $M = (E, \mathcal{C}_G)$ is the cycle matroid of an arbitrary (not just plane) cubic graph. In this case we have

$$P_M(4) = 2^{|V|/2 - 1} F_M(4). \tag{4.17}$$

Considering our discussion in Section 4.1, in particular (4.7), the proof reduces to showing that $|\mathcal{K}(C_1, C_2)| = 2^{|V|/2 - 1}$ holds for all $C_1, C_2 \in \mathcal{C}$ with $C_1 \cup C_2 = E$.

Specializing once more to plane graphs G we find $F_G(\lambda) = \frac{\chi_{G^*}(\lambda)}{\lambda}$ where $\chi_{G^*}(\lambda)$ is the chromatic polynomial of the dual graph G^*. Thus the formulas read (with the factor λ in $P_G(\lambda)$):

$$P_G(-2) = (-1)^{|F|} 2^{|V| - 2} \chi_{G^*}(4), \tag{4.18}$$

and, for cubic plane graphs,

$$P_G(4) = 2^{|V|/2 - 1} \chi_{G^*}(4). \tag{4.19}$$

Formula (3.22) can also be directly generalized to binary matroids. Considering the weightings $W : \alpha = \beta = -1, \gamma = 0$ and $W' : \alpha' = \beta' = 0, \gamma' = 1$; as before, we find

$$Q(S, W, \lambda) = (-1)^{|E|} T_M(1 + \lambda, 1 + \lambda)$$

and

$$Q(S, W', \lambda) = \lambda^{\dim(\mathcal{L} \cap \widehat{Z})}.$$

Now, $X(K) + Y(C)$ is in $\widehat{Z}$ if and only if $K = C$ holds, which means $\mathcal{L} \cap \widehat{Z} = Z(C \cap K)$. With (4.15) we thus obtain the result

$$(-1)^{\dim(C \cap K)} |C \cap K| = (-1^{|E|} T_M(-1, -1),$$

and in particular,

$$C \cap K = \{0\} \iff |T_M(-1, -1)| = 1.$$

By similar arguments it is possible to define the Penrose polynomial and transition polynomial for arbitrary isotropic systems (see [28]). Furthermore, one can look at arbitrary finite fields $GF(q)$ instead of $GF(2)$. In [20] one finds a number interesting results for q-linear matroids.

4.4 The Hopf algebra and the Penrose polynomial

Let us return to the Penrose polynomial $P_G(\lambda)$ of a plane graph G. In sections 3.1 and 3.2 we have seen an interpretation of $P_G(\lambda)$ for positive integers $\lambda = k$ and for the single negative value $\lambda = -2$. What then, if anything, does $P_G(\lambda)$ count for arbitrary negative numbers?

Look again at the chromatic polynomial $\chi_G(\lambda)$ of an arbitrary graph G. Stanley showed in [43] the remarkable result that $|\chi_G(-1)|$ enumerates the number of *acyclic orientations* of G, that is, without directed cycles. He gave also an interpretation for arbitrary negative numbers $\lambda = -k$.

It was maybe Gian-Carlo Rota who first suggested the technique of Hopf algebras to tackle problems of this type. Let us first collect the basic concepts.

A *Hopf algebra* over $\mathbb{C}$ is a bialgebra $\mathcal{A}$ with two operations, the product $\mu_{\mathcal{A}} : \mathcal{A} \otimes \mathcal{A} \longrightarrow \mathcal{A}$, the coproduct $\Delta_{\mathcal{A}} : \mathcal{A} \longrightarrow \mathcal{A} \otimes \mathcal{A}$, and two unitary maps $\eta_{\mathcal{A}} : \mathbb{C} \longrightarrow \mathcal{A}$, $\varepsilon_{\mathcal{A}} : \mathcal{A} \longrightarrow \mathbb{C}$ with the usual compatibility conditions (see [1]). In addition, $\mathcal{A}$ possesses a so-called *antipodal* map $S_{\mathcal{A}} : \mathcal{A} \longrightarrow \mathcal{A}$ which satisfies

$$\mu_{\mathcal{A}} \circ (S_{\mathcal{A}} \otimes \mathrm{id}_{\mathcal{A}}) \circ \Delta_{\mathcal{A}} = \eta_{\mathcal{A}} \circ \varepsilon_{\mathcal{A}} = \mu_{\mathcal{A}} \circ (\mathrm{id}_{\mathcal{A}} \otimes S_{\mathcal{A}}) \circ \Delta_{\mathcal{A}}.$$

The standard example of a Hopf algebra is the polynomial algebra $\mathbb{C}[x]$ with

$$\mu: \quad x^i \otimes x^j = x^{i+j}$$

$$\Delta: \quad x^n \longrightarrow \sum_{k=0}^{n} \binom{n}{k} x^k \otimes x^{n-k}$$

$$\eta: \quad c \longrightarrow c \cdot 1$$

$$\varepsilon: \quad x^n \longrightarrow \begin{cases} 1 & \text{for } n = 0 \\ 0 & \text{otherwise,} \end{cases}$$

and the antipode S: $f(x) \longrightarrow f(-x)$.

Given two Hopf algebras $\mathcal{A}$ and $\mathcal{B}$ over $\mathbb{C}$ with antipodes $S_\mathcal{A}$ and $S_\mathcal{B}$, the morphism $H : \mathcal{A} \longrightarrow \mathcal{B}$ is called a *Hopf map* if

$$S_\mathcal{B} \circ H = H \circ S_\mathcal{A}. \qquad (4.20)$$

The basic idea is now the following:

(i) define a Hopf algebra $\mathcal{A}$, freely generated by plane graphs G (or equivalently by 2-coloured medial graphs $\tilde{G}$);

(ii) interpret the antipode $S_\mathcal{A}$ in graph-theoretic terms;

(iii) show that the map $P : G \longrightarrow P_G(\lambda)$ which associates with every plane graph G its Penrose polynomial $P_G(\lambda)$ is a Hopf map from $\mathcal{A}$ into $\mathbb{C}[\lambda]$.

Then it follows from (4.20) that

$$P_G(-\lambda) = P_{S_\mathcal{A}(G)}(\lambda),$$

and we obtain a formula for negative integers $-k$ since we know what $P_{S_\mathcal{A}(G)}(\lambda)$ counts for $\lambda = k$, Theorem 3.1. A very interesting construction of such a Hopf algebra $\mathcal{A}$ which satisfies (i) to (iii) was described in [36]. For the details the reader is referred to [36]. We discuss just one main result which is particularly relevant in our situation.

Let $G = (V, E, F)$ be a plane graph and $\tilde{G} = (E, L)$ its 2-coloured medial graph. A partition $\pi = A_1|A_2|\ldots|A_{s(\pi)}$ of the edge-set L of $\tilde{G}$ is called a *Penrose partition* if all subgraphs $\tilde{G}|A_i$ are Eulerian and if for vertices in $\tilde{G}|A_i$ of degree 2 the two incident edges are both white or crossing. Let $\Pi(\tilde{G})$ be the set of Penrose partitions.

For a Penrose partition π of L we call $f : L \longrightarrow \{1, \ldots, k\}$ a *k-valuation* of π if for all i the map f induces in $\tilde{G}|A_i$ on the vertices of degree 4 a k-valuation in the sense of section 3.1, and assumes the same value on the two incident edges of vertices of degree 2.

Theorem 4.9 *For a plane graph G we have*

$$P_G(-k) = \sum_{\pi \in \Pi(\tilde{G})} (-1)^{s(\pi)}(s(\pi))! \cdot (\#k - valuations of \pi).$$

As an example consider $k = 1$. Then all vertices in $\widetilde{G}|A_i$ have degree 2, and we obtain

$$P_G(-1) = \sum_{A \subseteq E} (-1)^{c(A)} (c(A))!,$$

where A runs through all edge-sets whose trails are *circuits*.

The technique of a Hopf algebra was applied to other combinatorial questions, see, for example, [13,37,38,39].

In closing we mention two interesting problems. First we know that 1 is a root of any polynomial $P_M(\lambda)$. Is it true that 1 is always a *single* root, provided that M is connected? Note that this holds for the chromatic polynomial when the graph is 2-connected.

The second problem concerns the evaluation of $P_M(\lambda)$ at $\lambda = -1$. It is true that $P_M(-1) \neq 0$ for all binary matroids M with non-vanishing Penrose polynomial? In particular, does it hold for plane bridgeless graphs?

In connection with $P_M(-1)$ Szegedy proposed another attractive conjecture. Let G be a cubic plane connected graph. We have already seen that the number $F_G(4)$ of 3-edge colourings divides $P_G(3)$, $P_G(4)$ and $P_G(-2)$, and the available data led Szegedy to suggest that $F_G(4)$ also divides $P_G(-1)$. Together with $P_G(-1) \neq 0$ this would provide yet another proof of the 4-colour theorem.

5 The 4-colour theorem revisited

We already know that the Penrose polynomial $P_G(\lambda) = \sum a_i \lambda^i$ of a bridgeless plane graph $G = (V, E, F)$ has degree $|F|$. What about the coefficients a_i? Do they have a combinatorial meaning?

Consider again the related chromatic polynomial $\chi_G(\lambda) = \sum b_i \lambda^i$ of an arbitrary (as always connected) graph $G = (V, E)$. From the recursion

$$\chi_G(\lambda) = \chi_{G \backslash e}(\lambda) - \chi_{G/e}(\lambda)$$

we infer immediately that $\chi_G(\lambda)$ has degree $|V|$, and further that the coefficients $b_n, b_{n-1}, \ldots, b_1$ form an *alternating* sequence, that is,

$$b_n = 1, b_{n-1} < 0, b_{n-2} > 0, b_{n-3} < 0, \ldots, (-1)^{n-1} b_1 > 0.$$

Hence when $G = (V, E, F)$ is plane then the corresponding result holds for $\chi_{G^*}(\lambda)$: the polynomial $\chi_{G^*}(\lambda)$ has degree $|F|$, and the coefficients alternate in sign.

Now let us look at $P_G(\lambda) = \sum a_i \lambda^i$. We have $a_0 = 0$ since 0 is a root of $P_G(\lambda)$. Suppose $G = (V, E, F)$ possesses two faces R_1, R_2 with two common boundaries e_1, e_2. Then it is easily verified (see [4]) that $P_G(\lambda) = 2P_{G/e_1}(\lambda)$ holds. Hence we may assume that in G any two faces have at most one common boundary (equivalently: G^* has no multiple edges). Let us call them *normal* graphs. For bridgeless plane normal graphs $G = (V, E, F)$ the following was shown in [4]:

$$a_r = 1 \quad (r = |F|); \tag{5.1}$$

$$a_{r-1} = -|E| + k_4^*, \text{ where } k_4^* = \#subgraphs \ K_4 \text{ in } G^*. \tag{5.2}$$

Now, clearly, $a_{r-1} < 0$, and we also have $a_{r-2} > 0$. What's more, all examples studied possess the property "alternating coefficients". So why should it be so much more difficult to verify this property for all Penrose polynomials $P_G(\lambda)$ when the analogous result for $\chi_{G^*}(\lambda)$ is almost trivial? Here is the reason as was first noted in [4].

Theorem 5.1 *If for a plane graph $G = (V, E, F)$ the Penrose polynomial $P_G(\lambda)$ has alternating coefficients, then G is 4-face colourable.*

The proof follows immediately from Theorem 3.6. Indeed, if $P_G(\lambda)$ has alternating coefficients $a_r > 0, a_{r-1} < 0, a_{r-2} > 0, \dots$ where $r = |F|$, then, for positive λ,

$$(-1)^r P_G(-\lambda) = (-1)^r \sum_{i=0}^{r} (-1)^i a_i \lambda^i = \sum_{i=0}^{r} (-1)^{r-i} a_i \lambda^i > 0,$$

since $(-1)^{r-i} a_i > 0$ for all $i \geq 1$. In particular, this implies $P_G(-2) \neq 0$, whence G is 4-face colourable by Theorem 3.6.

So, the property "alternating coefficients" implies the 4-colour theorem, but it is not all clear whether these are equivalent conditions. The alternating property has been shown to hold (see [4,28]) for

• Eulerian graphs

• series-parallel graphs and hence, in particular, for

• outerplanar graphs.

Proceeding to arbitrary graphs and binary matroids, which are always considered without bridges, the property "alternating" need, of course, not hold anymore. The smallest example is the dual Fano matroid F^* where, as we have seen in Section 4.2, the Penrose polynomial is 0. There are two further minor-minimal examples: the *Petersen Graph* Pet (which is not planar) and the dual matroid K_5^* of the complete graph K_5. For the Penrose polynomials we compute (without the factor λ):

$$
\begin{aligned}
P_{\text{Pet}}(\lambda) &= \lambda^6 - 15\lambda^5 + 94\lambda^4 - 200\lambda^3 - 216\lambda^2 + 1040\lambda - 704 \\
&= (\lambda - 1)(\lambda - 2)(\lambda - 4)(\lambda + 2)(\lambda^2 - 10\lambda + 44); \\
P_{K_5^*}(\lambda) &= \lambda^4 - 5\lambda^3 + 0 \cdot \lambda^2 + 20\lambda - 16 \\
&= (\lambda - 1)(\lambda - 2)(\lambda - 4)(\lambda + 2).
\end{aligned}
$$

The three matroids F^*, Pet and K_5^* now lead directly to one of the most famous conjectures due to Tutte which reads in our terminology:

(5.3)   *Let $M = (E, C)$ be a bridgeless binary matroid. If $P_M(4) = 0$, then M contains F^*, Pet or K_5^* as minor.*

Tutte presented his conjecture in [47] in geometric terms (his so-called "tangential 2-block conjecture"). The truth of (5.3) would, in particular, imply for plane graphs the 4-colour theorem (since K_5 is not planar) and also the "snark" conjecture, recently shown by Robertson, Seymour and Thomas with massive computer calculations: *if a 3-regular bridgeless graph G has no 3-edge colouring, then G contains the Peterson graph as minor.*

The most important progress towards a proof of (5.3) (however, using the 4-colour theorem) is due to Seymour [41]: the only possible further obstructions in (5.3), apart from F^*, Pet, K_5, are all *graphic*, that is, matroids $M = (E, C_G)$, with G a graph.

So let us end with a question whose affirmative answer would imply all conjectures above, and in particular, the 4-colour theorem:

Are F^*, Pet and K_5^* the only bridgeless minor-minimal binary matroids whose Penrose polynomial does not have alternating coefficients?

References

[1] E. Abe, *Hopf Algebras*, Cambridge University Press, Cambridge, 1977.

[2] M. Aigner, *Combinatorial Theory*, Springer 1979 (reprinted in Classics of Mathematics, 1997).

[3] M. Aigner and J.J. Seidel, Knoten, Spin Modelle und Graphen, *Jahresber. Deutsch. Math.-Verein.* **97** (1995), 75–96.

[4] M. Aigner, The Penrose polynomial of a plane graph, *Math. Ann.* **307** (1997), 173–189.

[5] M. Aigner and H. Mielke, The Penrose polynomial of a binary matroid, *Monatsh. Math.*, **131** (2000), 1–13.

[6] K. Appel and W. Haken, Every planar map is four–colorable, *Bull. Amer. Math. Soc.* **82** (1976), 711–712.

[7] R. Arratia, B. Bollobás, G. Sorkin, The interlace polynomial: a new graph polynomial, preprint.

[8] A. Bouchet, Isotropic Systems, *European J. Combin.* **8** (1987), 231–244.

[9] A. Bouchet, Tutte–Martin polynomials and orienting vectors of isotropic systems, *Graphs Combin.* **7** (1991), 235–252.

[10] A. Bouchet, Multimatroids. I. Coverings by independent sets, *SIAM J. Discrete Math.* **10** (1997), 626–646.

[11] T. Brylawski and J.G. Oxley, The Tutte polynomial and its applications, *Matroid Applications,* Cambridge University Press, Cambridge, 1992, 123–225.

[12] H. Crapo, The Tutte polynomial, *Aequationes Math.* **3** (1969), 211–229.

[13] J. Ellis–Monaghan, New results for the Martin polynomial, *J. Combin. Theory Ser. B* **74** (1998), 326–352.

[14] C.F. Gauss, *Gesammelte Werke VIII* (pp. 272, 282–286), Teubner, Leipzig, 1900.

[15] M.C. Golumbic, *Algorithmic Graph Theory and Perfect Graphs,* Academic Press, New York, 1980.

[16] F. Jaeger, On the Penrose number of cubic diagrams, *Discrete Math.* **74** (1988), 85–97.

[17] F. Jaeger, On Tutte polynomials and link polynomials, *Proc. Amer. Math. Soc.* **103** (1988), 647–654.

[18] F. Jaeger, On Tutte polynomials and cycles of planar graphs, *J. Combin. Theory Ser. B* **44** (1988), 127–146.

[19] F. Jaeger, On edge–colorings of cubic graphs and a formula of Roger Penrose, *Ann. Discrete Math.* **41** (1989), 267–280.

[20] F. Jaeger, On Tutte polynomials of matroids representable over $GF(q)$. *European J. Combin.* **10** (1989), 247–255.

[21] F. Jaeger, On transition polynomials of 4-regular graphs, *Cycles and Rays,* Kluwer, 1990, 123–150.

[22] V. Jones, A polynomial invariant for knots via von Neumann algebras, *Bull. Amer. Math. Soc.* **12** (1985), 103–111.

[23] L. Kauffman, State models and the Jones polynomial, *Topology* **26** (1987), 297–309.

[24] M. Las Vergnas, Le polynôme de Martin d'un graph Eulérien, *Ann. Discrete Math.* **17** (1983), 397–411.

[25] W. Lickorish, Polynomials for links, *Bull. London Math. Soc.* **20** (1988), 558–588.

[26] L. Lovász and M. L. Marx, A forbidden subgraph characterization of Gauss codes, *Bull. Amer. Math. Soc.* **82** (1976), 121–122.

[27] P. Martin, *Enumeration euléréenne dans les multigraphes et invariants de Tutte-Grothendieck.* Thesis, Univ. Grenoble 1977.

[28] H. Mielke, *The Penrose Polynomial*, Diss., Freie Universität Berlin.

[29] J.G. Oxley, *Matroid Theory*. Oxford University Press, Oxford, 1992.

[30] R. Penrose, Applications of negative dimensional tensors. *Combinatorial Mathematics and its Applications*, Academic Press, 1971, 221-244.

[31] N. Robertson, D.P. Sanders, P.D. Seymour and R. Thomas, The four-colour theorem. *J. Combin. Theory Ser. B* **70** (1997), 2-44.

[32] P. Rosenstiehl, Bicycles et diagonales des graphes planaires, *Cahiers Centre Études Rech. Opér.* **17** (1975), 365-383.

[33] P. Rosenstiehl, Solution algébrique du problème de Gauss sur la permutation des points d' intersection d'une ou plusieurs courbes fermées du plan, *C. R. Acad. Sci. Paris Sér. A* **283** (1976), 551-553.

[34] P. Rosenstiehl, A new proof of the Gauss interlace conjecture, *Adv. in Appl. Math.* **23** (1999), 3-13.

[35] P. Rosenstiehl and R. Reed, On the principal edge tripartition of a graph, *Ann. Discrete Math.* **3** (1978), 195-226.

[36] I. Sarmiento, Hopf algebras and the Penrose polynomial, *European J. Combin.*, to appear.

[37] W. Schmitt, Antipodes and incidence coalgebras, *J. Combin. Theory Ser. A* **46** (1987), 264-290.

[38] W. Schmitt, Incidence Hopf algebras, *J. Pure Appl. Algebra* **96** (1994), 299-330.

[39] W. Schmitt, Hopf algebra methods in graph theory, *J. Pure Appl. Algebra* **101** (1995), 77-90.

[40] W. Schwärzler and D.J.A. Welsh, Knots, matroids and the Ising model, *Math. Proc. Cambridge Philos. Soc.* **113** (1993), 107-139.

[41] P.D. Seymour, On Tutte's extension of the four-colour problem, *J. Combin. Theory Ser. B* **31** (1981), 82-94.

[42] H. Shank, The theory of left-right paths, *Combinatorial Mathematics III*, Lecture Notes in Math. **452**, Springer, Berlin, 1975, 42-54.

[43] R.P. Stanley, Acyclic orientations of graphs, *Discrete Math.* **5** (1973), 171-178.

[44] C. Szegedy, On the number of 3-edge colorings of cubic graphs, preprint.

[45] W. Tutte, A ring in graph theory, *Proc. Cambridge. Philos. Soc.* **43** (1947), 26–40.

[46] W. Tutte, A contribution to the theory of chromatic polynomials, *Canad. J. Math.* **6** (1954), 80–91.

[47] W. Tutte, On the algebraic theory of graph colorings., *J. Combin. Theory* **1** (1966), 15–50.

[48] O. Veblen, An application of modular equations in Analysis Situs, *Ann. Math.* **14** (1912), 86–94.

[49] D.J.A. Welsh, *Complexity: Knots, Colourings and Counting,* London Math. Soc. Lecture Note Series **186**, Cambridge University Press, Cambridge, 1993.

[50] H. Whitney, A set of topological invariants for graphs, *Amer. J. Math.* **55** (1933), 231–235.

II. Mathematisches Institut
Freie Universität Berlin
D-14195 Berlin
Germany
aigner@math.fu-berlin.de

Some cyclic and 1-rotational designs

Ian Anderson

Abstract

A variety of combinatorial designs possessing 'cyclic' structure over the integers are discussed. In particular, details of recent developments in the study of cyclically resolvable cyclic Steiner 2-designs and $\mathbf{Z}$-cyclic triplewhist tournaments are presented.

1 Introduction

In this paper we discuss certain designs which possess cyclic structure. We make no attempt to be exhaustive, but choose topics of particular interest to us. Further details of developments in these and other areas can be found in the definitive *CRC Handbook of Combinatorial Designs* [40] and at the website http://www.emba.uvm.edu/~dinitz/hcd.html where corrections and updates can be found.

Before making any formal definitions we present a few illustrative examples.

Examples *The seven point plane.* Take the point set as $\{1,\dots,7\}$ and the lines as the 3-element sets $\{1,2,4\}$, $\{2,3,5\}$, $\{3,4,6\}$, $\{4,5,7\}$, $\{5,6,1\}$, $\{6,7,2\}$, $\{7,1,3\}$. Then any two distinct points are incident with precisely one line, and any two lines intersect in a single point. This finite projective plane is cyclic in the sense that each line is a translate of the first, being of the form $\{1+i, 2+i, 4+i\}$, arithmetic being performed modulo 7.

Examples *One-factorisation of K_6.* The sets

$$
\begin{array}{lll}
\{\infty, 0\} & \{1,4\} & \{2,3\} \\
\{\infty, 1\} & \{2,0\} & \{3,4\} \\
\{\infty, 2\} & \{3,1\} & \{4,0\} \\
\{\infty, 3\} & \{4,2\} & \{0,1\} \\
\{\infty, 4\} & \{0,3\} & \{1,2\}
\end{array}
$$

are precisely all of the 2-element subsets of $\{\infty, 0, \dots, 4\}$, arranged so that each row contains each element exactly once. The design could be used to schedule the games of a round-robin tournament involving six teams. Again, there is cyclic structure; each row is obtained by adding $1 \pmod 5$ to each entry of the previous row, leaving ∞ unaltered.

47

Examples   Z-*cyclic whist tournament* $Wh(8)$. The games

$$
\begin{array}{ll}
\infty, 0 \; v \; 4,5 & 1,3 \; v \; 2,6 \\
\infty, 1 \; v \; 5,6 & 2,4 \; v \; 3,0 \\
\infty, 2 \; v \; 6,0 & 3,5 \; v \; 4,1 \\
\infty, 3 \; v \; 0,1 & 4,6 \; v \; 5,2 \\
\infty, 4 \; v \; 1,2 & 5,0 \; v \; 6,3 \\
\infty, 5 \; v \; 2,3 & 6,1 \; v \; 0,4 \\
\infty, 6 \; v \; 3,4 & 0,2 \; v \; 1,5
\end{array}
$$

have the property that any two members of $\{\infty, 0, 1, \dots, 6\}$ partner each other once and oppose each other twice. The games are arranged in seven rounds, with each player involved in one game in each round. Again there is cyclic structure; each round is obtained from the previous one by adding 1 modulo 7.

We now make some formal definitions.

Definition   A (v, k, λ) *design* is a collection of k-element subsets (called *blocks*) of a v-element set $\mathcal{V}$ such that each pair of elements of $\mathcal{V}$ occur together in exactly λ of the blocks.

It is straightforward to show that each element then occurs in the same number r of blocks and that there are b blocks where

$$
bk = vr \quad \text{and} \quad \lambda(v - 1) = r(k - 1).
$$

For example, the seven point plane of Example 1.1 is a $(7, 3, 1)$ design with $b = 7$ and $r = 3$.

Definition   A $(v, k, 1)$ design is called a *Steiner 2-design*; in particular, a $(v, 3, 1)$ design is called a *Steiner triple system of order* v and is denoted by $STS(v)$.

Any $STS(v)$ has $b = \dfrac{1}{6} v(v - 1)$ blocks, and can exist only if $v \equiv 1$ or $3 \pmod 6$. It was shown by Kirkman [56] in 1847 that an $STS(v)$ exists for all such v.

Definition   A *finite projective plane* (FPP) *of order* n is a $(n^2 + n + 1, n + 1, 1)$ design.

Such designs exist for all prime power values of n; none is known to exist for any other values of n.

Definition Let $\mathcal{V}, \mathcal{B}$ denote the sets of elements, blocks of a (v, k, λ) design, and let σ be a permutation on $\mathcal{V}$. When $B = \{b_1, \ldots, b_k\}$ is a block, define $B^\sigma = \{b_1^\sigma, \ldots, b_k^\sigma\}$. If $\mathcal{B}^\sigma = \{B^\sigma : B \in \mathcal{B}\} = \mathcal{B}$, then σ is called an *automorphism* of the design, and if there exists an automorphism σ of order $v = |\mathcal{V}|$, the design is said to be *cyclic*. In this case we can identify $\mathcal{V}$ with $\mathbb{Z}_v$, the set of integers modulo v, and we can take $\sigma : i \mapsto i + 1 \, (\bmod \, v)$.

If $B = \{b_1, \ldots, b_k\}$ is a block in a cyclic design, each $B + i = \{b_1 + i, \ldots, b_k + i\}$ is called a translate of B, and the set of distinct translates of B is called the *orbit* of B. If the orbit of B has v distinct blocks it is called a *full* orbit; otherwise it is *short*. If $k|v$ and $B = \left\{0, \frac{v}{k}, 2\frac{v}{k}, \ldots, (k-1)\frac{v}{k}\right\}$, then the orbit of B is called the *regular short orbit*. It can be shown that a cyclic $(v, k, 1)$ design can exist only if $v \equiv 1$ or $k \bmod k(k-1)$, and that if $v \equiv 1 \bmod k(k-1)$ all orbits are full, whereas if $v \equiv k \bmod k(k - 1)$ one orbit is the regular short orbit and the remaining orbits are full.

Definition A (v, k, λ) design which has an automorphism of order $v - 1$ with one fixed point, is called 1-*rotational*.

A 1-rotational $(v, k, 1)$ design can exist only when $v \equiv k \bmod k(k - 1)$. The $(6, 2, 1)$ design in Example 1.2 is 1-rotational, as is the underlying $(8, 4, 3)$ design of Example 1.3. It is common to use ∞ to denote the element fixed by σ; the remaining elements can be taken to be the elements of $\mathbb{Z}_{v-1}$.

Definition A design is *resolvable* if the collection $\mathcal{B}$ of blocks can be partitioned into classes $\mathcal{R}_1, \ldots, \mathcal{R}_r$ so that every element of $\mathcal{V}$ is in precisely one block of each class. The classes $\mathcal{R}_i$ are called the *resolution classes*.

Clearly, a (v, k, λ) design can be resolvable only if $k|v$. A resolvable STS(v) is called a *Kirkman triple system* of order v, and is denoted by KTS(v); such a design can exist only if $v \equiv 3 \, (\bmod \, 6)$, and it is a consequence of the work of Lu [62] and Ray-Chaudhuri and Wilson [73] that such a design exists for all such v.

If $\mathcal{R}_i$ is a resolution class, let $\mathcal{R}_i^\sigma = \{B^\sigma : B \in \mathcal{R}_i\}$.

Definition A resolvable (v, k, λ) design is *cyclically resolvable* if it possesses a nontrivial automorphism σ of order v which preserves the resolution, that is, such that $\{\mathcal{R}_i^\sigma\} = \{\mathcal{R}_i\}$.

We shall discuss cyclically resolvable cyclic $(v, k, 1)$ designs in section 5; these can exist only when $v \equiv k \bmod k(k - 1)$.

Cyclic designs can be obtained from difference sets or difference families.

Definition A *(cyclic)* $(v, k, 1)$ *difference set* is a subset $\{d_1, \ldots, d_k\}$ of $\{0, \ldots, v - 1\}$ such that the differences $\pm(d_i - d_j)$, $i \neq j$, give each nonzero element of $\mathbb{Z}_v$ once.

Examples $\{1,2,4\}$ is a $(7,3,1)$ difference set in $\mathbb{Z}_7$, since the differences are ±1, ±2, $\pm3 \pmod 7$.

The translates of a $(v,k,1)$ difference set form the blocks of a $(v,k,1)$ design with $b = v$. Designs with $b = v$ are called *symmetric*.

Definition Let $D_i = \{d_{i1},\dots,d_{ik}\}$, $1 \le i \le s$, be subsets of $\{0,\dots,v-1\}$ such that the differences $\pm(d_{i\ell} - d_{ik})$ altogether give each nonzero element of $\mathbb{Z}_v$ once. Then $D_1,\dots,D_s$ are said to form a $(v,k,1)$ *(cyclic) difference family* (abbreviated to $(v,k,1)$DF).

The translates of the D_i then form a $(v,k,1)$ design, of which the D_i are called the *base blocks*.

Examples $\{1,2,5\}$ and $\{1,3,9\}$ give differences ±1, ±3, ±4; ±2, ±6, $\pm5 \pmod{13}$.

Since counting differences gives $sk(k-1) = v-1$, difference families can exist only when $v \equiv 1 \ \mathrm{mod}\, k(k-1)$. For $v \equiv k \ \mathrm{mod}\, k(k-1)$ we have the following.

Definition Suppose that $k|v$. Then subsets $E_i = \{e_{i1},\dots,e_{ik}\}$, $1 \le i \le s$, of $\{0,\dots,v-1\}$ whose differences give each element of $\mathbb{Z}_v - \left\{0, \dfrac{v}{k},\dots,(k-1)\dfrac{v}{k}\right\}$ exactly once, are said to form a *partial difference family*.

The translates of the E_i, along with the $\dfrac{v}{k}$ distinct translates of $\left\{0, \dfrac{v}{k},\dots,\right.$ $\left.(k-1)\dfrac{v}{k}\right\}$, then form a cyclic $(v,k,1)$ design.

Examples The blocks $\{0,1,4\}$ and $\{0,2,8\}$ give differences ±1, ±3, ±4; ±2, ±6, ±7 in $\mathbb{Z}_{15}$, that is, each element of $\mathbb{Z}_{15} - \{0,5,10\}$ exactly once. So the translates of $\{0,1,4\}$ and $\{0,2,8\}$, along with the five distinct translates of $\{0,5,10\}$, form a cyclic STS(15).

Constructing $(v,k,1)$ difference families is a major research area. By the following theorem of Wilson [76] which has recently been improved by Chen and Zhu [38], it is known that cyclic $(p,k,1)$ designs must exist for all sufficiently large primes $p \equiv 1 \ \mathrm{mod}\, k(k-1)$.

Theorem 1.1 [76] *If $p \equiv 1 \ \mathrm{mod}\, k(k-1)$ is prime, and $p > \left(\dfrac{k(k-1)}{2}\right)^{k(k-1)}$, then a $(p,k,1)$DF exists.*

For 1-rotational $(v,k,1)$ designs we require $v \equiv k \ \mathrm{mod}\, k(k-1)$. Such designs can be constructed from 1-rotational difference families.

Definition A 1-*rotational* $(v, k, 1)$ *difference family* is a set of k-subsets D_i of $\mathbb{Z}_{v-1}$ such that each element of $\mathbb{Z}_{v-1} - \left\{0, \dfrac{v-1}{k-1}, \dots, (k-2)\dfrac{(v-1)}{k-1}\right\}$ arises exactly once as a difference between two elements of some D_i.

The translates of the blocks D_i and of $\left\{\infty, 0, \dfrac{v-1}{k-1}, \dots, (k-2)\dfrac{v-1}{k-1}\right\}$ then form a 1-rotational $(v, k, 1)$ design.

Examples $\{1, 2, 4, 8\}$ gives differences ± 1, $\pm 2, \pm 3$, ± 4, ± 6, ± 7 (mod 15), that is, all elements of $\mathbb{Z}_{15} - \{0, 5, 10\}$ once. So the translates of $\{1, 2, 4, 8\}$ and $\{\infty, 0, 5, 10\}$, form a 1-rotational $\{16, 4, 1\}$ design.

For recursive constructions, the concept of a difference matrix is useful.

Definition A $(v, k, 1)$ *difference matrix* is a $k \times v$ matrix $A = (a_{ij})$ with each $a_{ij} \in \mathbb{Z}_v$ such that, for $r \neq s$, the differences $a_{rj} - a_{sj}$, $1 \leq j \leq v$, give all the elements of $\mathbb{Z}_v$ once.

In the case of $k = 4$, it is well known that the $4 \times v$ matrix with rows $R = (0, 1, \dots, v-1)$, $-R$, $2R$ and $-2R$ is a $(v, 4, 1)$ difference matrix provided g.c.d.$(v, 6) = 1$. Another useful difference matrix is given as follows.

Lemma 1.2 *If* g.c.d.$(v, (k-1)!) = 1$, *then the matrix* $D = (d_{ij})$, $d_{ij} \equiv ij \pmod{v}$, $0 \leq i \leq k-1$, $0 \leq j \leq v-1$, *is a* $(v, k, 1)$ *difference matrix*.

The use of difference matrices is illustrated in the following theorem.

Theorem 1.3 [52]

(i) *Let* $v \equiv 1 \mod k(k-1)$. *Suppose that a cyclic* $(v, k, 1)$ *design, a cyclic* $(u, k, 1)$ *design and a* $(u, k, 1)$ *difference matrix exist. Then a cyclic* $(uv, k, 1)$ *design exists.*

(ii) *Let* $v \equiv k \mod k(k-1)$. *Suppose that cyclic* $(v, k, 1)$ *and* $(ku, k, 1)$ *designs and a* $(u, k, 1)$ *difference matrix exist. Then a cyclic* $(uv, k, 1)$ *design exists.*

Proof of (i), when $u \equiv 1 \mod k(k-1)$. Let $D = (d_{ij})$ be the $(u, k, 1)$ difference matrix. For each base block $B = \{b_1, \dots, b_k\}$ of the cyclic $(v, k, 1)$ design, construct u blocks $B_i = \{b_1 + d_{1i}v, \dots, b_k + d_{ki}v\}$, $1 \leq i \leq u$, and for each base block $C = \{c_1, \dots, c_k\}$ of the $(u, k, 1)$ design form the block $\{vc_1, \dots, vc_k\}$. These blocks are the base blocks of a cyclic $(uv, k, 1)$ design. □

See Colbourn and Colbourn [39], Grannell and Griggs [48], Jimbo [51], Jimbo and Kuriki [52] and Jungnickel [54] for a variety of composition theorems of this form. These constructions are in fact all encompassed by a recent very general construction of Buratti [29].

For some applications we require our difference matrices to have no repetitions in any row. A $(v, k, 1)$ difference matrix is called *good* (or *homogeneous*) if each row contains each element of $\mathbb{Z}_v$ once. The $(v, 4, 1)$ difference matrix described above, with rows $\pm R$ and $\pm 2R$, is good. The following result is easily established.

Lemma 1.4 *A good $(v, k, 1)$ difference matrix exists if and only if a $(v, k + 1, 1)$ difference matrix exists.*

2 Early results

2.1 Steiner triple systems

The first block designs to receive much attention were the Steiner triple systems. It was Anstice [17, 18] who produced the first infinite family of cyclic STS.

Theorem 2.1 *Let p be a prime, $p \equiv 1 \pmod 6$; put $p = 3.2^{s+1}n + 1$ where n is odd. Let θ be a primitive root of p and let $\omega = \theta^{2^{s+1}n}$ (so that $\omega^3 = 1$). Then the sets*

$$\{1, \omega, \omega^2\} \times \theta^{2^{s+1}i+j}, \quad 0 \leq i \leq n - 1, \quad 0 \leq j \leq 2^s - 1$$

form a $(p, 3, 1)\,\mathrm{DF}$ whose translates form a cyclic $\mathrm{STS}(p)$.

Examples Take $p = 19$ and $\theta = 2$, so that $\omega = 7$. The sets $\{1, 7, 11\}$, $\{4, 9, 6\}$, $\{16, 17, 5\}$ form a $(19, 3, 1)\mathrm{DF}$ whose translates form a cyclic $\mathrm{STS}(19)$.

Anstice's method is a remarkably early use of primitive roots in the construction of combinatorial designs (see [6] for further comments on his papers). Later, Netto [69] gave another construction, taking sets $\{0, 1, \theta^{2^s n}\} \times \theta^i$, $0 \leq i \leq 2^s n - 1$. Then Bose, in his major contribution [21] to the development of design theory, gave a construction very similar to Anstice's, which was clearly unknown to him; he took the same set $\{1, \omega, \omega^2\}$, but different multiples of it. Bose also provided similar constructions for cyclic $(v, 4, 1)$ and $(v, 5, 1)$ designs, and we shall return to these in Section 3.

It is only in the case $k = 3$ that it has been determined precisely for which values of v a cyclic $(v, k, 1)$ design exists. The case $k = 3$ was dealt with by Peltesohn [71].

Theorem 2.2 *A cyclic $\mathrm{STS}(v)$ exists for all $v \equiv 1$ or $3 \pmod 6$, $v \geq 7$, $v \neq 9$.*

Peltesohn's proof involved considering the cases $v \equiv 1, 3, 7, 9, 13, 15 \pmod{18}$ separately. Heffter had observed that if $\{1, \ldots, 3n\}$ can be partitioned into triples of the form $\{a, b, \pm(a+b)\}$ in $\mathbb{Z}_{6n+1}$, then the sets $\{0, a, a+b\}$ will form a $(6n+1, 3, 1)$DF. When $v = 6n + 3$ it is the set $\{1, \ldots, 2n, 2n+2, \ldots, 3n+1\}$ that has to be partitioned into triples, so as to produce a partial difference family. Peltesohn provided a complete solution to Heffter's problem, thereby establishing the existence of cyclic STS in all possible cases except $v = 9$. As an illustration, here is her solution to Heffter's problem with $v = 18s + 1$, $s \geq 2$.

$$\{3r + 1, \; 4s - r + 1, \; 4s + 2r + 2\}, \quad 0 \leq r \leq s - 1$$
$$\{3r + 2, \; 8s - r, \; 8s + 2r + 2\}, \quad\;\; 0 \leq r \leq s - 1$$
$$\{3r + 3, \; 6s - 2r - 1, \; 6s + r + 2\}, \quad 0 \leq r \leq s - 2$$
$$\{3s, \; 3s + 1, \; 6s + 1\}.$$

2.2 Finite projective planes

Kirkman [57] showed in 1850 that a finite projective plane of order p exists for all primes p. A few years later, in a remarkable paper [58] which essentially presented the idea of a difference set, he obtained cyclic FPPs of orders $2, 3, 4, 5$ and 8. He remarked that it was a surprise to obtain such designs of orders 4 and 8, which are not prime, and he asserted that it was improbable that a FPP of order 6 exists.

It was not until 1938 that the existence of cyclic FPPs of all prime power orders was established by Singer.

Theorem 2.3 [74] *A cyclic FPP of order n exists whenever $n = p^\alpha$, p prime.*

The standard proof of this result involves a consideration of the projective geometry $PG(2, n)$; an alternative approach is presented by Bose and Chowla [23].

2.3 Whist tournaments

The first whist tournaments to be published were due to Mitchell, Safford and Whitfeld; they appeared in early issues of the journal *Whist*, in Mitchell's *Duplicate Whist* [67], and in Moore's paper [68]. Most of these designs were 1-rotational. See my recent article [9] on the methods used in their construction, and [5] for a history of the study of whist tournaments.

In his important 1896 paper [68] Moore constructed 1-rotational whist tournaments $Wh(3p + 1)$ for all primes $p \equiv 1 \pmod{4}$. We note in passing that Moore's paper included many remarkable early results, such as the existence of a complete set of mutually orthogonal latin squares of prime power order, but has nevertheless been almost ignored by the mathematical community. (Can you, for example, find any reference to it in the account of Moore's mathematical work in [70]?) Apart from a family of cyclic whist tournaments due to Watson [75], to which we shall refer in Section 7, few further examples of cyclic or 1-rotational whist tournaments were known when Baker and Wilson established the existence of $Wh(v)$ for all $v \equiv 0$ or $1 \pmod 4$. (See [20] and [8].) More recently there has been renewed interest in cyclic and 1-rotational $Wh(v)$, and we shall consider specifically cyclic and 1-rotational *triplewhist* tournaments in later sections. Note that cyclic and 1-rotational $Wh(v)$ together are usually called $\mathbb{Z}$-cyclic $Wh(v)$ in the literature.

3   Radical difference families

Let k be an odd integer, and let $p \equiv 1 \bmod k(k-1)$ be prime. Let H denote the multiplicative group of $\mathbb{Z}_p - \{0\}$. Then a $(p, k, 1)$DF is a *radical* difference family if each of its blocks is a coset of the set of kth roots of unity in H.

Thus, for example, Anstice's difference family in Theorem 2.1 is radical, ω being a cube root of unity. Radical difference families were also used by Bose in his construction of cyclic $(p, 5, 1)$ designs. For a prime $p = 20t + 1$ with primitive root θ, the blocks $\{1, \theta^{4t}, \theta^{8t}, \theta^{12t}, \theta^{16t}\} \times \theta^{2i}$, $0 \le i \le t - 1$, form a difference family provided $\theta^{4t} + 1$ is an odd power of θ.

When k is even, a difference family is radical if each block is a coset of the set consisting of the $(k - 1)$th roots of unity and 0. For example, when $k = 4$, Bose constructed cyclic $(p, 4, 1)$ designs for $p = 20t + 1$ by considering $\{0, 1, \theta^{4t}, \theta^{8t}\} \times \theta^{2i}$, $0 \le i \le t - 1$. These form a radical DF provided $\theta^{4t} - 1$ is an odd power of θ.

Recently, Buratti [24] has improved on these results, giving *necessary and sufficient* conditions for the existence of radical difference families when $k = 4$ or 5.

Theorem 3.1 [24]

(i) *Let* $p = 12t + 1$ *be a prime, with* 2^e *the largest power of 2 dividing* t. *Then a* $(p, 4, 1)$ *radical difference family exists* $\Leftrightarrow$ -3 *is not a* 2^{e+2}th *power in* $\mathbb{Z}_p$.

(ii) *Let* $p = 20t + 1$ *be a prime, with* 2^e *the largest power of 2 dividing* t. *Then a* $(p, 5, 1)$ *radical difference family exists* $\Leftrightarrow$ $(11 + 5\sqrt{5})/2$ *is not a* 2^{e+1}th *power in* $\mathbb{Z}_p$.

In a later paper [25], Buratti improved upon earlier work of Wilson [76] giving a sufficient condition for the existence of a $(p, k, 1)$ radical difference family which he proved to be necessary at least for $k \le 7$.

As well as producing many cyclic $(v, k, 1)$ designs, radical difference families have been used by Genma, Mishima and Jimbo [47] in their study of cyclically resolvable cyclic designs; we return to these in Section 5.

Although radical difference families provide a powerful method of constructing cyclic designs, it should not, of course, be thought that they are the only means. Buratti, in [26], presented a method of constructing $(p, 4, 1)$DFs for primes $p \equiv 1 \,(\mathrm{mod}\,12)$, showing that his method worked for all such $p < 10^6$. His method was to take as base blocks $\{0, 1, b, b^2\} \times \theta^{6i}$ for suitably chosen b. Similarly he obtained $(p, 5, 1)\mathrm{DF}$ for all primes $p \equiv 1 \,(\mathrm{mod}\,20)$, $p < 10^4$. More recently Chen and Zhu [36, 37] built on these and other results to show that the necessary condition $p \equiv 1 \,\mathrm{mod}\, k\,(k-1)$ is also sufficient for the existence of $(p, k, 1)\mathrm{DF}$ in the cases $k = 4, 5, 6$ (except for $p = 61$ in the case $k = 6$).

Theorem 3.2 [36, 37]

(i) *A* $(p, 4, 1)\mathrm{DF}$ *exists for all primes* $p \equiv 1 \,(\mathrm{mod}\,12)$.

(ii) *A* $(p, 5, 1)\mathrm{DF}$ *exists for all primes* $p \equiv 1 \,(\mathrm{mod}\,20)$.

(iii) *A* $(p, 6, 1)\mathrm{DF}$ *exists for all primes* $p \equiv 1 \,(\mathrm{mod}\,30), p \neq 61$.

4 Mainly concerning Steiner triple systems

4.1 1-rotational STS

Any 1-rotational STS(v) must contain a short orbit. For if $\{\infty, x, y\}$ is one of the triples, $\{\infty, y, 2y - x\}$ is also a triple, so $2y - x \equiv x \,(\mathrm{mod}\,v - 1)$, that is, $y = x \pm \frac{1}{2}(v - 1)$. So all triples involving ∞ must be of the form $\{\infty, x, x + \frac{1}{2}(v - 1)\}$.

Existence of 1-rotational STS was established by Phelps and Rosa.

Theorem 4.1 [72] *There exists a 1-rotational* STS$(v) \Leftrightarrow v \equiv 3 \text{ or } 9 \,(\mathrm{mod}\,24)$.

Examples Take $\{\infty, 0, 4\}$ and $\{0, 1, 3\}$ as base blocks. Their translates $(\mathrm{mod}\,8)$ form a 1-rotational STS(9). This STS is in fact a 1-rotational KTS(9); resolution classes are as follows, read horizontally.

$$
\begin{array}{lll}
\{\infty, 0, 4\} & \{2, 3, 5\} & \{6, 7, 1\} \\
\{\infty, 1, 5\} & \{3, 4, 6\} & \{7, 0, 2\} \\
\{\infty, 2, 6\} & \{4, 5, 7\} & \{0, 1, 3\} \\
\{\infty, 3, 7\} & \{5, 6, 0\} & \{1, 2, 4\}
\end{array}
$$

When do 1-rotational KTS(v) exist? Apart from some isolated examples such as for $v = 9$ and 33, the only known general result is the following, due to Buratti and Zuanni.

Theorem 4.2 [33] *If all prime factors p of v satisfy $p \equiv 1 \,(\mathrm{mod}\,12)$ then there exists a 1-rotational* KTS$(2v + 1)$.

Such KTS are of interest since, as shown by Buratti [30], if a 1-rotational KTS$(2v + 1)$ exists then a dicyclic $(4v, 4, 1)$ design exists. Here *dicyclic* means that the design admits the dicyclic group of order $4v$ as an automorphism group acting regularly on its elements.

4.2   2-rotational STS

A STS(v) is *2-rotational* if it admits an automorphism consisting of one fixed point and two cycles of length $\frac{1}{2}(v-1)$. Phelps and Rosa determined precisely when such STS exist.

Theorem 4.3 [72] *A 2-rotational* STS(v) *exists* $\Leftrightarrow$ $v \equiv 1, 3, 7, 9, 15$ *or* 19 (mod 24).

Two-rotational KTS(v) were constructed as early as 1852. Anstice's cyclic STS(p), described in Theorem 2.1, were obtained incidentally on the way to establishing the following result.

Theorem 4.4 [18] *A 2-rotational* KTS($2p+1$) *exists whenever p is a prime,* $p \equiv 1 \, (\mathrm{mod} \, 6)$.

Anstice's method was extended by Anderson as follows.

Theorem 4.5 [7] *A 2-rotational* KTS($2v+1$) *exists whenever all prime factors of v are* $\equiv 1 \, (\mathrm{mod} \, 6)$.

4.3   Cyclically resolvable KTS

The 1-rotational KTS($2v+1$) obtained by Buratti and Zuanni in Theorem 4.3 are in fact *cyclically resolvable* in the sense that each translate of a resolution class is also a resolution class. Recursive constructions for cyclically resolvable 1-rotational designs have been given by Jimbo and Vanstone.

Lemma 4.6 [53] *If there exist a cyclically resolvable 1-rotational $(v+1, k, 1)$ design, a cyclically resolvable 1-rotational $(u(k-1)+1, k, 1)$ design and a good $(u, k, 1)$ difference matrix, then a cyclically resolvable 1-rotational $(uv+1, k, 1)$ design also exists.*

This lemma was used in the case $k = 3$ in the proof of Theorem 4.3. We also note here that in the case $k = 4$ it can be used, along with the cyclically resolvable 1-rotational $(3p+1, 4, 1)$ designs obtained by Moore [68] for all primes p, $p \equiv 1 \, (\mathrm{mod} \, 4)$, to establish the existence of cyclically resolvable 1-rotational $(3v+1, 4, 1)$ designs whenever v is a product of primes p, $p \equiv 1 \, (\mathrm{mod} \, 4)$. See [33], [13], [55] for this result. However, an alternative approach due to Lu, not depending on Moore's result, will be discussed in Section 7.

5 Cyclically resolvable cyclic designs

These are cyclically resolvable designs in $\mathbb{Z}_v$, where the mapping $x \to x+1 \,(\text{mod}\, v)$ sends resolution classes into resolution classes.

The first families of such $(v, k, 1)$ designs were obtained by Genma, Mishima and Jimbo [47], who considered the case of k odd. For resolvability, we need $k|v$; they considered $v = pk$ where p is a prime, noting that a necessary condition is $p \equiv 1 \,\text{mod}\, k(k-1)$.

Theorem 5.1 [47] *If a $(p, k, 1)$ radical difference family exists with p prime and k odd, then a cyclically resolvable cyclic $(pk, k, 1)$ design exists.*

Since a $(p, 3, 1)$ radical difference family exists for all $p \equiv 1 \,(\text{mod}\, 6)$ (see Theorem 2.1), the following result is established.

Theorem 5.2 [47] *If $p \equiv 1 \,(\text{mod}\, 6)$ is prime, a cyclically resolvable cyclic $(3p, 3, 1)$ design exists.*

The following example is the one obtained in [47] for $p = 7$.

Examples A cyclically resolvable cyclic $(21, 3, 1)$ design

1,4,16	8,11,2	15,18,9	19,20,3	5,6,10	12,13,17	0,7,14
2,5,17	9,12,3	16,19,10	20,0,4	6,7,11	13,14,18	1,8,15
$\vdots$						
7,10,1	14,7,8	0,3,15	4,5,9	11,12,16	18,19,2	6,13,20

- -

1,11,9	4,14,12	7,17,15	10,20,18	13,2,0	16,5,3	19,8,6
2,12,10	5,15,13	8,18,16	11,0,19	14,3,1	17,6,4	20,9,7
3,13,11	6,16,14	9,19,17	12,1,20	15,4,2	18,7,5	0,10,8

Note that the design has base blocks $\{1, 4, 16\}$, $\{19, 20, 3\}$, $\{1, 11, 9\}$ and $\{0, 7, 14\}$, and observe how the translates of $\{1, 11, 9\}$ make up the last three resolution classes.

Using Theorem 3.1(ii), the following result was similarly obtained.

Theorem 5.3 [47] *If $p = 20t + 1$ is prime, if 2^e is the largest power of 2 dividing t, and if $(11 + 5\sqrt{5})/2$ is not a 2^{e+1}th power in $\mathbb{Z}_p$, then a cyclically resolvable cyclic $(5p, 5, 1)$ design exists.*

Since then, Buratti [28] has given a variation of the Genma-Mishima-Jimbo method which yields a cyclically resolvable cyclic $(5p, 5, 1)$ design at least for all $p \equiv 1 (\text{mod}\, 20)$, $p < 1000$.

The case of k even has recently been considered by Lam and Miao [59]. They presented a construction which is valid for both even and odd k, and they used Weil's theorem and computation to show that, when $k = 4$, the cyclotomic conditions required by the method are satisfied for all relevant $p \equiv 13 \,(\text{mod}\, 24)$.

Theorem 5.4 [59] *If $p = 12t + 1$ is prime and t is odd then a cyclically resolvable cyclic $(4p, 4, 1)$ design exists.*

A recursive construction in [47] can then be used to show that a cyclically resolvable cyclic $(3v, 3, 1)$ design exists whenever v is a product of primes $\equiv 1 \,(\mathrm{mod}\, 6)$ and that such a $(4v, 4, 1)$ design exists whenever v is a product of primes $\equiv 13 \,(\mathrm{mod}\, 24)$.

The recursive construction of Genma, Mishima and Jimbo works for cyclically resolvable $(vk, k, 1)$ designs which possess the following two properties (as do the designs of Theorems 5.2 – 5.4):

(P1): The translates of $D = \{0, v, \dots, (k-1)v\}$ lie in k different resolution classes;

(P2): If $\mathcal{R}$ is any resolution class not containing one of the translates of D, then the minimum positive integer t satisfying $\mathcal{R} + t = \mathcal{R}$ is k.

Lemma 5.5 [47] *If there exist*

(a) *cyclically resolvable cyclic $(kv, k, 1)$ and $(ku, k, 1)$ designs with properties (P1) and (P2), and*

(b) *a good $(u, k, 1)$ difference matrix,*

then there exists a cyclically resolvable cyclic $(kuv, k, 1)$ design which also has properties (P1) and (P2).

We note that Mishima and Jimbo [66] have recently provided another recursive construction which yields cyclically resolvable cyclic designs which cannot be obtained via this lemma.

Before leaving [47], we also note that Buratti has extracted from it the following useful result.

Theorem 5.6 [27] *If a $(v, k, 1)\,\mathrm{DF}$ exists, where every prime factor of v is $\equiv 1 \,(\mathrm{mod}\, k)$, then a partial $(kv, k, 1)\,\mathrm{DF}$ exists.*

It follows from this result and from Theorem 3.2(i) that a cyclic $(4p, 4, 1)$ design exists for all primes $p \equiv 1 \,(\mathrm{mod}\, 12)$ [31]. Compare this with the result of Check and Colbourn [35], that a cyclic $(4p, 4, 1)$ design exists for all $p \equiv 13 \,(\mathrm{mod}\, 24)$. Of course, in view of Theorem 5.5, it is now known that a *cyclically resolvable $(4p, 4, 1)$ design exists for all $p \equiv 13 \,(\mathrm{mod}\, 24)$.*

6 Whist tournaments

Definition A *whist tournament* $Wh(4n)$ is a schedule of games each involving four of $4n$ players, in which the games are arranged into $4n - 1$ rounds each of n games. In each game two players partner each other against two others, each player plays in one game in each round, and overall

(a) each player partners each other player exactly once;

(b) each player opposes each other player exactly twice.

The underlying structure, ignoring partners and opponents, is therefore that of a resolvable $(4n, 4, 3)$ design.

The study of $Wh(4n)$ requires the simultaneous study of corresponding tournaments $Wh(4n + 1)$ for $4n + 1$ players.

Definition A $Wh(4n+1)$ is a schedule of games each involving four of $4n+1$ players, the games being arranged into $4n + 1$ rounds each of n games. In each game two players oppose two others. Each player plays in one game in all but one of the rounds, and conditions (a) and (b) are again satisfied.

Most of the early $Wh(v)$ possessed cyclic structure. The $Wh(8)$ of Example 1.3 is 1-rotational; the following is an example of a cyclic $Wh(13)$.

Examples *A cyclic $Wh(13)$.*

Round 1:	$1, 12 \; v \; 8, 5$	$2, 11 \; v \; 3, 10$	$4, 9 \; v \; 6, 7$
Round 2:	$2, 0 \; v \; 9, 6$	$3, 12 \; v \; 4, 11$	$5, 10 \; v \; 7, 8$
$\vdots$			$\vdots$
Round 13:	$0, 11 \; v \; 7, 4$	$1, 10 \; v \; 2, 9$	$3, 8 \; v \; 5, 6$

To check that this works, we first note that in the initial round the partner differences are $\pm(12-1), \pm(8-5), \pm(11-2), \pm(10-3), \pm(9-4), \pm(7-6)$, i.e all the nonzero elements of $\mathbb{Z}_{13}$, each once. Similarly, the opponent differences yield every nonzero element of $\mathbb{Z}_{13}$ twice.

One of the earliest constructions was the following, explicitly stated by Baker [20] but implicit in Watson's paper [75].

Theorem 6.1 [20, 75] *If $p \equiv 1 \, (\mathrm{mod}\, 4)$ is prime, there exists a cyclic $Wh(p)$.*

Proof Let θ be a primitive root of the prime $p = 4t + 1$. Then the games $\theta^i, -\theta^i \; v \; \theta^{t+i}, -\theta^{t+i}$, $0 \le i \le t - 1$ form the initial round of a cyclic $Wh(p)$. $\qquad \square$

The existence of $Wh(4n)$ and $Wh(4n+1)$ for all $n \geq 1$ was established around 1970 (see [8], [20]). Twenty years later, the work of Anderson and Finizio ([11], [12], etc) renewed interest in the construction of cyclic or 1-rotational (that is, $\mathbb{Z}$-cyclic) whist designs. A typical early result was the following.

Theorem 6.2 [12] *Let $q \equiv 3 \pmod 4$ be prime, $q \geq 7$, and let $p_1, \ldots, p_n$ be primes, $p_i \equiv 1 \pmod 4$. Then if there exists a $\mathbb{Z}$-cyclic $Wh(q+1)$ there also exists a $\mathbb{Z}$-cyclic $Wh(qp_1^{\alpha_1} \ldots p_n^{\alpha_n} + 1)$ for all $\alpha_i \geq 0$.*

Many constructions of $\mathbb{Z}$-cyclic $Wh(v)$ have now been obtained, but the aim of showing that a $\mathbb{Z}$-cyclic $Wh(v)$ exists for $v \equiv 0$ or $1 \pmod 4$, $v \neq 9$, is still far from being achieved. Examples of v for which existence has been established include all v of the form $p_1^{\alpha_1} \ldots p_n^{\alpha_n}$ (each prime $p_i \equiv 1 \pmod 4$) [75], $v = q^2$ (q prime, $q \equiv 3 \pmod 4$) $7 \leq q < 500$) [60], $v = 3p_1^{\alpha_1} \ldots p_n^{\alpha_n} + 1$ [63], $v = 27p + 1$ (p prime, $p \equiv 1 \pmod{36}, p < 5000$) [45], as well as all those that can be deduced from the following two theorems of Anderson, Finizio and Leonard.

Theorem 6.3 [16] *If there exist $\mathbb{Z}$-cyclic $Wh(u)$ and $Wh(v)$, where $u, v \equiv 1 \pmod 4$, and if a good $(u, 4, 1)$ difference matrix exists, then a $\mathbb{Z}$-cyclic $Wh(uv)$ exists.*

Theorem 6.4 [16] *If $u \equiv 1 \pmod 4$ and $w \equiv 3 \pmod 4$, $w > 3$, where $\mathbb{Z}$-cyclic $Wh(u)$ and $Wh(w+1)$ and a good $(w, 4, 1)$ difference matrix exist, then a $\mathbb{Z}$-cyclic $Wh(uw+1)$ exists.*

Examples Since a $\mathbb{Z}$-cyclic $Wh(49)$ and $Wh(8)$ exist, and a good $(7, 4, 1)$ difference matrix exists, a $\mathbb{Z}$-cyclic $Wh(7^3 + 1)$, that is, a $\mathbb{Z}$-cyclic $Wh(344)$, exists.

For the remainder of this paper, we change our notation slightly. Consider the game $a, b\, v\, c, d$. It involves four players seated round a table, with partners opposite each other, so it makes sense to represent it by (a, c, b, d), giving the *cyclic* order of the players round the table. In this notation, the initial round games of Example 1.3 can be written as $(\infty, 4, 0, 5)$ and $(1, 2, 3, 6)$. We use this notation from now on.

7   Triplewhist tournaments

Of special interest are triplewhist tournaments, which were introduced by Moore [68].

Definition In a game (a, c, b, d) of a whist tournament, pairs $\{a, c\}$ and $\{b, d\}$ are called *opponent pairs of the first kind*, and pairs $\{a, d\}$ and $\{b, c\}$ are *opponent pairs of the second kind*.

Definition A *triplewhist tournament* $TWh(v)$ is a $Wh(v)$ in which each pair of players are opponents of the first kind exactly once (and hence are opponents of the second kind exactly once).

Examples *The $Wh(8)$ of Example 1.3 is a $TWh(8)$.* Its first round games are $(\infty, 4, 0, 5)$ and $(1, 2, 3, 6)$. Check the differences between opponent pairs of the first kind not involving ∞ in the initial round: these differences are $\pm(5-0), \pm(2-1), \pm(6-3)$, that is, $\pm 1, \pm 2, \pm 3 \pmod 7$. It is also clear that ∞ will be an opponent of the first kind of every other player once.

Triplewhist tournaments are so called because each gives rise to three different whist tournaments, since partner pairs can be interchanged with either kind of opponent pairs. It is now known (Lu, Zhu [64]) that $TWh(4n)$ exist for all n except possibly $n = 3, 14$, and $TWh(4n+1)$ exist for all n except possibly 13 values of $n \leq 38$. But what about $\mathbb{Z}$-cyclic triplewhist tournaments?

7.1 $\mathbb{Z}$-cyclic $TWh(3v+1)$

Moore [68] proved that $\mathbb{Z}$-cyclic $TWh(3p+1)$ exist for all primes $p \equiv 1 \pmod 4$ and that $\mathbb{Z}$-cyclic $TWh(2^n)$ exist for all *even* values of n. (Note that such 2^n are also of the form $3v+1$.) Apart from a few isolated examples, no other $\mathbb{Z}$-cyclic $TWh(v)$ were known to exist until it was proved in [11] that $\mathbb{Z}$-cyclic $TWh(3p^n+1)$ exist for all primes $p \equiv 1 \pmod 4$ and all $n \geq 1$. It was then shown [13] that $\mathbb{Z}$-cyclic $TWh(3p_1^{\alpha_1} \ldots p_n^{\alpha_n} + 1)$ exist whenever the p_i are primes $\equiv 1 \pmod 4$ which are *compatible* in the sense that each is of the form $p_i = 2^m t_i + 1$, t_i odd, *with the same m*. Since then the compatibility condition has been removed (see e.g. Buratti and Zuanni [33], Lu [63]). The compatibility problem arose because of number theoretic complications arising in the reduced set of residues $(\bmod\, p_1^{\alpha_1} \ldots p_n^{\alpha_n})$, but different (and simpler!) approaches avoid this problem. The approach of Lu is particularly elegant, so we now say a little about an adaptation of that approach.

The $Wh(13)$ of Example 6.1 has the property that the partner pairs in the initial round are of the form $\{x, -x\}$, $x \in \mathbb{Z}_{13} - \{0\}$. If v is an odd integer, the pairs $\{1, v-1\}, \{2, v-2\}, \ldots, \{\frac{1}{2}(v-1), \frac{1}{2}(v+1)\}$ form a difference family known as a *patterned starter*. Note that such a patterned starter in $\mathbb{Z}_5$ was used in the initial round games of the league schedule of Example 1.2. It was shown by Watson [75] that, if v is a product of primes $p_i \equiv 1 \pmod 4$, then there exists a $\mathbb{Z}$-cyclic $Wh(v)$ in which the initial round partner pairs are those given by the patterned starter. Such $Wh(v)$ are called *$\mathbb{Z}$-cyclic patterned starter whist tournaments* and are denoted by $ZCPSWh(v)$. Watson took as his initial round games those of the form $(x, \theta x, -x, -\theta x)$ where $\theta^2 \equiv -1 \pmod v$; note that this construction contains that of Theorem 6.2 as a special case.

Lu's result can be expressed as follows.

Theorem 7.1 [63] *If there exists a $ZCPSWh(v)$ where g.c.d.$(v, 6) = 1$, then there exists a $\mathbb{Z}$-cyclic $TWh(3v+1)$.*

Proof Corresponding to each initial round game $(x, y, -x, -y)$ of the $\mathbb{Z}CPSWh(v)$, form three games $(3x, -3x, 3y + v, -3y + v)$, $(3x + v, 3y + 2v, -3y + 2v, -3x + v)$ and $(3x + 2v, -3y, -3x + 2v, 3y)$. Take all of these, along with $(\infty, v, 0, 2v)$ as the initial round games of the required $TWh(3v + 1)$. □

Examples $(1, 2, 4, 3)$ is the initial round game of $\mathbb{Z}CPSWh(5)$. From that we obtain the three games $(3, 12, 11, 14)$, $(8, 1, 4, 2)$, $(13, 9, 7, 6)$, which along with $(\infty, 5, 0, 10)$ give the initial round games of a $\mathbb{Z}$-cyclic $TWh(16)$.

Using Watson's $Wh(v)$, we obtain the following corollary.

Theorem 7.2 *There exists a $\mathbb{Z}$-cyclic $TWh(3v + 1)$ whenever v is a product of primes $p \equiv 1 \,(\mathrm{mod}\, 4)$.*

Other $TWh(3v + 1)$ also follow. For example, Leonard's $Wh(q^2)$, q prime, $q \equiv 3 \,(\mathrm{mod}\, 4)$ [60] are $\mathbb{Z}CPSWh(q^2)$, so we obtain $\mathbb{Z}$-cyclic $TWh(3q^2 + 1)$ for such q.

Lu's result also leads immediately to the 1-rotational $(3v + 1, 4, 1)$ designs discussed in Section 4.3. Indeed, whenever a $\mathbb{Z}CPSWh(v)$ exists with g.c.d.$(v, 6) = 1$, form the block $\{3x, -3x, 3y + v, -3y + v\}$ corresponding to the initial round game $(x, y, -x, -y)$. These blocks form a 1-rotational difference family in $\mathbb{Z}_{3v}$ whose translates, along with those of $\{\infty, 0, v, 2v\}$, form a 1-rotational $(3v + 1, 4, 1)$ design.

Theorem 7.3 *If a $\mathbb{Z}CPSWh(v)$ exists with g.c.d.$(v, 6) = 1$, then a cyclically resolvable 1-rotational $(3v + 1, 4, 1)$ design exists.*

A further development relating to Theorem 7.2 has recently been achieved by Buratti and Zuanni [34]. This concerns the construction of triplewhist tournaments which can be decomposed into sub-triplewhist tournaments. By considering a $\mathbb{Z}$-cyclic $TWh(v)$ as a resolved perfect Cayley design, they have obtained the following result.

Theorem 7.4 [34]

(i) *If v is a product of primes $p_i \equiv 1 \,(\mathrm{mod}\, 4)$, then there exists a $\mathbb{Z}$-cyclic $TWh(3v + 1)$ which is decomposable into $TWh(4)$.*

(ii) *If v is a product of primes $p_i \equiv 1 \,(\mathrm{mod}\, 8)$, $p_i \neq 17, 89$, then there exists a $\mathbb{Z}$-cyclic $TWh(7v + 1)$ which is decomposable into $TWh(8)$.*

Use is made in (ii) of the existence of 1-rotational resolvable $(7p + 1, 8, 1)$ designs for all $p \equiv 1 \,(\mathrm{mod}\, 8)$, $p \neq 17, 89$. These were constructed by Greig [49] for all such $p < 5000$ and for all such $p > 5000$ by Buratti [34].

7.2   Z-cyclic $TWh(v), v \equiv 1 \pmod 4$

We now turn to the construction of $TWh(v)$ for $v \equiv 1 \pmod 4$. Note that the conditions for games (a_i, b_i, c_i, d_i), $1 \le i \le m$ to form the initial round of a $\mathbb{Z}$-cyclic $TWh(4m+1)$ are

(i) the pairs $\{a_i, b_i\}$, $\{c_i, d_i\}$ form a $(4m+1, 2, 1)\,\mathrm{DF}$;

(ii) the pairs $\{a_i, c_i\}$, $\{b_i, d_i\}$ form a $(4m+1, 2, 1)\,\mathrm{DF}$;

(iii) the pairs $\{a_i, d_i\}$, $\{b_i, c_i\}$ form a $(4m+1, 2, 1)\,\mathrm{DF}$.

Examples The games $(1, 19, 10, 9)$, $(24, 21, 8, 13)$, $(25, 11, 18, 22)$, $(20, 3, 26, 6)$, $(16, 14, 15, 28)$, $(7, 17, 12, 5)$, $(23, 2, 27, 4)$ form the first round of a $\mathbb{Z}$-cyclic $TWh(29)$. For example the pairs in (i) (the opponent pairs of the first kind) give differences $\pm 11, \pm 1, \pm 3, \pm 5, \pm 14, \pm 4, \pm 12, \pm 9, \pm 2, \pm 13, \pm 10, \pm 7, \pm 8, \pm 6$, that is, all members of $\mathbb{Z}_{29} - \{0\}$.

We first consider $TWh(p)$ with p prime. Finizio [41] was the first to obtain such designs, showing how to construct them for all primes $p \equiv 5 \pmod 8$, $29 \le p < 2000$. The result was extended to all such $p \ge 29$ by Anderson, Cohen and Finizio.

Theorem 7.5 [10] *If p is a prime, $p \equiv 5 \pmod 8$, $p \ge 29$, then a $\mathbb{Z}$-cyclic $TWh(p)$ exists.*

The proof involved showing that, for all such $p \ne 61$, we can always find a primitive root θ of p for which $\theta^2 \pm \theta + 1$ are both squares in $\mathbb{Z}_p$. Using a similar approach, the work of Liaw [61] and McNay [65] dealt with half of the remaining primes.

Theorem 7.6 [61] *If p is a prime, $p \equiv 9 \pmod{16}$, there exists a $\mathbb{Z}$-cyclic $TWh(p)$.*

Since then, Buratti [32] has completely solved the problem. He showed that the previous existence theorems (7.8 and 7.9) can be proved more easily by requiring only that θ is a non-square (rather than a primitive root). He then gave a new construction which dealt with all $p \equiv 1 \pmod 8$, $p \ne 17$, together.

Theorem 7.7 [32] *If p is a prime, $p \equiv 1 \pmod 4$, $p \ge 29$, a $\mathbb{Z}$-cyclic $TWh(p)$ exists.*

We can now apply the triplewhist version of Theorem 6.4 to establish the following result.

Theorem 7.8 *A $\mathbb{Z}$-cyclic $TWh(v)$ exists whenever v is a product of primes $p \equiv 1 \pmod 4$, $p \ge 29$.*

Note that Theorems 6.4 and 6.5 hold for $TWh(v)$ rather than just $Wh(v)$: if the input tournaments are triplewhist, then so is the resulting one.

Note too that the product result of Theorem 6.4 requires a good difference matrix which may not exist if u is a multiple of 3. But for triplewhist tournaments this problem does not arise, due to the following result.

Theorem 7.9 [16] *If $v \equiv 1 \pmod 4$ and a $\mathbb{Z}$-cyclic $TWh(v)$ exists, then a good $(v, 4, 1)$ difference matrix exists.*

Proof We construct a good difference matrix $A = (a_{ij})$, $1 \le i \le 4$, $1 \le j \le v$, as follows. Take $a_{iv} = 0$ for each i. For each $j < v$, take $a_{1j} = j$, $a_{2j} = j$'s initial round partner, $a_{3j} = j$'s initial round opponent of the first kind, $a_{4j} = j$'s initial round opponent of the second kind.   □

As a consequence, we obtain the following elegant result.

Theorem 7.10 *If there exist $\mathbb{Z}$-cyclic $TWh(u)$ and $TWh(v)$, with $u, v \equiv 1 \pmod 4$ then a $\mathbb{Z}$-cyclic $TWh(uv)$ exists.*

Many new families of $\mathbb{Z}$-cyclic $TWh(v)$ have been obtained in recent years. Some of these have related to the problem of constructing $\mathbb{Z}$-cyclic $TWh(v)$ where $v \equiv 1 \pmod 4$ contains some of the primes $5, 13, 17$ as factors; such designs cannot be obtained by product constructions based on Theorem 7.10. One such general result is due to Finizio and Merritt [44] whose construction gives a $\mathbb{Z}$-cyclic $TWh(v)$ whenever v is of the form $3qp_1^{\alpha_1} \ldots p_r^{\alpha_r}$ with $q \in \{7, 11\}$ and $p_i \equiv 1 \pmod 4$, $p_i \ge 5$. Another direction has been taken by Attinger and Leonard [19] who construct $\mathbb{Z}$-cyclic $TWh(q^2)$ for $q = 11, 19$ and 23. The case $q = 7$ has recently been solved by Ge and Zhu [46].

7.3 Directed triplewhist tournaments

Anderson and Finizio have recently studied $\mathbb{Z}$-cyclic triplewhist tournaments which are also Mendelsohn designs. A $Wh(v)$ is said to be *directed* if each player plays precisely one game with any other player on his left, and one game with that player on his right. Directed whist tournaments are equivalent to Mendelsohn designs with block size 4. The question is: are there triplewhist tournaments which are also directed whist tournaments?

Examples The $TWh(29)$ of Example 7.7 is also directed. To check this, observe that the cyclic differences between adjacent elements in the initial round games are $\pmod{29}$ $1 - 19 = 11$, $19 - 10 = 9$, $10 - 9 = 1$, $9 - 1 = 8$, $24 - 21 = 3$, etc; check that each nonzero element of $\mathbb{Z}_{29}$ arises once.

It was shown in [14] that directed triplewhist tournaments $TWh(v)$ exist for all sufficiently large $v \equiv 1 \pmod 4$. Further, it is shown in [14] that $\mathbb{Z}$-cyclic examples exist for all primes $p \equiv 5 \pmod 8$, $p \ge 29$, and in [15] existence is shown for all primes $p \equiv 1 \pmod 8$, $p < 10,000$ (with at most nine exceptions). The directed triplewhist version of Theorem 6.4 can be used to obtain examples for all products of such primes.

8 Generalised whist tournaments

Recently the following variation on the idea of a whist tournament was introduced.

Definition A *pitch tournament* for $8n$ players, PTCH($8n$), is a schedule of games each involving 8 players, 4 playing against 4 others, the games being played in $8n - 1$ rounds each of n games, each player playing once in each round, and such that

(a) each player partners every other player 3 times;

(b) each player opposes every other player 4 times.

A similar definition applies to a PTCH($8n + 1$), which has $8n + 1$ rounds each of n games, with one player sitting out in each round.

Pitch tournaments were introduced by Finizio and Lewis in [42], and $\mathbb{Z}$-cyclic examples were the subject of their subsequent paper [43]. Each game can be represented by $(a, b, c, d; e, f, g, h)$, where a, b, c, d are partners playing against e, f, g, h.

Examples $(1, 13, 16, 4; 9, 15, 8, 2)$ and $(3, 5, 14, 12; 10, 11, 7, 6)$ are the initial round games of a $\mathbb{Z}$-cyclic PTCH(17).

Examples $(\infty, 4, 0, 5; 1, 2, 3, 6)$ is the initial round games of a $\mathbb{Z}$-cyclic PTCH(8).

These can be checked by considering partner differences and opponent differences.

Corresponding to Theorem 6.2, there is the following result.

Theorem 8.1 [43] *If $p = 8m + 1$ is prime and θ is a primitive root of p, the games $(1, \theta^{2m}, \theta^{4m}, \theta^{6m}; \theta^m, \theta^{3m}, \theta^{5m}, \theta^{7m}) \times 1, \theta, \dots, \theta^{m-1}$ form the initial round of a $\mathbb{Z}$-cyclic PTCH(p).*

The underlying structure of a PTCH(v) is clearly that of a $(v, 8, 7)$ design which is resolvable when $v \equiv 0 \,(\text{mod}\, 8)$ and nearly resolvable when $v \equiv 1 \,(\text{mod}\, 8)$. In two recent papers, Abel, Greig, Finizio and Lewis [1, 2] have shown that PTCH($8n$) exist for all $n > 1615$ and PTCH($8n + 1$) exist for all $n > 224$. In so doing, they reduce the number of values of n, for which the existence of a resolvable $(8n, 8, 7)$ design is undecided, to 21, and the number of n, for which the existence of nearly resolvable $(8n + 1, 8, 7)$ designs is undecided, to 5.

Whist and pitch tournaments are special cases of what are called generalized whist tournaments.

Definition A *generalized whist tournament design* (t, k)-$GWhD(v)$ is a schedule of games, each involving k of v players arranged into k/t teams of t players. The games are arranged into $v-1$ or v rounds according as $v \equiv 0$ or $1 \pmod k$, and altogether each pair of players play together as teammates in exactly $t-1$ games and oppose each other in exactly $k - t$ games.

Thus, for example, a $(2,4)$-$GWhD(v)$ is a $Wh(v)$ and a $(4,8)$-$GWhD(v)$ is a $PTCH(v)$.

Generalized whist tournaments were defined and discussed in [3], where many examples, some of which are $\mathbb{Z}$-cyclic, were constructed.

Examples *A* $\mathbb{Z}$-*cyclic* $(3,9)$-$GWhD(27)$. For initial round games take $\{\infty, 0, 13; 1, 3, 9; 14, 16, 22\}$, $\{2, 12, 21; 6, 10, 11; 4, 7, 18\}$ and $\{15, 25, 8; 17, 5, 20; 19, 23, 24\}$.

The basic idea behind generalized whist tournaments seems to have first appeared, however, in a paper of Bose and Cameron [22] which effectively dealt with not-necessarily-resolvable $(t, 2t)$-$GWhD(v)$. They consider all such possible designs with $v \le 13$.

Examples $(\infty, 2, 5; 6, 10, 4)$ and $(1, 9, 0; 3, 8, 7)$ can be taken as the initial round games of a $\mathbb{Z}$-cyclic $(3, 6)$-$GWhD(12)$.

Bose and Cameron give the following general construction (cf. Theorems 6.2 and 8.3).

Theorem 8.2 [22] *Let* $p = 6m + 1$ *be prime and let* θ *be a primitive root of* p. *Then the games* $(1, \theta^{2m}, \theta^{4m}; \theta^m, \theta^{3m}, \theta^{5m}) \times 1, \theta, \ldots, \theta^{m-1}$ *form the initial round of a* $(3, 6)$-$GWhD(p)$.

In a recent paper [4], $(2, 6)$-$GWhD(v)$ have been thoroughly studied. The underlying designs are $(v, 6, 5)$ designs, and as a result of this study some new resolvable $(6n, 6, 5)$ and nearly resolvable $(6n + 1, 6, 5)$ designs have been obtained. As a result, it is now known that a resolvable $(6n, 6, 5)$ design exists for all n except possibly for $n = 29$.

9   Greig's log table method

In this final section we discuss a powerful method of constructing many $\mathbb{Z}$-cyclic designs.

It was shown by Moore [68] that a $\mathbb{Z}$-cyclic $TWh(2^n)$ exists whenever n is even. In the case of odd n, $\mathbb{Z}$-cyclic $TWh(8)$ and $TWh(32)$ have been known for some time, and I have recently pointed out [9] how Mitchell's method of constructing a $TWh(32)$ can be used to construct a $\mathbb{Z}$-cyclic $TWh(128)$. However, no real progress on the case of n odd had been made until Greig [50] recently described a method which establishes the existence of $\mathbb{Z}$-cyclic $TWh(2^n)$ for *all* $n \ge 2$. His log table method is described in the following example.

Examples Consider the finite field GF(32) generated by the primitive element θ satisfying $\theta^5 = \theta^3 + 1$. Each power of θ is a polynomial in θ of degree at most 4, and hence can be represented by a 5-digit binary sequence. For example, $\theta^5 = \theta^3 + 1$ is represented by 01001, $\theta^7 = \theta^5 + \theta^2 = \theta^3 + \theta^2 + 1$ by 01101.

i	representation of θ^i		i	representation of θ^i
0	0 0 0 0 1		16	0 1 1 0 0
1	0 0 0 1 0		17	1 1 0 0 0
2	0 0 1 0 0		18	1 1 0 0 1
3	0 1 0 0 0		19	1 1 0 1 1
4	1 0 0 0 0		20	1 1 1 1 1
5	0 1 0 0 1		21	1 0 1 1 1
6	1 0 0 1 0		22	0 0 1 1 1
7	0 1 1 0 1		23	0 1 1 1 0
8	1 1 0 1 0		24	1 1 1 0 0
9	1 1 1 0 1		25	1 0 0 0 1
10	1 0 0 1 1		26	0 1 0 1 1
11	0 1 1 1 1		27	1 0 1 1 0
12	1 1 1 1 0		28	0 0 1 0 1
13	1 0 1 0 1		29	0 1 0 1 0
14	0 0 0 1 1		30	1 0 1 0 0
15	0 0 1 1 0			

If we label the rows of an 8×4 array by the initial 3-digit parts of the 5-digit representations, and the columns by the final 2-digit parts, and place i in the position corresponding to the representation of θ^i (putting ∞ in the 00000 position), we obtain the following table.

	00	01	10	11
000	∞	0	1	14
001	2	28	15	22
010	3	5	29	26
011	16	7	23	11
100	4	25	6	10
101	30	13	27	21
110	17	18	8	19
111	24	9	12	20

The rows then give the initial blocks of a $\mathbb{Z}$-cyclic resolvable $(32, 4, 3)$ design. But, further, these rows actually give the initial round games of a $\mathbb{Z}$-cyclic $TWh(32)$ if we take as partners entries in columns whose labels agree in their second digit; as opponents of the first kind, entries in columns whose labels agree in their first digit; and as opponents of the second kind, entries in columns whose labels differ in both digits.

This method works in general [50].

Theorem 9.1 *The rows of the* $2^{n-2} \times 4$ *log table of* $\mathrm{GF}(2^n)$, $n \geq 2$, *with* ∞ *as the log of zero, form the initial round games of a* $\mathbb{Z}$-*cyclic* $TWh(2^n)$.

We could arrange the logs differently.

Examples The rows of the log table below form the initial resolution class of a 1-rotational resolvable $(32, 8, 7)$ design.

	000	001	010	011	100	101	110	111
00	∞	0	1	14	2	28	15	22
01	3	5	29	26	16	7	23	11
10	4	25	6	10	30	13	27	21
11	17	18	8	19	24	9	12	20

Further, if we take as partners those entries in columns whose labels have the same initial digit, we obtain the initial round games of a $\mathbb{Z}$-cyclic $\mathrm{PTCH}(32)$: $\{\infty, 0, 1, 14; 2, 28, 15, 22\}$, $\{3, 5, 29, 26; 16, 7, 23, 11\}$, $\{4, 25, 6, 10; 30, 13, 27, 21\}$, $\{17, 18, 8, 19; 24, 9, 12, 20\}$. Indeed, this $\mathrm{PTCH}(32)$ satisfies the internal structure conditions that are a straight-forward generalization of the triplewhist conditions for whist tournaments.

As a final example of Greig's method, we show how the $(3, 9)$-$GWhD(27)$ of Example 8.4 is obtained.

Examples Consider $\mathrm{GF}(27)$ with primitive element θ satisfying $\theta^3 = \theta + 2$. Powers of θ are represented by 3-digit binary sequences as follows.

i	sequence representing θ^i	i	sequence representing θ^i
0	0 0 1	13	0 0 2
1	0 1 0	14	0 2 0
2	1 0 0	15	2 0 0
3	0 1 2	16	0 2 1
4	1 2 0	17	2 1 0
5	2 1 2	18	1 2 1
6	1 1 1	19	2 2 2
7	1 2 2	20	2 1 1
8	2 0 2	21	1 0 1
9	0 1 1	22	0 2 2
10	1 1 0	23	2 2 0
11	1 1 2	24	2 2 1
12	1 0 2	25	2 0 1

Arranging in a 3×9 array we obtain the following log table.

	00	01	02	:	10	11	12	:	20	21	22
0	∞	0	13	:	1	9	3	:	14	16	22
1	2	21	12	:	10	6	11	:	4	18	7
2	15	25	8	:	17	20	5	:	23	24	19

The rows give the $(3,9)$-$GWhD(27)$ of Example 8.4.

References

[1] R.J.R. Abel, N.J. Finizio, M. Greig and S.J. Lewis, Pitch tournament designs and other BIBDs — existence results for the case $v = 8n$, *J. Combin. Des.*, to appear.

[2] R.J.R. Abel, N.J. Finizio, M. Greig and S.J. Lewis, Pitch tournament designs and other BIBDs — existence results for the case $v = 8n + 1$, *Congr. Numer.* **138** (1999), 175–192.

[3] R.J.R. Abel, N.J. Finizio, M. Greig and S.J. Lewis, Generalized whist tournament designs, preprint.

[4] R.J.R. Abel, N.J. Finizio, M. Greig and S.J. Lewis, $(2,6)$-$GWhD(v)$-existence results and some $\mathbb{Z}$-cyclic solutions, *Congr. Numer.*, to appear.

[5] I. Anderson, A hundred years of whist tournaments, *J. Combin. Math. Combin. Comput.* **19** (1995), 129–150.

[6] I. Anderson, Cyclic designs in the 1850s; the work of Rev R.R. Anstice, *Bull. Inst. Combin. Appl.* **15** (1995), 41–46.

[7] I. Anderson, Some 2-rotational and cyclic designs, *J. Combin. Des.* **4** (1996), 247–254.

[8] I. Anderson, *Combinatorial designs and tournaments*, Oxford University Press, Oxford, 1997.

[9] I. Anderson, Early whist tournaments of Whitfeld and Mitchell, and a hidden treasure, *Bull. Inst. Combin. Appl.* **29** (2000), 89–93.

[10] I. Anderson, S.D. Cohen and N.J. Finizio, An existence theorem for cyclic triplewhist tournaments, *Discrete Math.* **138** (1995), 31–42.

[11] I. Anderson and N.J. Finizio, A generalisation of a construction of E.H. Moore, *Bull. Inst. Combin. Appl.* **6** (1992), 39–46.

[12] I. Anderson and N.J. Finizio, Many more $\mathbb{Z}$-cyclic whist tournaments, *Congr. Numer.* **94** (1993), 123–129.

[13] I. Anderson and N.J. Finizio, Cyclically resolvable designs and triplewhist tournaments, *J. Combin. Des.* **1** (1993), 347–358.

[14] I. Anderson and N.J. Finizio, Triplewhist tournaments that are also Mendelsohn designs, *J. Combin. Des.* **5** (1997), 397–406.

[15] I. Anderson and N.J. Finizio, On the construction of directed triplewhist tournaments, *J. Combin. Math. Combin. Comput.*, **35** (2000), 107–115.

[16] I. Anderson, N.J. Finizio and P.A. Leonard, New product theorems for $\mathbb{Z}$-cyclic whist tournaments, *J. Combin. Theory Ser. A* **88** (1999), 162–166.

[17] R.R. Anstice, On a problem in combinations, *Cambridge Dublin Math. J.* **7** (1852), 279–292.

[18] R.R. Anstice, On a problem in combinations, *Cambridge Dublin Math. J.* **8** (1853), 149–152.

[19] L. Attinger and P.A. Leonard, New families of $\mathbb{Z}$-cyclic triplewhist tournaments, *Congr. Numer.* **133** (1998), 163–169.

[20] R.D. Baker, Whist tournaments, *Congr. Numer.* **14** (1975), 89–100.

[21] R.C. Bose, On the construction of balanced incomplete block designs, *Ann. Engenics* **9** (1939), 353–399.

[22] R.C. Bose and J.M. Cameron, Calibration designs based on solutions to the tournament problem, *J. Res. Nat. Bur. Standards Sect. B* **71B** (1967), 149–160.

[23] R.C. Bose and S. Chowla, Theorems in the additive theory of numbers, *Comment. Math. Helvet.* **37** (1962), 141–147.

[24] M. Buratti, Improving two theorems of Bose on difference families, *J. Comb. Des.* **3** (1995), 15–24.

[25] M. Buratti, On simple radical difference families, *J. Comb. Des.* **3** (1995), 161–168.

[26] M. Buratti, Constructions of $(q, k, 1)$ difference families with q a prime power and $k = 4, 5$, *Discrete Math.* **138** (1995), 169–175.

[27] M. Buratti, From a $(G, k, 1)$ to a $(C_k + G, k, 1)$ difference family, *Des. Codes Cryptogr.* **11** (1997), 5–9.

[28] M. Buratti, On resolvable difference families, *Des. Codes Cryptogr.* **11** (1997), 11–23.

[29] M. Buratti, Recursive constructions for difference matrices and relative difference families, *J. Combin. Des.* **6** (1998), 165–182.

[30] M. Buratti, 1-rotational Kirkman triple systems generate dicyclic Steiner 2-designs with block size 4, *Bull. Inst. Combin. Appl.* **26** (1999), 91–95.

[31] M. Buratti, Some regular Steiner 2-designs with block size 4, *Ars. Combin.* **55** (2000), 135–137.

[32] M. Buratti, Existence of $\mathbb{Z}$-cyclic triplewhist tournaments for a prime number of players, *J. Combin. Theory Ser. A* **90** (2000), 315–325.

[33] M. Buratti and F. Zuanni, G-invariantly resolvable Steiner 2-designs that are 1-rotational over G, *Bull. Belg. Math. Soc. Simon Stevin* **5** (1998), 221–235.

[34] M. Buratti and F. Zuanni, Perfect Cayley designs as generalisations of perfect Mendelsohn designs, *Des. Codes Cryptogr.*, to appear.

[35] P.L. Check and C.J. Colbourn , Concerning difference families with block size four, *Discrete Math.* **133** (1994), 97–103.

[36] K. Chen and L. Zhu, Existence of $(q, 6, 1)$ difference families with q a prime power, *Des. Codes Cryptogr.* **15** (1998), 167–173.

[37] K. Chen and L. Zhu, Existence of $(q, k, 1)$ difference families with q a prime power and $k = 4, 5$, *J. Combin. Des.* **7** (1999), 21–31.

[38] K. Chen and L. Zhu, Improving Wilson's bound on difference families, *Utilitas Math.* **55** (1999), 189–200.

[39] C.J. Colbourn and M.J. Colbourn, On cyclic block designs, *C.R. Math. Rep. Acad. Sci. Canada* **2** (1980), 95–98.

[40] C.J. Colbourn and J.H. Dinitz (eds), *CRC Handbook of Combinatorial Designs*, CRC Press, Baton Rouge, 1996.

[41] N.J. Finizio, $\mathbb{Z}$-cyclic triplewhist tournaments when the number of players involves primes of the form $8n + 5$, *J. Combin. Des.* **2** (1994), 31–40.

[42] N.J. Finizio and S.J. Lewis, Pitch tournaments, *Congr. Numer.* **130** (1998), 19–27.

[43] N.J. Finizio and S.J. Lewis, Some specializations of pitch tournament designs, *Util. Math.* **56** (1999), 33–52.

[44] N.J. Finizio and A.J. Merritt, $\mathbb{Z}$-cyclic triplewhist tournaments — some exceptional cases, *Congr. Numer.* **131** (1998), 19–34.

[45] N.J. Finizio and A.J. Merritt, Some new $\mathbb{Z}$-cyclic whist tournaments, *Discrete Appl. Math.* **101** (2000), 115–130.

[46] G. Ge and L. Zhu, Frame construction for Z-cyclic triple whist tournaments, preprint.

[47] M. Genma, M. Mishima and M. Jimbo, Cyclic resolvability of cyclic Steiner 2-designs, *J. Combin. Des.* **5** (1997), 177–187.

[48] M.J. Grannell and T.S. Griggs, Product constructions for cyclic block designs II. Steiner 2-designs, *J. Combin. Theory Ser. A* **42** (1986), 179–183.

[49] M. Greig, Some group divisible design constructions, *J. Combin. Math. Combin. Comput.* **27** (1998), 33–52.

[50] M. Greig, Resolvable balanced incomplete designs with composite prime power block sizes, preprint.

[51] M. Jimbo, Recursive constructions for cyclic BIB designs and their generalizations, *Discrete Math.* **116** (1993), 79–95.

[52] M. Jimbo and S. Kuriki, On a composition of 2-cyclic designs, *Discrete Math.* **46** (1983), 249–255.

[53] M. Jimbo and S.A. Vanstone, Recursive constructions for resolvable and doubly resolvable 1-rotational Steiner 2-designs, *Util. Math.* **26** (1984), 45–61.

[54] D. Jungnickel, Composition theorems for difference families and regular planes, *Discrete Math.* **23** (1978), 151–158.

[55] S. Kageyama and Y. Miao, Note on a paper "1-rotational designs with block size 4", *Bull. Inst. Combin. Appl.* **20** (1997), 82–84.

[56] T.P. Kirkman, On a problem in combinations, *Cambridge Dublin Math. J.* **2** (1847), 191–204.

[57] T.P. Kirkman, Note on an unanswered prize question, *Cambridge Dublin Math. J.* **5** (1850), 255–262.

[58] T.P. Kirkman, On the perfect r-partitions of $r^2 - r + 1$, *Trans. Historical Soc. Lancs. Cheshire* **9** (1857), 127–142.

[59] C. Lam and Y. Miao, On cyclically resolvable cyclic Steiner 2-designs, *J. Combin. Theory Ser. A* **85** (1999), 194–207.

[60] P.A. Leonard, Some new Z-cyclic whist tournaments, *Util. Math.* **49** (1996), 223–232.

[61] Y.S. Liaw, Construction of Z-cyclic whist tournaments, *J. Combin. Des.* **4** (1996), 219–233.

[62] Jiaxi Lu, *Collected works of Lu Jiaxi on combinatorial designs*, Inner Mongolia People's Press, Huhhot, China, 1990.

[63] Y. Lu, Existence of $\mathbb{Z}$-cyclic $TWh(3p_1^{\alpha_1} \ldots p_n^{\alpha_n} + 1)$ for primes $p_i \equiv 1 \,(\mathrm{mod}\,4)$, preprint.

[64] Y. Lu and L. Zhu, On the existence of triplewhist tournaments $TWh(v)$, *J. Combin. Des.* **5** (1997), 249–256.

[65] G. McNay, Cohen's sieve with quadratic conditions, *Util. Math.* **49** (1996), 191–201.

[66] M. Mishima and M. Jimbo, Recursive constructions for cyclic quasiframes and cyclically resolvable cyclic Steiner 2-designs, *Discrete Math.* **211** (2000), 135–152.

[67] J.T. Mitchell, *Duplicate whist*, McClurg, Chicago, (1891); second edition 1897.

[68] E.H. Moore, Tactical Memoranda I–III, *Amer. J. Math.* **18** (1896), 264–303.

[69] E. Netto, Zur Theorie der Tripelsysteme, *Math. Ann.* **42** (1893), 143–152.

[70] K.H. Parshall and D.E. Rowe, *The emergence of the American Mathematical Research Community, 1876–1900: J.J. Sylvester, Felix Klein and E.H. Moore*, AMS/LMS History of Mathematics Vol.8, Providence, 1991.

[71] R. Peltesohn, Eine Lösung der beiden Heffterschen Differenzenprobleme, *Compositio Math.* **6** (1939), 251–257.

[72] K.T. Phelps and A. Rosa, Steiner triple systems with rotational automorphisms, *Discrete Math.* **33** (1981), 57–66.

[73] D.K. Ray-Chaudhuri and R.M. Wilson, Solution to Kirkman's schoolgirls problem, *Proc. Symp. Math.* **19** (1971), 187–203.

[74] J. Singer, A theorem in projective geometry and some applications to number theory, *Trans. Amer. Math. Soc.* **43** (1938), 377–385.

[75] G.L. Watson, Bridge problem, *Math. Gazette* **38** (1954), 129–130.

[76] R.M. Wilson, Cyclotomy and difference families in elementary abelian groups, *J. Number Theory* **4** (1972), 17–47.

Department of Mathematics
University of Glasgow
Scotland
United Kingdom
ia@maths.gla.ac.uk

Orthogonal designs and third generation wireless communication

A. R. Calderbank and A. F. Naguib

Abstract

This paper connects the practice of wireless communication with the mathematics of quadratic forms developed by Radon and Hurwitz about a hundred years ago. Orthogonal designs, known as space-time block codes in the communications literature, provide the bridge between the two subjects. The columns of the design represent different time slots, the rows represent different transmit antennas, and the entries are the symbols to be transmitted. Multiple transmit antennas provide independent paths from the base station to the mobile terminal, and in effect this creates a single channel that is more reliable than any constituent path. The mathematics developed by Hurwitz and Radon is used to derive fundamental limits on transmission rates. The algebraic structure of the 2×2 space-time block code (a representation of Hamilton's biquaternions) is used to suppress interference from a second space-time user, when a second antenna is available at the mobile terminal.

1 Introduction

Classical coding theory is concerned with the representation of information that is to be transmitted over some noisy channel. This general framework includes the algebraic theory of error correcting codes, where codewords are strings of symbols taken from some finite field, and it includes data transmission over Gaussian channels, where codewords are vectors in Euclidean space. Fifty years of information theory and coding has led to a number of consumer products that make essential use of coding to improve reliability; for example, compact disk players, hard disk drives and wireline modems. The discovery of turbo codes by Berrou, Glavieux, and Thitmajshima [3] has led to the construction of codes that essentially achieve the Shannon capacity of the Gaussian channel. This means that fifty years from now it will be disappointing if the focus of coding theory is simply point to point communication in the presence of noise. Wireless communication presents a new opportunity for coding theory, where the objective is to provide immunity to multiple channel impairments. The focus in this paper is the combination of noise, fading signal strength, and interference from other users sharing the same channel.

We set the stage with a quotation from Marconi:

It is dangerous to put limits on wireless.

This is an opinion expressed in 1932, but it holds true today given the gold rush to provide wireless Internet access. The difficulty in providing attractive

wireless data rates is that band-limited wireless channels are narrow pipes that
do not readily accommodate rapid flow of data. However these pipes can be
broadened by deploying multiple transmit and receive antennas. Foschini [7]
and Telatar [29] provide outage capacity curves under the assumption that
fading is quasistatic, that is constant over a data frame, and then changing
in an independent manner. Figure 1 shows the potential gain on narrowband
30kHz TDMA channels (IS-136) employed by AT&T Wireless Services - with
only two antennas at the base station and the mobile, there is the potential to
increase the achievable data rate by a factor of 6.

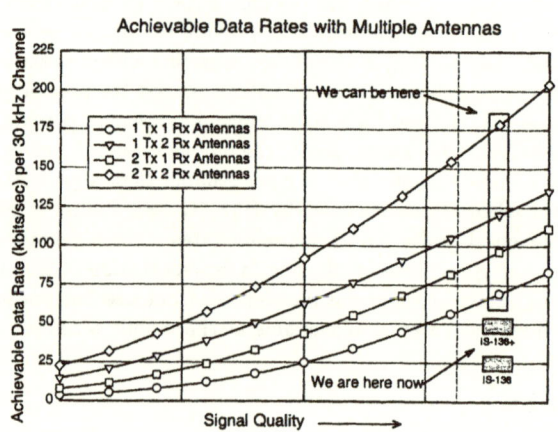

Figure 1: Potential data rates on the IS-136 narrowband TDMA channels employed by AT&T Wireless Services. Here 10% outage capacity is the transmission rate that can be supported 90% of the time, and retransmission protocols
on the radio link are used to correct frames or packets that are received in
error.

Two antennas at the base station provide two independent paths from the
base station to the mobile, and by spreading information across the two paths,
and by appropriate signal processing at the receiver, we see, in effect a single
channel that is better than either path. This link level advantage, over a single
path from base station to mobile, is called *diversity*. Reducing the variation in
signal strength at the mobile terminal also allows smoother and more efficient
power control, since the base station is continually adjusting transmit power
on the basis of reported signal strength at the mobile. This means that the
base station requires less power to support existing users, or that more users
can be supported for a given constraint on radiated signal power at the base
station. System studies by Parkvall et al. [19] for Wideband Code Division
Multiple Access or W-CDMA show up to 100% increase in the number of users
that can be supported on the downlink from the base station to the mobile

terminal.

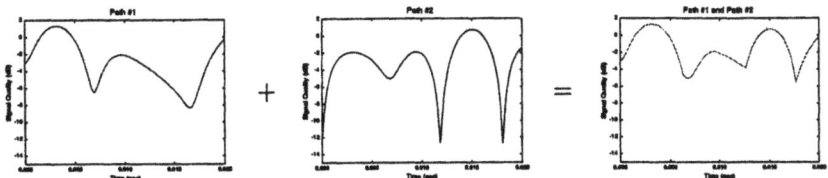

Figure 2: Diversity means that two paths are better than one!

The earliest form of transmit diversity is the delay diversity scheme proposed by Wittneben [34], where a signal is transmitted from the second antenna, then delayed one time slot, and transmitted from the first antenna. The focus of this paper is providing transmit diversity through orthogonal designs, which are known as block space-time codes in the communications literature. The name captures the fact that information is distributed across space (multiple transmit antennas), as well as over time as in classical coding theory. The simplest example is the 2×2 code

$$(c_1, c_2) \rightarrow \begin{pmatrix} c_1 & -\bar{c}_2 \\ c_2 & \bar{c}_1 \end{pmatrix}$$

discovered by Alamouti [1]. The columns of this matrix represent different time slots, the rows represent different antennas, and the entries are the symbols to be transmitted. Typically these are drawn from a small constellation of equally spaced points on the unit circle. The signals r_1, r_2 received over two consecutive symbol periods are given by

$$\begin{pmatrix} r_1 \\ -\bar{r}_2 \end{pmatrix} = \begin{pmatrix} h_1 & h_2 \\ -\bar{h}_2 & \bar{h}_1 \end{pmatrix} \begin{pmatrix} c_1 \\ c_2 \end{pmatrix} + \begin{pmatrix} n_1 \\ n_2 \end{pmatrix} \qquad (1.1)$$

where h_1, h_2 are the *path gains* from the two transmit antennas to the mobile, and the noise samples n_1, n_2 are independent samples of a zero-mean complex Gaussian random variable with power spectral density $N_0/2$ per dimension. The signal strength at the mobile terminal depends on the path taken by the data through air, hence the term *path gain*. Decoding is remarkably simple, provided that the path gains are known at the mobile (this is accomplished in practice through some sacrifice in rate by inserting pilot tones into the data frame for channel estimation). We form

$$\begin{pmatrix} \bar{h}_1 & h_2 \\ \bar{h}_2 & -h_1 \end{pmatrix} \begin{pmatrix} r_1 \\ r_2^* \end{pmatrix} = (|h_1|^2 + |h_2|^2) \begin{pmatrix} c_1 \\ c_2 \end{pmatrix} + \begin{pmatrix} n_1' \\ n_2' \end{pmatrix},$$

and observe that the new noise samples remain independent of each other; the vector (n_1', n_2') has zero mean and covariance $(|h_1|^2 + |h_2|^2)I_2$. We may then decode c_1, c_2 separately, rather than decoding the pair (c_1, c_2) jointly, which is far more complex. Decoding fails only if $h_1 = h_2 = 0$, when the antenna at the mobile experiences a deep fade. This all works because the columns of the matrix

$$H = \left(\begin{array}{cc} h_1 & h_2 \\ -\overline{h}_2 & \overline{h}_1 \end{array} \right) \tag{1.2}$$

are orthogonal, regardless of the actual values of the path gains h_1, h_2. If we were to view these path gains as indeterminates, then we have

$$HH^* = (|h_1|^2 + |h_2|^2)I_2,$$

which is the condition that defines a *complex orthogonal design.*

Section 2 provides more examples of orthogonal designs, as part of a survey of the algebraic theory that includes the connection to normed algebras and sums of squares. We derive fundamental limits on transmission rates, and construct optimal designs that meet these upper bounds using an extraspecial 2-group that plays a central role in the mathematics of quadratic and alternating bilinear forms. Our treatment follows the work of Wolfe [35], on amicable pairs of real orthogonal designs, and is elementary in the sense that it emphasizes sets of anti-commuting real matrices rather than representations of Clifford algebras.

Section 3 teases apart *diversity gain*, that results from multiple transmit antennas, and *coding gain*, that results from how symbols are correlated across transmit antennas. We consider the squared distance between two codewords c and c' at the output of the wireless channel. We introduce a matrix B with rows indexed by transmit antennas and columns indexed by time slots, where the it^{th} entry is the difference $c_{it} - c'_{it}$. Squared distance turns out to be proportional to

$$\sum_{j=1}^{M} h_j BB^* h_j^*,$$

where h_j is the vector of path gains from the different transmit antennas to the jth receive antenna. The vector h_j varies over time, and when it finds the null space of the matrix BB^* the jth receive antenna experiences a deep fade. *Diversity gain* is then just the minimum rank of the matrix B, where the minimization is over all pairs of codewords. Coding gain depends on the product of the nonzero eigenvalues, and again there is a minimization over all pairs of codewords. This analysis leads to two design criteria for space-time codes, and when the number of antennas is small, the trellis codes presented in Tarokh et al. [28] come close (within 2.5 dB) to the outage capacity. The block space-time codes considered here provide diversity gain but no coding gain. However they can easily be concatenated with block or convolutional outer codes if additional immunity to fading is required.

Section 4 describes a decoding algorithm that is able to suppress interference from a second space-time user, when a second antenna is available at the mobile. When the signal power of the interferer is equal to that of the desired signal, performance is the same as that of a system employing two transmit and only one receive antenna. When there is no interference the second antenna provides increased immunity to fading (diversity gain). The decoding algorithm exploits the algebraic structure of the space-time block code. For the 2×2 code the structure is that of a division algebra; the rule for multiplying matrices H corresponds to the rule for multiplying biquaternions (Hamilton's construction of quaternions using pairs of complex numbers). Our treatment adds geometric insight to the original paper by Naguib and Seshadri [17], and is more general. The impact of interference suppression is a doubling of system capacity, since it is now possible to support twice as many users in the same spectrum. What may happen in the future is co-location of two users on the same channel; there would be four antennas at the base station, two antennas at the mobile, and the terminal would support twice the standard data rate (around 1 Mb/s for the evolution of GSM called EDGE).

2 The algebra of orthogonal designs

Orthogonal designs have been studied fairly extensively in the mathematics literature; see, for example, Geramita and Seberry [8]. We begin by considering small examples.

2.1 Square designs in dimensions 2, 4, and 8

Let $u_0, u_1, \ldots, u_{s-1}$ be positive integers, and let $x_0, x_1, \ldots, x_{s-1}$ be commuting indeterminates. A *real orthogonal design* of type $(u_0, u_1, \ldots, u_{s-1})$ and size N is an $N \times N$ matrix X with entries $0, \pm x_0, \pm x_1, \ldots, \pm x_{s-1}$ satisfying

$$XX^T = \left(\sum_{j=0}^{s-1} u_j x_j^2 \right) I_N .$$

In the application to wireless communications, there are s information bits and N time slots, so we define the *rate* R of a real orthogonal design by $R = s/N$.

A *complex orthogonal design* of size N and type $(u_0, \ldots, u_{s-1}; v_1, \ldots, v_t)$ is a matrix $Z = X + iY$, where X, Y are real orthogonal designs of type $(u_0, \ldots, u_{s-1})$ and $(v_1, \ldots, v_t)$ respectively, and where

$$ZZ^* = \left(\left(\sum_{j=0}^{s-1} u_j x_j^2 \right) + \left(\sum_{j=1}^{t} v_j y_j^2 \right) \right) I_N$$

Since

$$\begin{aligned} ZZ^* &= (X + iY)(X^T - iY^T) \\ &= (XX^T + YY^T) + i(YX^T - XY^T) \end{aligned}$$

it follows that $XY^T = YX^T$. Note that if $t = s + 1$, then the entries of $X + iY$ are linear combinations of the complex indeterminates $z_j = x_j + iy_{j+1}$ and their conjugates $\bar{z}_j$, $j = 0, \dots, s - 1$. In fact the definition of a complex orthogonal design found in Tarokh et al. [27] is given in terms of these complex indeterminates. The *rate* of a complex design is $R = (s + t)/2N$.

$N = 2$, $R = 1$: This is the representation of the complex numbers $\mathbb{C}$ as a 2×2 matrix algebra over the real numbers $\mathbb{R}$, where the complex number $x_0 + ix_1$ corresponds to the matrix

$$\begin{pmatrix} x_0 & x_1 \\ -x_1 & x_0 \end{pmatrix}$$

$N = 4$, $R = 1$: This is the representation of the quaternions $\mathbb{Q}$ as a 4×4 matrix algebra over $\mathbb{R}$, where the quaternion $x_0 + ix_1 + jx_2 + kx_3$ corresponds to the matrix

$$\begin{bmatrix} x_0 & x_1 & x_2 & x_3 \\ -x_1 & x_0 & -x_3 & x_2 \\ -x_2 & x_3 & x_0 & -x_1 \\ -x_3 & -x_2 & x_1 & x_0 \end{bmatrix} = x_0 I + x_1 \begin{bmatrix} & 1 & & \\ -1 & & & \\ & & & -1 \\ & & 1 & \end{bmatrix}$$

$$+ x_2 \begin{bmatrix} & & 1 & \\ & & & 1 \\ -1 & & & \\ & -1 & & \end{bmatrix} + x_3 \begin{bmatrix} & & & 1 \\ & & -1 & \\ & 1 & & \\ -1 & & & \end{bmatrix}. \tag{2.1}$$

We may also view quaternions as pairs of complex numbers, where the product of quaternions (a, b) and (a', b') is given by

$$(a, b)(a', b') = (aa' - \bar{b}'b, ab' + \bar{a}'b)$$

These are Hamilton's Biquaternions, and we refer the reader to van der Waerden [30] for further details. We may associate the pair (a, b) with the 2×2 complex matrix

$$\begin{pmatrix} a & b \\ -\bar{b} & \bar{a} \end{pmatrix}$$

and observe that matrix multiplication coincides with the rule for multiplying biquaternions. Note that this matrix of complex indeterminates is an example of a 2×2 complex orthogonal design. In general, a complex design of size N with $t = s + 1$ determines a real design of size $2N$ through the substitutions

$$a = x_0 + ix_1 \longrightarrow \begin{pmatrix} x_0 & x_1 \\ -x_1 & x_0 \end{pmatrix}$$

$$\bar{a} = x_0 - ix_1 \longrightarrow \begin{pmatrix} x_0 & -x_1 \\ x_1 & x_0 \end{pmatrix}$$

$$ia = -x_1 + ix_0 \longrightarrow \begin{pmatrix} -x_1 & x_0 \\ -x_0 & -x_1 \end{pmatrix}.$$

N = 8, R = 1: This is the representation of the octonions or Cayley numbers as an 8-dimensional algebra over $\mathbb{R}$. This algebra was discovered first by Graves in 1843 (see [9]), and then independently by Cayley [5]. We view octonions as 4-tuples of complex numbers, where the product $c = ab$ of octonions $a = (a_0, a_1, a_2, a_3)$ and $b = (b_0, b_1, b_2, b_3)$ is given by

$$
\begin{aligned}
c_0 &= a_0 b_0 - \overline{b_1} a_1 - \overline{b_2} a_2 - \overline{a_3} b_3 \\
c_1 &= b_1 a_0 + a_1 \overline{b_0} - a_3 \overline{b_2} + b_3 \overline{a_2} \\
c_2 &= b_2 a_0 - \overline{a_1} b_3 + a_2 \overline{b_0} + \overline{b_1} a_3 \\
c_3 &= b_3 \overline{a_0} + a_1 b_2 - b_1 a_2 + a_3 b_0
\end{aligned}
$$

We may view octonions as vectors in $\mathbb{R}^8$, and we may realize multiplication by the complex number $x_0 + ix_1$ by applying the 2×2 matrix $\begin{pmatrix} x_0 & x_1 \\ -x_1 & x_0 \end{pmatrix}$. Given octonions x, y the product xy is given by $xy = xR(y)$, where

$$
R(y) = \begin{bmatrix}
y_0 & y_1 & y_2 & y_3 & y_4 & y_5 & y_6 & y_7 \\
-y_1 & y_0 & -y_3 & y_2 & -y_5 & y_4 & y_7 & -y_6 \\
-y_2 & y_3 & y_0 & -y_1 & -y_6 & -y_7 & y_4 & y_5 \\
-y_3 & -y_2 & y_1 & y_0 & -y_7 & y_6 & -y_5 & y_4 \\
-y_4 & y_5 & y_6 & y_7 & y_0 & -y_1 & -y_2 & -y_3 \\
-y_5 & -y_4 & y_7 & -y_6 & y_1 & y_0 & y_3 & -y_2 \\
-y_6 & -y_7 & -y_4 & y_5 & y_2 & -y_3 & y_0 & y_1 \\
-y_7 & y_6 & -y_5 & -y_4 & y_3 & y_2 & -y_1 & y_0
\end{bmatrix}
$$

This matrix is an example of an 8×8 real orthogonal design, but it is not derived from a 4×4 complex design. In fact we will prove in Section 2.3 that a complex orthogonal design of size 4 cannot involve more than 6 real indeterminates; see Tarokh et al. [27], where nonexistence of complex designs with $t = s + 1 = 4$ is proved in the Appendix.

N = 8, R = 3/4: Observe that in the rules for multiplying octonions, there is one term $\overline{a_i}$ in each expression and it is paired with b_3 or $\overline{b_3}$. It follows that right multiplication of an octonion a by octonions of the form $b = (b_0, b_1, b_2, 0)$ can be represented as $aR(b_0, b_1, b_2, 0)$, where

$$
R(b_0, b_1, b_2, 0) = \begin{pmatrix}
b_0 & b_1 & b_2 & 0 \\
-\overline{b_1} & \overline{b_0} & 0 & b_2 \\
-\overline{b_2} & 0 & \overline{b_0} & -b_1 \\
0 & -\overline{b_2} & \overline{b_1} & b_0
\end{pmatrix}
$$

Since this is a matrix of complex indeterminates with orthogonal columns, $R(b_0, b_1, b_2, 0)$ is an orthogonal design. The matrix $U = I_2 \oplus \left(\frac{1}{\sqrt{2}} \begin{pmatrix} 1 & 1 \\ 1 & -1 \end{pmatrix} \right)$

is unitary, so that

$$
U^*R(b_0, b_1, b_2, 0)U = \left| \begin{array}{cc|cc} b_0 & b_1 & b_2/\sqrt{2} & b_2/\sqrt{2} \\ -\bar{b}_1 & \bar{b}_0 & b_2/\sqrt{2} & -b_2/\sqrt{2} \\ \hline -\bar{b}_2/\sqrt{2} & -\bar{b}_2/\sqrt{2} & \frac{(b_0+\bar{b}_0-b_1+\bar{b}_1)}{2} & \frac{(\bar{b}_0-b_0-b_1-\bar{b}_1)}{2} \\ -\bar{b}_2/\sqrt{2} & \bar{b}_2/\sqrt{2} & \frac{(\bar{b}_0-b_0-b_1-\bar{b}_1)}{2} & \frac{(b_0+\bar{b}_0+b_1-\bar{b}_1)}{2} \end{array} \right| \tag{2.2}
$$

provides a different representation of this design. Left multiplication of (2.2) by diag$[1, 1, -1, -1]$ gives the orthogonal design discovered by Tarokh et al. [27], that is derived from a particular pair of *amicable* orthogonal designs. Orthogonal designs X and Y are said to be *amicable* if $X + iY$ is a complex orthogonal design, that is, if and only if $XY^T = YX^T$. For $N = 4$, it easy to verify that

$$
X = \left[\begin{array}{cc|cc} x_0 & x_1 & x_2 & x_2 \\ -x_1 & x_0 & x_2 & -x_2 \\ \hline x_2 & x_2 & -x_0 & -x_1 \\ x_2 & -x_2 & x_1 & x_0 \end{array} \right] \qquad Y = \left[\begin{array}{cc|cc} y_0 & y_1 & y_2 & y_2 \\ y_1 & -y_0 & y_2 & -y_2 \\ \hline -y_2 & -y_2 & y_1 & y_0 \\ -y_2 & y_2 & y_0 & -y_1 \end{array} \right]
$$

are a pair of amicable orthogonal designs. The complex orthogonal design discovered by Tarokh et al. [27] is obtained by expressing the entries of $X + iY$ in terms of the indeterminates $b_j = x_j + iy_j$ and $\bar{b}_j$ for $j = 0, 1$, and 2. This connection between the original design and the octonions was made by Schmidt and Wright [23].

Remarks The smallest example of a pair of amicable orthogonal designs is given by

$$
X = \begin{pmatrix} x_0 & x_1 \\ -x_1 & x_0 \end{pmatrix}, \qquad Y = \begin{pmatrix} y_0 & y_1 \\ y_1 & -y_0 \end{pmatrix}.
$$

If we represent the complex number $a = a_0 + ia_1$ by the pair (a_0, a_1), then

$$
(a_0, a_1) \longrightarrow (a_0, a_1) \begin{pmatrix} x_0 & x_1 \\ -x_1 & x_0 \end{pmatrix} \quad \text{represents} \quad a \to (x_0 + ix_1)a ,
$$

$$
(a_0, a_1) \longrightarrow (a_0, a_1) \begin{pmatrix} x_0 & -x_1 \\ x_1 & x_0 \end{pmatrix} \quad \text{represents} \quad a \to \overline{(x_i + ix_1)a} , \text{ and}
$$

$$
(a_0, a_1) \longrightarrow (a_0, a_1) \begin{pmatrix} x_0 & x_1 \\ x_1 & -x_0 \end{pmatrix} \quad \text{represents} \quad a \to (x_0 + ix_1)\bar{a} .
$$

The identity $XY^T = YX^T$ expresses the equivalence of the maps $a \to xa \to y\overline{x}\overline{a}$ and $a \to y\bar{a} \to \bar{x}y\bar{a}$. . Amicable pairs of orthogonal designs serve as building blocks in the construction of larger designs. Given a real orthogonal design of type (p, q) on the variables z_0, z_1, simply replace z_0 by $\begin{pmatrix} x_0 & x_1 \\ -x_1 & x_0 \end{pmatrix}$, and replace z_1 by $\begin{pmatrix} y_0 & y_1 \\ y_1 & -y_0 \end{pmatrix}$, to obtain a new orthogonal design of type (p, p, q, q) on the variables x_0, x_1, y_0, y_1. For more information about amicable orthogonal designs, we refer the reader to Chapter 5 of Geramita and Seberry [8].

2.2 Normed algebras and sums of squares

We develop this connection between normed algebras and sums of squares through an example. Define bilinear forms $\Phi_i = \Phi_i(x_0, x_1, x_2, x_3; y_0, y_1, y_2, y_3)$ $i = 0, 1, 2, 3$ using the matrix (2.3) as follows:

$$
\begin{bmatrix} \Phi_0 \\ \Phi_1 \\ \Phi_2 \\ \Phi_3 \end{bmatrix} = \begin{bmatrix} x_0 & -x_1 & -x_2 & -x_3 \\ x_1 & x_0 & x_3 & -x_2 \\ x_2 & -x_3 & x_0 & x_1 \\ x_3 & x_2 & -x_1 & x_0 \end{bmatrix} \begin{bmatrix} y_0 \\ y_1 \\ y_2 \\ y_3 \end{bmatrix}.
$$

The celebrated Four Square Identity

$$(x_0^2 + x_1^2 + x_2^2 + x_3^2)(y_0^2 + y_1^2 + y_2^2 + y_3^2) = \Phi_0^2 + \Phi_1^2 + \Phi_2^2 + \Phi_3^2$$

expresses the fact that the rows of the matrix (2.3) are orthogonal. We may use the bilinear forms determined by the real orthogonal design to define a multiplication on vectors in $\mathbb{R}$. The product of the vectors (x_0, x_1, x_2, x_3) and (y_0, y_1, y_2, y_3) is $(\Phi_0, \Phi_1, \Phi_2, \Phi_3)$. Note that the vector $(1, 0, 0, 0)$ is a 2-sided multiplicative identity. We arrive at the standard multiplication of quaternions, since the matrices

$$
\begin{bmatrix} & -1 & & \\ 1 & & & \\ & & & 1 \\ & & -1 & \end{bmatrix}, \quad \begin{bmatrix} & & -1 & \\ & & & -1 \\ 1 & & & \\ & 1 & & \end{bmatrix}, \quad \text{and} \quad \begin{bmatrix} & & & -1 \\ & & 1 & \\ & -1 & & \\ 1 & & & \end{bmatrix}
$$

describe left multiplication by **i**, **j**, and **k** respectively. This multiplication turns $\mathbb{R}^4$ into a 4-dimensional algebra with the property that

$$(xy, xy) = (x, x)(y, y)$$

for all vectors x, y. This is the defining property of a *normed algebra*.

In general, a real orthogonal design with rate 1 determines an identity

$$(x_0^2 + x_1^2 + \cdots + x_{N-1}^2)(y_0^2 + y_1^2 + \cdots + y_{N-1}^2) = (\Phi_0^2 + \Phi_1^2 + \cdots + \Phi_{N-1}^2)$$

where each $\Phi_i = \Phi_i(x_0, \ldots, x_{N-1}; y_0, \ldots, y_{N-1})$ is a bilinear form. Conversely an N Square Identity determines a real orthogonal design. Multiplication of vectors in $\mathbb{R}^N$ according to the rule

$$xy = (\Phi_0(x; y), \ldots, \Phi_{N-1}(x; y))$$

then defines a normed algebra of dimension N over $\mathbb{R}$ with respect to the standard inner product; that is

$$(xy, xy) = (x, x)(y, y)$$

for all vectors x, y in $\mathbb{R}^N$.

Rate 1 complex designs determine normed algebras over $\mathbb{C}$. The complex bilinear forms Φ_0, Φ_1 determined by the 2×2 design are given by

$$\begin{pmatrix} \Phi_0 \\ \Phi_1 \end{pmatrix} = \begin{pmatrix} a_0 & -\bar{a}_1 \\ a_1 & \bar{a}_0 \end{pmatrix} \begin{pmatrix} b_0 \\ b_1 \end{pmatrix},$$

and the identity

$$(|a_0|^2 + |a_1|^2)(|b_0|^2 + |b_1|^2) = |\Phi_0|^2 + |\Phi_1|^2$$

reduces to the classical Four Square Identity. Note that the vector $(1, 0)$ acts as a 2-sided multiplicative identity for the normed complex algebra.

The problem of finding all N Square Identities reduces to finding all normed algebras, and writing down the multiplication table for each of these algebras relative to all orthonormal bases. The following classification theorem, proved by Hurwitz [12] in 1896, resolves most of the difficulties.

Theorem 2.1 (Hurwitz) *Every normed real algebra with an identity is isomorphic to one of the following four algebras: the real numbers, the complex numbers, the quaternions, or the Cayley numbers.*

Furthermore, every normed real algebra can be obtained from a normed algebra with an identity by introducing a new rule (○) for multiplication given by

$$u \circ v = A(u)B(v)$$

where A, B are orthogonal (norm preserving) transformations. For proofs, we refer the reader to Chapters 17 and 18 of Kantor and Solodovnikov [13]. Hence

Theorem 2.2 *The only values of N for which there exist*

(1) *an N Square Identity,*

(2) *a rate 1 real orthogonal design*

are $N = 1, 2, 4,$ and 8.

Remarks We may relax the definition of a real orthogonal design X to allow entries that are linear combinations of indeterminates, and to allow identities of the form

$$X^T X = \text{diag}[d_{i0}x_1^2 + \cdots + d_{iN-1}X_{N-1}^2] = \sum_{i=0}^{N-1} x_i^2 D_i$$

with positive coefficients d_{ij}. We write $X = \sum_{i=0}^{N-1} x_i A_i$ so that

$$
\begin{aligned}
A_i^T A_i &= D_i, \quad \text{for} \quad i = 0, 1, \ldots, N-1 \\
A_i^T A_j &= -A_j^T A_i, \quad \text{for} \quad 0 \le i < j \le N-1.
\end{aligned}
$$

If $D_i^{1/2}$ is the diagonal matrix defined by $D_i = D_i^{1/2} D_i^{1/2}$, and if $B_i = A_i D_i^{-1/2}$, then we have

$$
\begin{aligned}
B_i^T B_i &= I, \quad \text{for} \quad i = 0, 1, \ldots, N-1 \\
B_i^T B_j &= B_j^T B_i, \quad \text{for} \quad 0 \le i < j \le N-1.
\end{aligned}
$$

Setting $Q = \sum_{i=0}^{N-1} x_i B_i$, then

$$Q^T Q = (x_0^2 + \cdots + x_{N-1}^2) I_N.$$

The orthogonal design Q determines a normed algebra as described above. The classification of normed algebras implies that this relaxation of the definition does not extend the range of values of N for which there exists a rate 1 real orthogonal design. A similar analysis applies to complex designs.

2.3 Hurwitz–Radon families of matrices

In this section we derive fundamental limits on the rate of real and complex orthogonal designs. We also construct optimal designs that meet these upper bounds using an extraspecial 2-group that plays a central role in the mathematics of quadratic and alternating bilinear forms. We begin by describing this group, which also enters into the construction of certain quantum error correcting codes; for more details about this connection, see Calderbank et al. [4].

Let k be a fixed positive integer, let $N = 2^k$, and let V denote the vector space $\mathbb{Z}_2^k$. Label the standard basis of $\mathbb{R}^N$ as e_v, $v \in V$. For $a, b \in V$, define the permutation matrix $X(a)$ and diagonal matrix $Z(b)$ as follows:

$$X(a) : e_v \to e_{v+a} \quad \text{and} \quad Z(b) = \text{diag}[(-1)^{b \cdot v}].$$

The groups $X(V) = \{X(a) | a \in V\}$ and $\{z(b) | b \in V\}$ are isomorphic to the additive group V. Next we verify that elements in the group E generated by $X(V)$ and $Z(V)$ either commute or anti-commute: for all $a, b \in V$

$$X(a) Z(b) = (-1)^{a \cdot b} Z(b) X(a), \tag{2.3}$$

since

$$e_v(X(a)^{-1}Z(b)^{-1}X(a)Z(b)) = \left[(-1)^{b\cdots(v+a)}e_{v+a}\right]X(a)Z(b)$$
$$= (-1)^{b\cdot(v+a)}(-1)^{b\cdot v}e_v.$$

It is easy to verify that every element of E takes the form $\pm X(a)Z(b)$, that the centre $\Xi(E) = \langle -I_N \rangle$, and that the quotient $E/\Xi(E)$ is elementary abelian of order 2^{2k}. This is the defining property of an extraspecial 2-group; see Huppert [11, pp. 355–357], Aschbacher [2, p. 109], or Suzuki [25, pp. 97–98], for more information about extraspecial 2-groups and quadratic forms.

It follows from (2.3) that

$$(X(a)Z(b))^2 = (-1)^{a\cdot b}I_N,$$

and that $X(a)Z(b)$ commutes with $X(a)Z(b)$ if and only if

$$a\cdot b' + a'\cdot b = 0.$$

If we represent $\pm X(a)Z(b)$ by the pair of binary vectors $(a|b)$, then the scalar product $Q(a|b) = a\cdot b$ defines a quadratic form on the quotient $E/\Xi(E)$, and the associated bilinear form is

$$Q(a+a'|b+b') - Q(a|b) - Q(a'|b') = a\cdot b' + a'\cdot b.$$

It is often useful to express the group elements in E as tensor products of the 2×2 Pauli matrices

$$\sigma_x = \begin{pmatrix} 0 & 1 \\ 1 & 0 \end{pmatrix}, \quad \sigma_z = \begin{pmatrix} 1 & 0 \\ 0 & -1 \end{pmatrix}, \quad \text{and} \quad \sigma_y = \sigma_x\sigma_z = \begin{pmatrix} 0 & -1 \\ 1 & 0 \end{pmatrix}.$$

The group E consists of all tensor products $\pm\omega_1 \otimes \cdots \otimes \omega_k$, where each ω_j is one of I_2, σ_x, σ_y, or σ_z. The correspondence is given by

$$(a_1\cdots a_k|b_1\cdots b_k) \leftrightarrow \pm\omega_1 \otimes \cdots \otimes \omega_k$$

where

$$\omega_j = \begin{cases} I_2, & \text{if } a_j = b_j = 0, \\ \sigma_x, & \text{if } a_j = 1 \text{ and } b_j = 0, \\ \sigma_z, & \text{if } a_j = 0 \text{ and } b_j = 1, \\ \sigma_y, & \text{if } a_j = b_j = 1. \end{cases}$$

Lemma 2.3 The matrix $(a_1\cdots a_k|b_1\cdots b_k)$ is symmetric or skew-symmetric according as the number of pairs (a_jb_j) equal to $(1,1)$ is even or odd.

Proof $(\omega_1 \otimes \cdots \otimes \omega_k)^T = \omega_1^T \otimes \cdots \otimes \omega_k^T.$ $\qquad\qquad\square$

This binary shorthand makes it easy to verify properties of families of matrices.

Lemma 2.4 *The group E contains the following families of matrices:*

(a) *for $k = 1$, a single skew-symmetric 2×2 matrix;*

(b) *for $k = 2$, a family of 3 skew-symmetric 4×4 matrices that pairwise anti-commute;*

(c) *for $k = 3$, a family of 7 skew symmetric 8×8 matrices that pairwise anti-commute.*

Proof Take

$(1\|1)$	$(10\|10)$	$(010\|010)$
	$(11\|01)$	$(011\|001)$
(a)	$(01\|11)$	$(001\|111)$
		$(101\|011)$
	(b)	$(110\|101)$
		$(111\|100)$
		$(100\|110)$

(c)

Let X be a real orthogonal design of size N and type $(u_0, \dots, u_{s-1})$. We write $X = \sum_{i=0}^{s-1} x_i A_i$, so that

$$A_i A_i^T = u_i I_N, \quad \text{for} \quad i = 0, 1, \dots, s-1, \quad \text{and}$$
$$A_i A_j^T = -A_j A_i^T, \quad \text{for} \quad 0 \le i < j \le s-1.$$

Setting $B_i = \frac{1}{(u_0 u_i)^{1/2}} A_i A_0^T$ provides a representation of this design with respect to a different orthogonal basis. Now we have $X = \sum_{i=0}^{s-1} x_i B_i$, where $B_0 = I_N$,

$$
\begin{aligned}
B_i^T &= -B_i, \quad \text{for} \quad i = 1, \dots, s-1, \\
B_i B_i^T &= I_N, \quad \text{for} \quad i = 1, \dots, s-1, \quad \text{and} \\
B_i B_j^T &= -B_j B_i^T, \quad \text{for} \quad 1 \le i < j \le s-1.
\end{aligned}
\tag{2.4}
$$

Equations (2.4) are the defining conditions for a *Hurwitz-Radon (H-R) family* of matrices. The next theorem, proved by Radon [22] in 1922, can be interpreted as a fundamental upper bound on the rate of a real orthogonal design.

Theorem 2.5 (Radon) *Given $N = 2^{4a+b} N_0$, where N_0 is odd, define $\rho(N) = 8a + 2^b$. Then*

(1) *The size s of a Hurwitz-Radon family of real $N \times N$ matrices is at most $\rho(N) - 1$, and*

(2) *There exists a Hurwitz-Radon family containing exactly $s = \rho(N) - 1$ integer $N \times N$ matrices.*

Here we only give the construction that achieves the bound. Note that if $N = 2^t N_0$, where N_0 is odd, then $\rho(N) = \rho(2^t)$. Since an H-R family of $2^t \times 2^t$ matrices B_i extends to an H-R family of $N \times N$ matrices $B_i \otimes I_{N_0}$, it follows that it is sufficient to provide a construction for integers N of the form $N = 2^t$.

Lemma 2.4 provides extremal H-R families of matrices for $N = 2, 4$, and 8. The general case is proved by induction. Let $N = 2^{4a+3}$, and suppose that $(a^i|b^i)$, $i = 1, \ldots, \rho(N) - 1$ is an H-R family of integer $N \times N$ matrices contained in the extraspecial group E. There are 4 cases to consider.

$\mathbf{N'} = 2^{4(a+1)}$, $\rho(\mathbf{N'}) = \rho(\mathbf{N}) + 1$: The new extremal family is $(10|10)$ and $(0a^i|1b^i)$, for $i = 1, 2, \ldots, \rho(N) - 1$.

$\mathbf{N'} = 2^{4(a+1)+1}$, $\rho(\mathbf{N'}) = \rho(\mathbf{N}) + 2$: The new extremal family is $(100|100)$, $(110|010)$, and $(00a^i|10b^i)$, for $i = 1, 2, \ldots, \rho(N) - 1$.

$\mathbf{N'} = 2^{4(a+1)+2}$, $\rho(\mathbf{N'}) = \rho(\mathbf{N}) + 4$: The new extremal family is $(1000|1000)$, $(1p0|1q0)$, where $(p|q) = (10|10)$, $(11|01)$, $(01|11)$ is an extremal H-R family of 4×4 matrices, and $(000a^i|100b^i)$ for $i = 1, 2, \ldots, \rho(N) - 1$.

$\mathbf{N'} = 2^{4(a+1)+3}$, $\rho(\mathbf{N'}) = \rho(\mathbf{N}) + 8$: The new extremal family is $(1000|1000)$, $(1p0|1q0)$, where $(p|q)$ is drawn from an extremal H-R family of 8×8 matrices, and $(0000a^i|1000b^i)$, for $i = 1, 2, \ldots, \rho(N) - 1$.

The last case is just the inductive hypothesis for $N = 2^{4(a+1)+3}$ so the proof is complete.   □

Following Wolfe [35] we now consider amicable pairs X, Y of real orthogonal designs of size N, where X has type $(u_0, \ldots, u_s)$ and Y has type $(v_1, \ldots, v_t)$. We write $X = \sum_{i=0}^{s} x_i A_i$, and $Y = \sum_{i=1}^{t} y_i B_i$ so that

$$A_i A_i^T = u_i I_N, \quad \text{for } i = 0, 1, \ldots, s; \quad B_i B_i^T = v_i I_N, \quad \text{for } i = 1, \ldots, t,$$

$$A_i A_j^T = -A_j A_i^T, \quad \text{for } 0 \leq i < j \leq s; \quad B_i B_j^T = -B_j B_i^T, \quad \text{for } 1 \leq i < j \leq t,$$

$$\text{and } A_i B_j^T = B_j A_i^T, \quad \text{for } i = 0, 1, \ldots, s \text{ and } j = 1, \ldots, t.$$

Setting

$$\alpha_i - \frac{1}{(u_o u_i)^{1/2}} A_i A_0^T, \quad \text{for } i = 0, 1, \ldots, s;$$

$$B_j = \frac{1}{(u_o v_j)^{1/2}} B_j A_0^T, \quad \text{for } j = 1, \ldots, t,$$

provides a representation of this amicable pair with respect to a different orthogonal basis. Now we have $\alpha_0 = I_N$,

$$\alpha_i = -\alpha_i^T, \quad \text{for } i = 1, \ldots, s; \quad \beta_i = \beta_i^T, \quad \text{for } i = 1, \ldots, t, \quad (2.5)$$

$$\alpha_i^2 = -I_N, \quad \text{for } i = 1, \ldots, s; \quad \beta_i^2 = I_N, \quad \text{for } i = 1, \ldots, t. \quad (2.6)$$

$$\alpha_i\alpha_j + \alpha_j\alpha_i = 0, \quad \text{for } 1 \le i < j \le s; \quad \beta_k\beta_l + \beta_l\beta_k = 0, \quad \text{for } 1 \le k < l \le t, \tag{2.7}$$

$$\text{and} \quad \alpha_i\beta_j = -\beta_j\alpha_i, \quad \text{for } i = 1,\dots,s \text{ and } j = 1,\dots,t. \tag{2.8}$$

A family of real $N \times N$ matrices $\{\alpha_i, 1 \le i \le s; \beta_j, 1 \le j \le t\}$ satisfying (2.5)–(2.8) will be called a *Hurwitz-Radon family of type (s,t)* or simply an *H-R (s,t) family of order N*. Equations (2.6)–(2.8) are a form of the relations found by Clifford [6] in his attempt to generalize the quaternions. Formal algebras over $\mathbb{R}$ which satisfy these relations are called Clifford algebras, and were used by Kawada and Iwahori [14] in their analysis of sets of anti-commuting real and complex matrices.

A *Clifford algebra $C^{s,t}$* of type (s,t) is an algebra over $\mathbb{R}$ with generators ϵ, $a_1,\dots,a_s,b_1,\dots,b_t$ and fundamental relations

$$\epsilon^2 = \epsilon, \quad \epsilon a_i = a_i\epsilon = a_i \text{ for } 1 \le i \le s; \quad \epsilon b_j = b_j\epsilon = b_j \text{ for } 1 \le j \le t,$$

$$a_i^2 = -\epsilon, \quad \text{for } i = 1,\dots,s; \quad b_i^2 = \epsilon, \quad \text{for } i = 1,\dots,t,$$

$$a_i a_j = -a_j a_i, \quad \text{for } 1 \le i < j \le s; \quad b_k b_l = -b_l b_k \text{ for } 1 \le k < l \le t,$$

and
$$a_i b_j = -b_j a_i, \quad \text{for } i = 1,\dots,s \text{ and } j = 1,\dots,t.$$

We refer the reader to Witt [33], Porteous [20], or Lam [15] for more information about quadratic forms and the representation theory of Clifford algebras. Existence of an H-R (s,t) family of order N implies that the algebra $C^{s,t}$ has a matrix representation of degree N over $\mathbb{R}$. The converse requires proof, since the images of the generators need not be symmetric, skew-symmetric, or even orthogonal. We begin with a result of Kawada and Iwahori that describes the irreducible matrix representations of $C^{s,t}$ over $\mathbb{R}$ with minimal degree.

Theorem 2.6 *Let d be the minimal degree (greater than 1) of an irreducible matrix representation of $C^{s,t}$. Then d varies with the parity of $s+t$ as follows. For $s+t = 2k$,*

(1) *if $t - k = 0$ or $1 \pmod 4$, then $d = 2^k$;*

(2) *if $t - k = 2$ or $3 \pmod 4$, then $d = 2^{k+1}$.*

For $s+t = 2k+1$,

(3) *if $t - k = 0, 2,$ or $3 \pmod 4$, then $d = 2^{k+1}$;*

(4) *if $t - k = 1 \pmod 4$, then $d = 2^k$.*

Remarks Let e_i, $i = 1, \ldots, s+t$ be generators for the Clifford algebra $C^{s,t}$, and for each subset I of $\{1, \ldots, s+t\}$, with the natural ordering, let $e_I = \Pi_{i \in I} e_i$. Clearly every element of the Clifford algebra is a linear combination of the elements e_I. In fact it turns out that the dimension of the linear space $\langle e_I \rangle$ is 2^{s+t} or 2^{2+t-1}, the lower value being a possibility only if $s - t - 1$ is divisible by 4, in which case $s + t$ is odd and $e_1 \cdots e_{s+t} = \pm 1$. We refer the reader to Chapter 15 of Porteous [20] for a proof. As an example, consider the quaternions, a 4-dimensional Clifford algebra of type $(0, 3)$, and note that the product $\mathbf{ijk} = -1$. It is clear from Theorem 2.6 that the maximum rate of a complex design of size N will grow as $\log N$.

If $s + t = 8h + p$, with $0 \le p < 8$, then $d = 2^{4h+\delta}$, where δ is given by Table 1.

t (mod 4) \ ρ	0	1	2	3	4	5	6	7
0	0	1	2	2	3	3	3	3
1	0	0	1	2	3	3	4	4
2	1	1	1	1	2	3	4	4
3	1	1	2	2	2	2	3	4

Table 1: Values of δ.

Given t symmetric, anti-commuting orthogonal matrices of size N, let $\rho_t(N) - 1$ be the number of skew-symmetric, anti-commuting orthogonal matrices of size N that anti-commute with the given t matrices. Thus

$$\rho_t(N) = \max\{s \mid C^{s-1,t} \text{ has an irreducible matrix representation over } \mathbb{R} \text{ of degree } N\}.$$

By Theorem 2.5, we have $\rho_0(N) = \rho(N)$ By the remark following Theorem 2.6, the dimension of a Clifford algebra is always a power of 2. Hence $\rho_t(2^a b) = \rho_t(2^a)$, for b odd, and it is sufficient to consider integers N of the form $N = 2^a$.

Let $N = 2^{4a+b}$, where $0 \le b < 4$. Then $\rho_t(N) - 1 = 8a - t + \lambda$ where λ is given by Table 2. The first step in calculating λ is to use Table 1 to find the largest p with $d = 2^{4a+b}$.

Lemma 2.7 We have

(1) $\rho_1(2) - 1 = 1$, $\rho_2(2) - 1 = 1$, $\rho_2(8) - 1 = 0$;

(2) $\rho_t(2N) = \rho_{t-1}(N) + 1$;

t (mod 4) \ b	0	1	2	3
0	0	1	3	7
1	1	2	3	5
2	-1	3	4	5
3	-1	1	5	6

Table 2: Values of λ.

(3) $\rho_t(N) = \rho_{t+8}(2^4 N)$.

Proof This is done by verification using Table 2. For example, in part (2), take $N = 2^{4a}$, and take $t = 0$ (mod 4). Then $t - 1 = 3$ (mod 4), and Table 2 gives $\rho_{t-1}(N) = 8a - (t - 1) - 1$. Now $2N = 2^{4a+1}$ and Table 2 gives $\rho_t(2n) = 8a - t + 1$, as required. The other cases follow similarly. □

Now the quantities $\rho_t(N)$ are completely determined by the quantities $\rho_0(N')$ and by the above lemma. Small values are collected in the following table.

N \ t	0	1	2	3	4	5	6	7	8	9	10	11	12
2	t	1	t										
2^2	3	2	2	2									
2^3	7	4	3	3	3	0							
2^4	8	8	5	4	4	4	1	0	0	0			
2^5	9	9	9	6	5	5	5	2	1	1	1		
2^6	11	10	10	10	7	6	6	6	3	2	2	2	
2^7	15	12	11	11	11	8	7	7	7	4	3	3	3
2^8	16	16	13	12	12	12	9	8	8	8	5	4	4

Table 3: The quantities $\rho_t(N) - 1$ for $t = 0, 1, \ldots, 12$ and for $N = 2^h$, $h = 1, \ldots, 8$. Note the 8-periodicity: the dotted line marks the beginning of the second period.

The next two lemmas combine to show there exists an H-R $(\rho_t(N) - 1, t)$ family of size N.

Lemma 2.8 (Slide Lemma) *Existence of an H-R (s,t) family of order N implies existence of an H-R $(s+1, t+1)$ family of order $2N$.*

Proof Suppose that $(a^i|b^i)$, $i = 1, \ldots, s$; $(c^j|d^j)$, $j = 1, \ldots, t$ is an H-R (s,t) family of $N \times N$ integer matrices contained in the extraspecial group E. The H-R $(s+1, t+1)$ family is

$$(1a^i|0b^i), \quad i = 1, \ldots, s \quad \text{and} \quad (10|10), \quad \text{together with}$$
$$(1c^j|0d^j), \quad j = 1, \ldots, t \quad \text{and} \quad (00|10).$$

Again these are integer matrices contained in the appropriate extraspecial group. □

Lemma 2.9 (Jump Lemma) *Existence of an H-R (s,t) family of order N implies existence of an H-R $(s, t+8)$ family of order $16N$.*

Proof Suppose that $(a^i|b^i)$, $i = 1, \ldots, s$; $(c^j|d^j)$, $j = 1, \ldots, t$ is an H-R (s,t) family of $N \times N$ integer matrices contained in E. The H-R $(s, t+8)$ family is

$$(1110a^i|1011b^i), \quad i = 1, \ldots, s, \quad \text{together with}$$
$$(1110c^j|1011d^j), \quad j = 1, \ldots, t, \quad \text{and}$$
$$(\mathbf{p}0|\mathbf{q}0) \text{ where } (\mathbf{p}|\mathbf{q}) \text{ range over the 8 pairs}$$

(0000|1111)	(0001|0000)
(1000|0001)	(1011|0111)
(0100|1001)	(0111|0011)
(0010|1101)	(1101|0101)

Again these are integer matrices contained in the appropriate extraspecial group. □

Wolfe [35] proves there exists an H-R $(\rho_t(N) - 1, t)$ family of integer matrices by using Lemma 2.8 to slide down the diagonal in Table 3, and by using Lemma 2.9 to jump from period to period.

Theorem 2.10 (Wolfe) *There exists an amicable pair X, Y of real orthogonal designs of size N, where X has type $(1, \ldots, 1)$ on variables $x_0, x_1, \ldots, x_s$, and Y has type $(1, \ldots, 1)$ on variables $y_1, \ldots, y_t$ if and only if $s \leq \rho_t(N) - 1$.*

Remarks In fact there exist pairs X, Y where the entries of X are 0, $\pm x_0, \ldots, \pm x_s$ and the entries of Y are $0, \pm y_1, \ldots, \pm y_t$. This is because we may construct an H-R (s,t) family using matrices from the extraspecial group E that have nonzero entries $0, \pm 1$ in different positions.

Theorem 2.11 (Wolfe) *Let X, Y be an amicable pair of real orthogonal designs of size $N = 2^h N_0$, where N_0 is odd. Then the total number of variables in X and Y is $2h + 2$, and this bound is achieved by designs X, Y that each involve $h + 1$ variables.*

Proof Define $\tau(N) = \max_{t \geq 0}\{\rho_t(N) + t\}$. Setting $h = 4a + b$, we obtain $\tau(N)$ as a function of b using Table 2.

b	0	1	2	3
$\tau(N)$	$2(4a) + 2$	$2(4a + 1) + 2$	$2(4a + 2) + 2$	$2(4a + 3) + 2$

Hence $\tau(N) = 2h + 2$ as required. Existence of amicable pairs X, Y, where each of X, Y involves $h + 1$ variables follows from Lemma 2.8 (the Slide Lemma). The starting point is the 2×2 complex orthogonal design that involves 2 complex variables. $\qquad\qquad\square$

Theorem 2.12 *The rate of a complex orthogonal design of order $N = 2^h$ is bounded above by $(2h + 2)/N$, and there are designs that meet the upper bound with equality.*

Remarks The following table shows that the 4×4 complex design discovered by Tarokh et al. [27] is optimal in the sense that it maximizes the total number of real variables.

Number of Variables in Y	Number of Variables in x
t	$\rho_t(4)$
4	0
3	3
2	3
1	3
0	4

This is the tip of a mathematical iceberg. Inspired by Wolfe, Shapiro [24] has abstracted existence questions for amicable pairs of real orthogonal designs to a more general setting, where he provides a complete solution.

3 Space-time codes for wireless communication

3.1 The system model

This section presents a mathematical model of a communications system with N transmit antennas and M receive antennas. The space-time encoder transforms the input data at time l into N code symbols $c_1(l), \ldots, c_N(l)$ that are transmitted simultaneously from the different transmit antennas. We assume the signal constellation is scaled so that the average energy is equal to 1. When this signal constellation consists of equally spaced points on the unit circle, no scaling is necessary. If E_s is the total energy radiated by all transmit antennas in a given time slot, then E_s/N is the energy per code symbol.

The signal $r_j(l)$ received by antenna j at time l is given by

$$r_j(l) = \sqrt{E_s/N} \sum_{i=1}^{N} h_{ij}(l)c_i(l) + \eta_j(l), \quad j = 1, \ldots, M, \qquad (3.1)$$

where the noise $\eta_j(l)$ at time l is modelled as independent samples of a zero-mean complex Gaussian random variable with variance $N_0/2$ per dimension. Note that we assume the noise processes at different receive antennas are independent. The coefficients $h_{ij}(l)$ model the effect of fading on signals from transmit antenna i to receive antenna j. We assume that signals transmitted from different antennas are subject to independent fades, and note that this can be achieved by separating transmit antennas by more than half the underlying wavelength, or by using antennas with different polarizations. We also assume that the path gains $h_{ij}(l)$ are constant during a data frame $(h_{ij}(1) = \cdots = h_{ij}(L))$, and that they vary from one frame to another. This variation is modelled as independent samples of a complex Gaussian variable, possibly with nonzero mean, and with variance 0.5 per dimension.

We rewrite (3.1) in matrix form as

$$r(l) = \sqrt{E_s/N}\,H(l)c(l) + \eta(l)\,, \tag{3.2}$$

where $r(l) = (r_1(l), \ldots, r_M(l))^T$, $H(l) = [h_{ij}(l)]$, $c(l) = (c_1(l), \ldots, c_N(l))^T$, and $\eta(l) = (\eta_1(l), \ldots, \eta_M(l))^T$. The *signal to noise ratio (SNR)* per receive antenna is given by E_s/N_0.

Next we consider the squared distance $d^2(c, c')$ between two codewords c and c' at the output of the wireless channel. We have

$$
\begin{aligned}
d^2(c, c') &= \sum_{j=1}^{N}\sum_{l=1}^{L} |h_{ij}(c_1(l) - c_1'(l)) + \cdots + h_{Nj}(c_N(l) - c_N'(l))|^2 \\
&= \sum_{j=1}^{M} h_j B B^* h_j^*\,,
\end{aligned}
$$

where $h_j = (h_{1j}, \ldots, h_{Nj})$ is the vector of path gains from the different transmit antennas to the jth receive antenna, and

$$B_{il} = [c_i(l) - c_i'(l)]\,.$$

If the path gains are known at the mobile, then standard methods (see Proakis [21]) show that the probability $P(c, c')$ of transmitting c and deciding in favour of c' satisfies

$$P(c, c') \leq \left(\prod_{i=1}^{R} \lambda_i\right)^{-M} (E_s/4N_0)^{-RM}\,, \tag{3.3}$$

where R is the rank of the matrix B, and λ_i, $i = 1, \ldots, R$ are the nonzero eigenvalues of BB^*. The first term on the right side measures the coding gain achieved by the space-time code, and the second term measures the diversity gain that results from multiple paths between base station and mobile. The coding gain is an approximate measure of the performance improvement over an uncoded system operating with the same diversity gain.

The first objective in designing a space-time code is to maximize diversity gain, as given by the rank of the error matrix B for all pairs of codewords c, c'. For a given diversity gain, the objective of code design is to maximize the product of the nonzero eigenvalues of B, again for all pairs of codewords c, c'.

3.2 Space-time trellis codes

The simplest form of transmit diversity is the delay diversity scheme proposed by Wittneben [34] for two transmit antennas, where a signal is transmitted from the second antenna, then delayed one time slot, and transmitted from the first antenna. Figure 3A shows an 8-state trellis representation of delay diversity for the 8-PSK constellation, and Figure 3B shows an 8-state space-time code for 8-PSK.

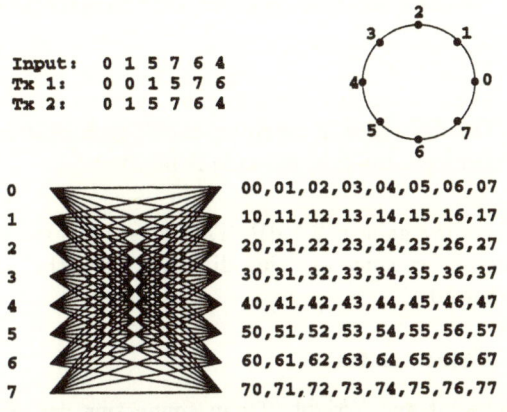

8-PSK 8-State Delay Diversity Code with 2 Transmit Antennas

Figure 3A. Delay diversity. The edge label xy means that symbol x is transmitted from the first antenna and symbol y from the second antenna. Labels on edges leaving a given state disagree in the second position. Labels on edges entering a given state disagree in the first position. Hence the diversity gain is 2.

The input data is encoded as a path through the *trellis* obtained by concatenating copies of the encoder state diagram. The decoder has a copy of this trellis. It processes the noisy samples and tries to find the path taken by the data. The decoding algorithm is dynamic programming, known to communication theorists as the Viterbi algorithm. Every time slot the decoder calculates and stores the most likely path terminating in a given state. The decoder also calculates the path metric, which measures distance from the received signals to the codeword corresponding to the most likely path. Both schemes provides diversity gain ($R = 2$), but with the space-time code there

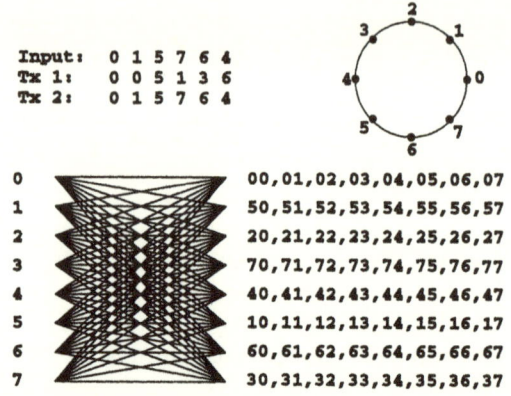

Input:	0 1 5 7 6 4
Tx 1:	0 0 5 1 3 6
Tx 2:	0 1 5 7 6 4

0	00,01,02,03,04,05,06,07
1	50,51,52,53,54,55,56,57
2	20,21,22,23,24,25,26,27
3	70,71,72,73,74,75,76,77
4	40,41,42,43,44,45,46,47
5	10,11,12,13,14,15,16,17
6	60,61,62,63,64,65,66,67
7	30,31,32,33,34,35,36,37

8-PSK 8-State Space-Time Code with 2 Transmit Antennas

Figure 3B. Space-time code. Labelling of even numbered states is identical to delay diversity. The difference in labelling of odd numbered states is that the symbol transmitted from the first antenna is negated.

is an additional coding gain of 2.5 dB. This means the space-time code will provide the same performance as delay diversity when the transmit power is reduced by a factor $10^{1/4}$.

3.3   Space-time block codes

We follow Schmidt and Wright [23] in connecting decoding of the 2×2 block space-time code with the algebra of quaternions. Observe that the map φ given by

$$\varphi \begin{pmatrix} r_1 \\ r_2 \end{pmatrix} = \begin{pmatrix} r_1 \\ -\bar{r}_2 \end{pmatrix}$$

is an isometry of complex space and that φ^2 is the identity. Define

$$Q(x_1, x_2) = \begin{pmatrix} x_1 & x_2 \\ -\bar{x}_2 & \bar{x}_1 \end{pmatrix}$$

so that

$$\varphi(Q(h)c) = Q(c)h \tag{3.4}$$

for all vectors $h = (h_1, h_2)^T$ and $c = (c_1, c_2)^T$ in $\mathbb{C}^2$. We may write (1.1) as

$$\varphi(r) = Q(h)c + \eta, \tag{3.5}$$

or as

$$r = Q(c)h + \varphi(\eta). \tag{3.6}$$

Decoding is possible, even when path gains are not known at the mobile, by using equations (3.5) and (3.6) in combination. Transmission begins with a known codeword c, and (3.6) is used to estimate the path gains h. The estimated path gains are then used to decode the next codeword via (3.5), and this codeword is used to provide a fresh estimate of the path gains via (3.6). This is called *incoherent detection* and it performs within 3 dB of *coherent detection* where perfect channel state information is available to the receiver; see [26] for details.

Next we interpret vectors in $\mathbb{C}^2$ as biquaternions, and recall that left multiplication by $Q(h)$ corresponds to right multiplication by the biquaternion $h^+ = (h_1, -\overline{h}_2)^T$. The decoding problem is that of finding the biquaternion codeword α that minimizes $\|Q(h)\alpha - \varphi(r)\|^2$. Now

$$\|Q(h)\alpha - \varphi(r)\|^2 = \|\alpha \circ h^+ - r\|^2$$

where $\circ$ denotes multiplication of biquaternions. If $\hat{h} = (\overline{h}_1, \overline{h}_2)$ then $h^+\hat{h} = \|h\|^2$, so that

$$\|Q(h)\alpha - \varphi(r)\|^2 = \left\| \alpha - \frac{r \circ \hat{h}}{\|h\|^2} \right\|^2 \|h\|^2. \qquad (3.7)$$

The first term in (3.7) provides a scaling of the received signal r to the energy of the constellation, and the second term represents the diversity gain.

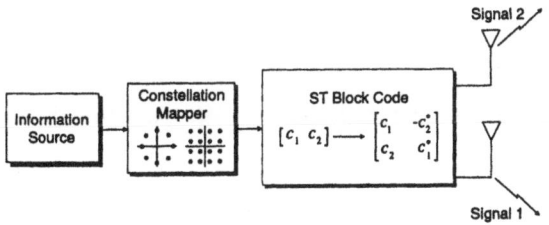

Figure 4: Space-time block coding with 2 transmit antennas.

4 Interference suppression with space-time block codes

4.1 The zero forcing solution

We follow Naguib and Seshadri [16, 17] in describing how a mobile terminal with two receive antennas uses the second antenna to separate two synchronous co-channel users, each employing the 2×2 block space-time code. Figure 5 is a representation of the system architecture. Define vectors r_1, r_2 where the entries of the vector r_i are the signals received at antenna i over two consecutive symbol periods. If $c = (c_1, c_2)$ and $s = (s_1, s_2)$ are the codewords transmitted by the first and second users respectively, then

$$
\begin{aligned}
r_1 &= H_1 \cdot c + G_1 \cdot s + \eta_1 \\
r_2 &= H_2 \cdot c + G_2 \cdot s + \eta_2
\end{aligned}
\tag{4.1}
$$

where the vectors η_1 and η_2 are complex Gaussian random variables with zero mean and covariance $N_0 I_2$. The matrices H_1 and H_2 capture the path gains from the first user to the first and second receive antennas respectively. The matrices G_1 and G_2 capture the path gains from the second user to the first and second receive antennas respectively. What is important is that the structure of the matrices H_1, H_2, G_1 and G_2 is identical, and this structure is given by (1.1).

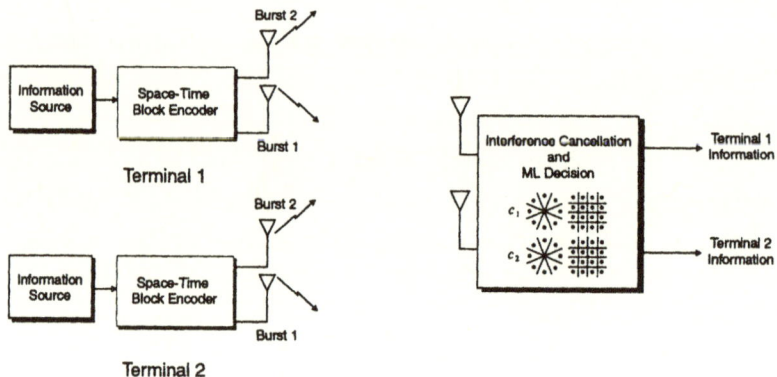

Figure 5: Interference cancellation with space-time block codes.

We now rewrite (4.1) in matrix form as

$$
r = \begin{bmatrix} r_1 \\ r_2 \end{bmatrix} = \begin{bmatrix} H_1 & G_1 \\ H_2 & G_2 \end{bmatrix} \begin{bmatrix} c \\ s \end{bmatrix} + \begin{bmatrix} \eta_1 \\ \eta_2 \end{bmatrix},
\tag{4.2}
$$

and we set

$$
H = \begin{bmatrix} H_1 & G_1 \\ H_2 & G_2 \end{bmatrix}
$$

The zero-forcing decoder employs a linear combination of received symbols to remove interference between the two users without any regard for noise enhancement. Naguib and Seshadri show that, if

$$
W = \begin{bmatrix} I_2 & -G_1 G_2^{-1} \\ -H_2 H_1^{-1} & I_2 \end{bmatrix},
\tag{4.3}
$$

then

$$W \begin{bmatrix} r_1 \\ r_2 \end{bmatrix} = \begin{bmatrix} \tilde{r}_1 \\ \tilde{r}_2 \end{bmatrix} = \begin{bmatrix} \tilde{H} & 0 \\ 0 & \tilde{G} \end{bmatrix} \begin{bmatrix} c \\ s \end{bmatrix} + \begin{bmatrix} \tilde{\eta}_1 \\ \tilde{\eta}_2 \end{bmatrix}. \tag{4.4}$$

The algebraic structure of the block space-time code (closure under addition, multiplication, and taking inverses) implies that the matrices $\tilde{H}$ and $\tilde{G}$ have the same structure as the matrices H_1, H_2, G_1 and G_2. The matrix W transforms the problem of joint detection of two co-channel users into separate detection of two individual space-time users. Essentially it reduces the joint detection problem to two instances of a previously solved problem.

Remarks In the zero forcing solution, it is the noise that is *forced* to be zero. Note that the matrix W is used to invert the channel, even when the power of the desired or interfering path is small, and the result is magnification of the noise in those circumstances. The Minimum Mean Squared Error (MMSE) solution described below is designed to limit this magnification.

The matrix H plays the role of the correlation matrix in multi-user detection in Code Division Multiple Access (CDMA) systems so that

$$\begin{bmatrix} H_1 & G_1 \\ H_2 & G_2 \end{bmatrix}^{-1} = \begin{bmatrix} \tilde{H}^{-1} & 0 \\ 0 & \tilde{G}^{-1} \end{bmatrix} \begin{bmatrix} I_2 & -G_1 G_2^{-1} \\ -H_2 H_1^{-1} & I_2 \end{bmatrix}$$

plays the role of the decorrelating detector. We refer the reader to Verdu [31] for more information about multi-user detection.

Given K synchronous co-channel users, each employing a space-time block code designed for N transmit antennas, there will be NK interfering signals arriving at the receiver. A substantial body of work by Winters and colleagues (see [32]) has shown that $N(K-1)+1$ antennas at the receiver are able to suppress $N(K-1)$ interfering signals and provide diversity order N to the desired user. It is important to note that this assumes no correlation between the interfering signals. Naguib and Seshadri [16, 17, 18] have used correlation of space-time signalling across time and space to show that only K receive antennas are needed to suppress $K-1$ space-time users, while providing diversity order N to the desired user. If there are M antennas at the receiver, where $M \geq K$, then diversity order $N(M-K+1)$ is provided to the desired user.

4.2 The minimum mean squared error solution

Define column vectors h_1, h_2, g_1 and g_2 by

$$[h_1, h_2] = \begin{bmatrix} H_1 \\ H_2 \end{bmatrix} \quad \text{and} \quad [g_1, g_2] = \begin{bmatrix} G_1 \\ G_2 \end{bmatrix}.$$

Let $\mathcal{H}$ be the subspace spanned by h_1 and h_2, and let $\mathcal{G}$ be the subspace spanned by g_1 and g_2. Let $\Gamma = Es/N_0$ be the Signal to Noise Ratio (SNR). Then the covariance matrix M of the received signal r is given by

$$M = E[rr^*] = HH^* + \frac{1}{\Gamma}I_4.$$

An estimate of the correlation matrix M is the input to the MMSE solution. Note that M includes the effects of any interfering signals.

Lemma 4.1 *If $i \neq j$, then*

$$h_i^* M h_j = g_i^* M g_j = 0.$$

More generally, let $v_1, v_2 \in \mathcal{H}$, $x_1, x_2 \in \mathcal{G}$, where

$$[v_1 v_2] = \begin{bmatrix} V_1 \\ V_2 \end{bmatrix}, \quad [x_1, x_2] = \begin{bmatrix} X_1 \\ X_2 \end{bmatrix},$$

and V_1, V_2, X_1 and X_2 have the structure of the 2×2 block space-time code. If $i \neq j$, then

$$v_i^* M v_j = v_i^* \begin{pmatrix} G_1 \\ G_2 \end{pmatrix} (G_1^* G_2^*) v_j = 0,$$

and

$$x_i^* M x_j = x_i^* \begin{pmatrix} G_1 \\ G_2 \end{pmatrix} (G_1^* G_2^*) x_j = 0.$$

Proof Observe that

$$\begin{aligned} M &= \begin{pmatrix} H_1 & G_1 \\ H_2 & G_2 \end{pmatrix} \begin{pmatrix} H_1^* & H_2^* \\ G_1^* & G_2^* \end{pmatrix} + \frac{1}{\Gamma}I_4 \\ &= \begin{pmatrix} H_1 \\ H_2 \end{pmatrix} (H_1^* H_2^*) + \begin{pmatrix} G_1 \\ G_2 \end{pmatrix} (G_1^* G_2^*) + \frac{1}{\Gamma}I_4, \end{aligned}$$

where the first term represents orthogonal projection onto $\mathcal{H}$, and the second term represents orthogonal projection onto $\mathcal{G}$. If $i \neq j$, then

$$v_i^* M v_j = v_i^* \begin{pmatrix} G_1 \\ G_2 \end{pmatrix} (G_1^* G_2^*) v_j.$$

The algebraic structure of the space-time block code implies

$$[V_1^* V_2^*] \begin{bmatrix} G_1 \\ G_2 \end{bmatrix} = \begin{pmatrix} a & b \\ -\bar{b} & \bar{a} \end{pmatrix}$$

for some $a, b \in \mathbb{C}$. Since the rows of this matrix are orthogonal, we have $v_i^* M v_j = 0$, for $i \neq j$. The proof that $x_i^* M x_j = 0$ is entirely similar, and we omit the details. $\qquad \square$

Theorem 4.2 *If* $i \neq j$, *then*

$$h_i^* M^k h_j = g_i^* M^k g_j = 0$$

for all integer exponents k.

Proof The proof for nonnegative exponents k is by induction. For the general case, note that M is the sum of a positive semidefinite matrix HH^*, and a positive definite matrix I_4/Γ, so it is nonsingular. Since M satisfies its characteristic polynomial, it follows that M^{-1} is a polynomial in M, and so the result follows for negative exponents.

Let $v_1, v_2 \in \mathcal{H}$, $x_1, x_2 \in \mathcal{G}$ where

$$[v_1, v_2] = \begin{bmatrix} V_1 \\ V_2 \end{bmatrix}, \quad [x_1, x_2] = \begin{bmatrix} X_1 \\ X_2 \end{bmatrix}$$

and V_1, V_2, X_1 and X_2 have the structure of the 2×2 block space-time code. By Lemma 4.1, we have

$$v_1^* M v_2 = x_1^* M x_2 = 0$$

and the inductive hypothesis is that for $0 \leq i < l$ we have

$$v_1^* M^i v_2 = x_1^* M^i x_2 = 0.$$

Now

$$v_1^* M^l v_2 = v_1^* M^{l-1} \left(\begin{pmatrix} H_1 \\ H_2 \end{pmatrix} (H_1^* H_2^*) + \begin{pmatrix} G_1 \\ G_2 \end{pmatrix} (G_1^* G_2^*) + \tfrac{1}{\Gamma} I_4 \right) v_2$$

$$= v_1^* M^{l-1} \begin{pmatrix} G_1 \\ G_2 \end{pmatrix} (G_1^* G_2^*) v_2$$

$$= v_1^* \left(\begin{pmatrix} H_1 \\ H_2 \end{pmatrix} (H_1^* H_2^*) + \begin{pmatrix} G_1 \\ G_2 \end{pmatrix} (G_1^* G_2^*) + \tfrac{1}{\Gamma} I_4 \right) M^{l-2} \begin{pmatrix} G_1 \\ G_2 \end{pmatrix} (G_1^* G_2^*) v_2.$$

By induction,

$$v_1^* \begin{pmatrix} G_1 \\ G_2 \end{pmatrix} (G_1^* G_2^*) M^{l-2} \begin{pmatrix} G_1 \\ G_2 \end{pmatrix} (G_1^* G_2^*) v_2 = 0,$$

so that

$$v_1^* M^l v_2 = \lambda v_1^* M^{l-2} \begin{pmatrix} G_1 \\ G_2 \end{pmatrix} (G_1^* G_2^*) v_2$$

for some constant λ. We now iterate, peeling off powers of M to obtain

$$v_1^* M^l v_2 = \lambda' v_1^* \begin{pmatrix} G_1 \\ G_2 \end{pmatrix} (G_1^* G_2^*) v_2,$$

for some constant λ', and this is zero, by Lemma 4.1. The proof that $x_1^* M^l x_2 = 0$ is entirely similar, and we omit the details. This completes the proof. $\square$

The MMSE receiver looks for a linear combination of received signals r_i, $i = 1, \ldots, 4$ that minimizes the mean squared error (due to co-channel interference and noise) with respect to a linear combination of the transmitted signals c_1 and c_2. We define the error cost function

$$\mathcal{T}(\alpha, \beta) = \|\alpha^* r - \beta^* c\|^2 = \left\| \sum_{i=1}^{4} \overline{\alpha}_i r_i - (\overline{\beta}_1 c_1 + \overline{\beta}_2 c_2) \right\|^2,$$

where $\alpha = (\alpha_1, \alpha_2, \alpha_3, \alpha_4)^T$, $r = (r_1, r_2, r_3, r_4)^T$, and $\beta = (\beta_1, \beta_2)^T$. The weights α and β are chosen to minimize the expectation $E[\mathcal{T}(\alpha, \beta)]$, but one of the coefficients β_1 and β_2 is set equal to one to avoid the zero solution. There are two cases to consider.

$\boldsymbol{\beta_1 = 1}$: Set $\alpha_1 = (\alpha_{11}, \alpha_{12}, \alpha_{13}, \alpha_{14}, -\beta_2)^T$ and $\tilde{r}_1 = (r_1, r_2, r_3, r_4, c_2)$, so that the error cost function is given by

$$\mathcal{T}(\alpha_1) = \|\alpha_1^* \tilde{r}_1 - c_1\|^2.$$

We need to find tap weights α_1 that minimize

$$
\begin{aligned}
E[\|\alpha_1^* \tilde{r}_1 - c_1\|^2] &= E[(\alpha_1^* \tilde{r}_1 - c_1)(\tilde{r}_1^* \alpha_1 - \overline{c}_1)] \\
&= \alpha_1^* E[\tilde{r}_1 \tilde{r}_1^*] \alpha_1 + E_s - \alpha_1^* E[\tilde{r}_1 \overline{c}_1] - E[c_1 \tilde{r}_1^*] \alpha_1. \quad (4.5)
\end{aligned}
$$

Recall that

$$
\tilde{r}_1 \left(\frac{r}{c_2} \right) = \left(\begin{array}{c|c} H & 0 \\ \hline 0 & 1 \end{array} \right) \left(\frac{c}{\frac{s}{c_2}} \right) + \left(\frac{\eta_1}{\frac{\eta_2}{0}} \right),
$$

that the data symbols are drawn from the unit circle, and that the noise samples are independent of the data and of each other. If E_s is the signal power and N_0 is the noise power, then

$$E[\tilde{r}_1 \tilde{r}_1^*]$$

$$
= E_s \left(\begin{array}{c|c} H & 0 \\ \hline 0 & 1 \end{array} \right) \left(\begin{array}{c|c} I_4 & \begin{array}{c} 0 \\ 1 \\ 0 \\ 0 \end{array} \\ \hline 0\,1\,0\,0 & 1 \end{array} \right) \left(\begin{array}{c|c} H^* & 0 \\ \hline 0 & 1 \end{array} \right) + N_0 \left(\begin{array}{c|c} I_4 & 0 \\ \hline 0 & 0 \end{array} \right)
$$

$$
= E_s \left(\begin{array}{c|c} HH^* + \frac{1}{\Gamma} I_4 & h_2 \\ \hline h_2^* & 1 \end{array} \right), \qquad (4.6)
$$

where h_2 is the second column of H. Now

$$E[\tilde{r}_1 c_1] = E_s \left(\begin{array}{c|c} H & 0 \\ \hline 0 & 1 \end{array} \right) \left(\begin{array}{c} 1 \\ 0 \\ 0 \\ 0 \\ \hline 0 \end{array} \right) = E_s \left(\begin{array}{c} h_1 \\ \hline 0 \end{array} \right), \qquad (4.7)$$

where h_1 is the first column of H. We minimize $\mathcal{T}(\alpha_1)$ by setting equal to zero the partial derivatives with respect α's in (4.5). Applying (4.6) and (4.7) we obtain

$$\left(\begin{array}{c|c} HH^* + \frac{1}{\Gamma}I_4 & h_2 \\ \hline h_2^* & 1 \end{array} \right) \alpha_1 = \left(\begin{array}{c} h_1 \\ \hline 0 \end{array} \right),$$

so that

$$\begin{array}{rcll} (\alpha_{11}, \alpha_{12}, \alpha_{13}, \alpha_{14})^T & = & (M - h_2 h_2^*)^{-1} h_1, & \text{and} \qquad (4.8) \\ \beta_2 & = & h_2^* (M - h_2 h_2^*)^{-1} h_1. & \qquad (4.9) \end{array}$$

Applying the Matrix Inversion Lemma (see p. 18 of Horn and Johnson [10]) gives

$$\begin{array}{rcl} (M - h_2 h_2^*)^{-1} & = & M^{-1} + \dfrac{M^{-1} h_2 h_2^* M^{-1}}{1 - h_2^* M^{-1} h_2}, \quad \text{and} \\[2ex] \beta_2 & = & \dfrac{h_2^* M^{-1} h_1}{1 - h_2^* M^{-1} h_1}. \end{array}$$

By Theorem 4.2, $h_2^* M^{-1} h_1 = 0$, so that $\beta_2 = 0$ and $(\alpha_{11}, \alpha_{12}, \alpha_{13}, \alpha_{14})^T = M^{-1} h_1$.

$\beta_2 = 1$: Set $\alpha_2 = (\alpha_{21}, \alpha_{22}, \alpha_{23}, \alpha_{24}, -\beta_1)^T$ and $\tilde{r}_2 = (r_1, r_2, r_3, r_4, c_1)^T$ so that the error cost function is given by

$$\mathcal{T}(\alpha_2) = \|\alpha_2^* \tilde{r}_2 - c_2\|^2.$$

By symmetry (interchanging the roles of h_1 and h_2 in (4.8) and (4.9)) the tap weights α_2 that minimize the expected value $E[\|\alpha_2^* \tilde{r}_2 - c_2\|^2]$ are given by

$$\begin{array}{rcl} (\alpha_{21}, \alpha_{22}, \alpha_{23}, \alpha_{24})^T & = & (M - h_1 h_1^*)^{-1} h_2, \quad \text{and} \\[2ex] \beta_1 & = & \dfrac{h_1^* M^{-1} h_2}{1 - h_1^* M^{-1} h_2}. \end{array}$$

Again it follows from Theorem 4.2 that $\beta_1 = 0$ and $(\alpha_{21}, \alpha_{22}, \alpha_{23}, \alpha_{24})^T = M^{-1} h_2$.

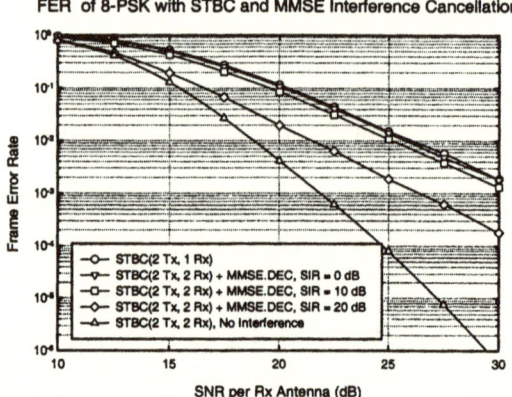

Figure 6: Performance (Frame Error Rate or FER) of MMMSE interference cancellation as a function of Signal to Noise Ratio (SNR) and Signal to Interference Ratio (SIR). When the SIR is 10 dBB (20 dB) the signal power P is 10 (100) times that of the interference power P' (SIR= $10 \log_{10} P/P'$).

Remarks Setting $\alpha_i = (\alpha_{i1}, \alpha_{i2}, \alpha_{i3}, \alpha_{i4})^T$ for $i = 1, 2$, we observe that

$$\alpha_1^* \alpha_2 = h_2^* \left(HH^* + \frac{1}{\Gamma} I_4 \right)^{-2} h_1 .$$

It follows from Theorem 4.2 that $\alpha_1^* \alpha_2 = 0$. This means that errors in decoding c_1 do not influence the decoding of c_2 and vice versa. The MMSE interference canceller maintains the separate detection feature of space-time block codes.

Note that this method of interference cancellation does not require explicit knowledge of whether or not an interferer is present. All that is required is an estimate for the covariance matrix M of the received signal vector (which includes implicitly the effects of any interfering signals).

Figure 6 shows the performance of MMSE interference cancellation as a function of Signal to Noise Ratio (SNR) and Signal to Interference Ratio (SIR), for two co-channel users, each employing the 2×2 space-time block code. When the power of the interferer equals that of the desired user, performance is identical to that obtained by the combination a single space-time user and a mobile terminal employing a single receive antenna. Performance improves as the power of the interferer decreases, and the limiting performance is that obtained by the combination of a single space-time user and a mobile terminal employing two receive antennas.

The generalization to M receive antennas is straightforward and requires calculation of $2M$ tap weights for α_1 and α_2.

Acknowledgements

The authors would like to thank the referee and J.C. Lagarias for helpful suggestions regarding organization of this paper.

References

[1] S. Alamouti, Space-block coding: A simple transmitter diversity technique for wireless communications, *IEEE J. Selected Areas Comm.* **16** (1998), 1451–1458.

[2] M. Aschbacher, Finite Group Theory, Cambridge University Press, Cambridge, 1986.

[3] C. Berrou, A. Glavieux, and P. Thitmajshima, Near Shannon limit error correcting coding and decoding: Turbo codes, *Proc. IEEE ICC'93*, 1993, 1063–1070.

[4] A.R. Calderbank, E.M. Rains, P.W. Shor, and N.J.A. Sloane, Quantum error correction via codes over GF(4), *IEEE Trans. Inform. Theory* **44** (1998), 1–19.

[5] A. Cayley, On certain results relating to quaternions, *Philos. Mag.* **26** (1845), 141–145.

[6] W.K. Clifford, Applications of Grassman's extensive algebra, *Amer. J. Math.* **1** (1878), 350–358.

[7] G.J. Foschini, Jr. and M.J. Gans, On limits of wireless communication in a fading environment using multiple antennas, Wireless Personal Communication, *Wireless Communications Magazine* **6** (1998), 311–335.

[8] A.V. Geramita and J. Seberry, *Orthogonal Designs, Quadratic Forms and Hadamard Matrices, Lecture Notes in Pure and Applied Mathematics* **43**, Marcel Dekker, New York, 1979.

[9] J. Graves, Note by Professor Sir W.R. Hamilton respecting the researches of John T. Graves, *Trans. Irish Acad.* **21** (1848), 338–341.

[10] R.A. Horn and C.R. Johnson, *Matrix Analysis*, Cambridge University Press, Cambridge, 1985.

[11] B. Huppert, *Endliche Gruppen I*, Springer, Berlin, 1967.

[12] A. Hurwitz, Uber die Komposition quadratischer Formen von beliebig vielen Variabeln, *Nachr. Gesell. Wiss. Göttingen*, (1898), 309–316.

[13] I.L. Kantor and A.S. Solodovnikov, *Hypercomplex Numbers, An Elementary Introduction to Algebras*, translated from the Russian by A. Shenitzer, Springer, New York, 1989.

[14] Y. Kawada and N. Iwahori, On the structure and representations of Clifford algebra, *J. Math. Soc. Japan* 2 (1950), 34–43.

[15] T. Y. Lam, *The Algebraic Theory of Quadratic Forms*, Mathematics Lecture Note Series, Benjamin, Reading, Mass., 1973.

[16] A.F. Naguib and N. Seshadri, Combined interference cancellation and ML decoding of space-time block codes, *Proc. Comm. Theory Miniconference*, held in conjunction with Globecomm '98 (Sydney, Australia), 1998, 7–15.

[17] A.F. Naguib and N. Seshadri, Combined interference cancellation and maximum likelihood decoding of space-time block codes, *IEEE J. Selected Areas Comm.*, to appear.

[18] A.F. Naguib and N. Seshadri, Combined interference cancellation and ML decoding of space-time block codes II: The general case, submitted.

[19] S. Parkvall, M. Karlsson, M. Samuelsson, L. Hedlund, and B. Goransson, Transmit diversity in WCDMA: Link and system level results, *Proc. IEEE VTC'00* (Osaka, Japan).

[20] I.R. Porteous, *Clifford Algebras and the Classical Groups*, Cambridge Studies in Advanced Mathematics 50, Cambridge University Press, Cambridge, 1995.

[21] J. G. Proakis, *Digital Communications*, McGraw-Hill, New York, 1989.

[22] J. Radon, Lineare Scharen orthogonaler Matrizen, *Abh. Math. Sem. Hamburg* 1 (1922), 1–14.

[23] S.E. Schmidt and P.E. Wright, Low complexity block space-time codes for high speed data transmission: I, preprint, 2000.

[24] D. Shapiro, Spaces of similarities IV - (s, t)-families, *Pacific J. Math.* 69 (1977), 223–244.

[25] M. Suzuki, *Group Theory II*, Springer, Berlin, 1986.

[26] V. Tarokh and H. Jafarkhani, A differential detection scheme for transmit diversity, *IEEE J. Selected Areas Comm.*, July, 2000.

[27] V. Tarokh, H. Jafarkhani, and A.R. Calderbank, Space-time block codes from orthogonal designs, *IEEE Trans. Inform. Theory*, **45** (1999), 1456–1467.

[28] V. Tarokh, N. Seshadri and A.R. Calderbank, Space-time codes for high data rate wireless communications: Performance criterion and code construction, *IEEE Trans. Inform. Theory* **44** (1998), 744–765.

[29] E. Telatar, Capacity of multi-antenna Gaussian channels, *AT&T Bell Labs Internal Technical Memorandum*, June, 1995.

[30] B.L. van der Waerden, *A History of Algebra, from al-Khwarzimi to Emmy Noether*, Springer, New York, 1985.

[31] S. Verdu, Multiuser Detection, Cambridge University Press, Cambridge, 1998.

[32] J.H. Winters, J. Salz, and R.D. Gitlin, The impact of antenna diversity on the capacity of wireless communications systems, *IEEE Trans. Comm.*, **42** (1994), 1740–1751.

[33] E. Witt, Theorie der quadratischen Formen in beliebigen Körpern, *J. Reine Angew. Math.* **176** (1937), 31–44.

[34] A. Wittneben, Base station modulation diversity for digital SIMULCAST, *Proc. IEEE VTC'91* (St. Louis, USA) **1**, 1991, 848–853.

[35] W. Wolfe, Amicable orthogonal designs - Existence, *Canad. J. Math.* **28** (1976), 1006–1020.

Information Sciences Research
AT&T Labs Research
Florham Park, NJ 07932
USA
rc@research.att.com

Morphics Technology Inc.
Campbell, CA 95008
USA
naguib@morphics.com

Computation in permutation groups: counting and randomly sampling orbits

Leslie Ann Goldberg

Abstract

Let Ω be a finite set and let G be a permutation group acting on Ω. The permutation group G partitions Ω into orbits. This survey focuses on three related computational problems, each of which is defined with respect to a particular input set $\mathcal{I}$. The problems, given an input $(\Omega, G) \in \mathcal{I}$, are (1) count the orbits (exactly), (2) approximately count the orbits, and (3) choose an orbit uniformly at random. The goal is to quantify the computational difficulty of the problems. In particular, we would like to know for which input sets $\mathcal{I}$ the problems are tractable.

1 Introduction

Let Ω be a finite set and let G be a permutation group acting on Ω. The permutation group G partitions Ω into *orbits*: Two elements of Ω are in the same orbit if and only if there is a permutation in G which maps one element to the other. This survey focuses on three related computational problems, each of which is defined with respect to a particular input set $\mathcal{I}$:

1. Given an input $(\Omega, G) \in \mathcal{I}$, count the orbits.

2. Given an input $(\Omega, G) \in \mathcal{I}$, approximately count the orbits.

3. Given an input $(\Omega, G) \in \mathcal{I}$, choose an orbit uniformly at random.

The goal is to quantify the computational difficulty of the problems. In particular, we would like to know for which input sets $\mathcal{I}$ the problems are tractable.

Many interesting orbit-counting problems come from the setting of "Pólya theory". In this setting, Σ is a fixed alphabet of size at least two. For every (infinite) set $\mathcal{G}$ of permutation groups, we get an input set $\mathcal{I}(\mathcal{G})$. In particular, the group $G \in \mathcal{G}$ corresponds to the input $(\Sigma^m, \widehat{G}) \in \mathcal{I}(\mathcal{G})$, where m is the degree of G, Σ^m is the set of length-m words over alphabet Σ, and $\widehat{G}$ is a permutation group acting on Σ^m which is induced by G. We will later give a precise definition of $\widehat{G}$ (as a function of G), but the rough idea is as follows: Every permutation $g \in G$ induces exactly one permutation $\hat{g} \in \widehat{G}$. The image of a word $\alpha \in \Sigma^m$ under $\hat{g}$ is the word β that is obtained from α by permuting the m alphabet symbols in α (applying g to the "positions" of the symbols). We use the notation $\mathcal{P}$ to denote the set of *all* permutation groups, so $\mathcal{I}(\mathcal{P})$ is the input set corresponding to all inputs in the Pólya-theory framework (over a fixed alphabet Σ).

Problems 1–3 are studied in Sections 3, 4 and 5 respectively. In those sections, we will give the precise complexity-theory definitions which we will need to formally capture the notion of "tractability". We will also give appropriate references, tracing the development of ideas. In this section, we introduce the area by giving a very high-level sketch, stating (roughly) what is known about each of the three problems, without giving details, definitions, or references.

There are natural input sets for which exact orbit-counting is tractable. Here is an example. Suppose that Ω_n is the set of all n-vertex trees and that G_n is the permutation group acting on Ω_n which is induced by vertex permutations. The orbits of Ω_n under G_n correspond to isomorphism classes. That is, there is one orbit for each *unlabelled* n-vertex tree. Unlabelled trees can be counted quickly using classical methods based on generating functions and dynamic programming. Thus, the orbit-counting problem is tractable for the input set $\mathcal{I} = \{(\Omega_n, G_n)\}$.

Perhaps surprisingly, the orbit-counting problem becomes intractable if we make the input set slightly more complicated: Suppose that each input corresponds to a tree T, and orbits correspond to unlabelled subtrees of T. It turns out that this orbit-counting problem is #P-complete. We will discuss the complexity class #P in Section 3. To give a rough idea, counting problems which are complete in #P are equivalent in difficulty to counting the number of satisfying assignments of a Boolean formula, and this is believed to be very demanding computationally. Thus, it is unlikely that there is a polynomial-time algorithm for counting unlabelled subtrees of a tree T. What happens when we make the input set still more complicated? Suppose now that each input corresponds to a graph H, and orbits correspond to unlabelled subtrees of H. By the previous result, the new problem (counting unlabelled subtrees of a graph) is at least as difficult as #P. But perhaps it is more difficult? Whether or not it is more difficult is unknown, but a complexity-theoretic result of Toda implies that it cannot be much more difficult. The new problem is complete in the complexity class $FP^{\#P}$, which, as we will see later, is not so different from #P. The problem of counting orbits in the Pólya-theory setting (with input set $\mathcal{I}(\mathcal{P})$) is also complete in $FP^{\#P}$. There are several natural problems for which the complexity of orbit-counting is unresolved. For example, there is no known efficient algorithm for exactly counting unlabelled n-vertex graphs. It seems plausible that this problem is computationally difficult, but we seem to lack the complexity-theory machinery to quantify the difficulty of such problems. These problems are members of the complexity class $\#P_1$, which will be discussed in Section 3.

There are many known examples of counting problems which are #P-complete to solve exactly but for which good (efficient) approximation algorithms exist. Interestingly, we do not know any natural[1] orbit-counting

---

[1] Technically, every counting problem without orbits can be expressed as an orbit-counting problem in which the relevant permutation group contains only the identity, but this is not what we mean by a "natural" orbit-counting problem.

problems which have this property, though perhaps this is just because orbit-counting problems have not been sufficiently studied.

There are examples of problems for which the complexity of exact orbit-counting is unresolved, but approximate orbit-counting is known to be tractable. For example, there is an efficient algorithm for approximately counting unlabelled n-vertex graphs. It is worth pointing out we do not know any efficient algorithms for this problem using traditional methods such as generating functions and dynamic programming. The only algorithm which is known relies upon a reduction from almost-uniform sampling, which we will discuss presently (see also Section 5.6).

As we mentioned before, some #P-complete counting problems have efficient approximation algorithms. For other #P-complete counting problems, the corresponding approximation problem can be shown to be "complete" in a formal sense which implies that there is no efficient approximation algorithm under standard complexity-theoretic assumptions. It would be very interesting to know whether the approximate orbit-counting problem is complete in this sense for the set of Pólya-theory inputs $\mathcal{I}(\mathcal{P})$. This seems to be a very difficult question. At present, the most that can be said is that the related question of approximately counting orbits over a coset (rather than over a group) is complete.

For many computational counting problems, there is a close connection between the complexity of approximately counting and the complexity of sampling uniformly at random. However, with a few exceptions, which we will discuss, these connections break down for orbit-counting problems. Thus, sampling must be studied separately from approximate counting.

Most of the work on sampling orbits has focused on three methods: inductive sampling, the orbit-sampling process, and rejection sampling. "Inductive sampling" can be used when generating functions for enumerating orbits can be efficiently evaluated. The idea is to find a recurrence for the coefficients of the generating function, which can be computed by dynamic programming. The recurrence should show how to express outputs corresponding to a given instance of the problem in terms of outputs to smaller instances. Sampling is then done recursively: The coefficients are used to probabilistically select the appropriate (recursive) sub-problem. For example, unlabelled n-vertex trees can be sampled in this way.

The "orbit-sampling process" is inspired by the orbit-counting lemma of Cauchy, Frobenius and Burnside. This lemma says that each orbit is represented $|G|$ times in the set $\Upsilon(\Omega, G)$, which is the set of pairs (α, g) such that $\alpha \in \Omega$, $g \in G$ and g fixes α (that is, g maps α to itself). The orbit-sampling process is a Markov chain with state space Ω. To make a transition from the state α, the chain first picks uniformly-at-random a permutation $g \in G$ which fixes α and then chooses the new state uniformly-at-random from the subset of Ω which is fixed by g. The orbit-sampling lemma can be used to show that the stationary distribution of this process is uniform on orbits. If the input set is chosen such that (1) each transition can be implemented in polynomial time, and (2) the Markov chain is rapidly mixing, then the orbit-process gives an efficient sampler for orbits.

Not too much is known about the mixing-time of the orbit-sampling process, even in the Pólya-theory setting. It is not rapidly-mixing for the complete set of inputs $\mathcal{I}(\mathcal{P})$. There is also an infinitely-large subset of $\mathcal{I}(\mathcal{P})$ for which the transitions can be implemented in polynomial time, but the process is still not rapidly mixing. The process is known to be rapidly mixing if the set $\mathcal{G}$ of permutation groups is either (1) the set of all symmetric groups S_n, or (2) the set of all cyclic groups (all permutation groups with a single generator), but nothing else is known. In particular, it is not even known whether the orbit-sampling process is rapidly-mixing for the situation in which orbits correspond to unlabelled n-vertex graphs.

As we mentioned earlier, there is an alternative efficient algorithm for sampling unlabelled n-vertex graphs. This algorithm uses the "rejection sampling" method, which will be explained in Section 5.6. The algorithm can be extended to the general orbit-sampling framework. Whether or not it leads to an efficient sampling algorithm depends upon the input set $\mathcal{I}$. What is clear is that it will not work unless the identity permutation accounts for a sufficiently large fraction of the pairs in $\Upsilon(\Omega, G)$. That is, a typical member of Ω must not be fixed by too many permutations in G.

An example of an input set which does not have this property is as follows: Each input corresponds to a degree-sequence in which every degree is bounded. Orbits correspond to unlabelled connected multigraphs with the given degree sequence. Here the degree sequence may contain many vertices of degree 1 or 2, in which case a typical member of Ω is a multigraph which is fixed by many automorphisms. The ideas which have been described so far can be combined to give an efficient sampling algorithm for this problem. Nevertheless, we lack good general techniques for orbit-sampling, especially when the objects in Ω have many symmetries.

The final section of the survey is devoted to a problem which is related to that of sampling and counting orbits – namely, the problem of listing orbits. Not too much is known about the problem, and the section gives pointers to some recent work in the area.

2   Definitions and preliminaries

Let Ω be a finite set and let G be a permutation group acting on Ω. If $\alpha \in \Omega$ and $g \in G$, we write α^g to denote the image of α under g. We write G_α to denote the subgroup of G consisting of the permutations in $\{g \in G \mid \alpha^g = \alpha\}$. We define the relation $\sim$ on Ω in which $\alpha \sim \beta$ if and only if there is a permutation $g \in G$ such that $\alpha^g = \beta$. The relation $\sim$ partitions Ω into equivalence classes, which are called *orbits*. We use the notation α^G to denote the orbit $\{\alpha^g : g \in G\}$ containing α and the notation $\Phi(\Omega, G)$ to denote the set of orbits. For each permutation $g \in G$, we let $\mathrm{fix}(g)$ denote the set $\{\alpha \in \Omega \mid \alpha^g = \alpha\}$. We let $\Upsilon(\Omega, G)$ denote the set

$$\Upsilon(\Omega, G) = \{(\alpha, g) \mid \alpha \in \Omega \text{ and } g \in G \text{ and } \alpha \in \mathrm{fix}(g) \}.$$

The following lemma was known to Cauchy and Frobenius (see [40]) but is often called "Burnside's Lemma". Following Cameron [7], we call it the "orbit-counting" lemma.

Lemma 2.1 The orbit-counting lemma   *Let G be a permutation group on the finite set Ω. Then for each orbit $\Delta \in \Phi(\Omega, G)$ we have*

$$|\{(\alpha, g) \in \Upsilon(\Omega, G) \mid \alpha \in \Delta\}| = |G|;$$

so

$$|\Upsilon(\Omega, G)| = |\Phi(\Omega, G)|\,|G|.$$

Examples Let Ω_n be the set of all n-vertex graphs and let G_n be the permutation group acting on Ω_n which is induced by vertex permutations. That is, G_n has $n!$ permutations — one for each permutation of the n vertices. If π is a permutation of the vertices and α is a graph in Ω_n then the image of α under the permutation corresponding to π is the graph obtained from α by applying π to the vertices. The orbits of Ω_n under G_n correspond to isomorphism classes. Thus, we can think of the orbits as representing the set of *unlabelled* n-vertex graphs.

We will let $\mathcal{U}_{\Omega,G}$ denote the uniform distribution on orbits. That is, for each orbit Δ in $\Phi(\Omega, G)$, the probability of Δ in $\mathcal{U}_{\Omega,G}$ (denoted $\mathcal{U}_{\Omega,G}(\Delta)$) is $1/|\Phi(\Omega, G)|$.

We will measure the distance between two probability distributions D_1 and D_2 over the discrete sample space Ψ using the *total variation distance* metric. Namely,

$$d_{\mathrm{tv}}(D_1, D_2) = \max_{A \subseteq \Psi} |D_1(A) - D_2(A)| = \frac{1}{2} \sum_{x \in \Psi} |D_1(x) - D_2(x)|.$$

2.1 A special case of Pólya's theorem

Many of the orbit-counting problems which we will consider come from the setting of Pólya theory. We will not be using the fully general version of Pólya's theorem. Instead, we will restrict our attention to the special case of the theorem that we define in this section. Suppose that $\Sigma = \{0, \ldots, k-1\}$ is a finite alphabet and that G is a group of permutations of the set $\{0, \ldots, m-1\}$, which we denote $[m]$. For every permutation $g \in G$, let $c(g)$ denote the number of cycles in g. Let Ω be the set Σ^m of length-m words over alphabet Σ. The group G has a natural action on Ω which is induced by permuting the "positions" $0, \ldots, m - 1$ of the alphabet symbols in the words. In particular, if $\alpha = a_0 a_1 \ldots a_{m-1}$ is a word in Ω then the image of α under the induced action of g is the word $\beta = b_0 b_1 \ldots b_{m-1}$, in which, for all $j \in [m]$, b_j is $a_{j g^{-1}}$ That is, b_j is the element a_i such that $i^g = j$. To avoid confusion, we use the symbol $\widehat{G}$ to denote the permutation group on Ω which is induced by G and we use the symbol $\hat{g}$ to denote the permutation of Ω which is induced by permutation $g \in G$.

Now fix($\hat{g}$) has $k^{c(g)}$ elements. In particular, if a word α is in fix($\hat{g}$), then all of the positions which form a single cycle of g must have the same alphabet symbol in α. There are k possible symbols which can be chosen. Thus the orbit-counting lemma (Lemma 2.1) gives us the following special case of Pólya's theorem.

Lemma 2.2 Pólya's theorem   *If $\Sigma = [k]$ is a finite alphabet, $\Omega = \Sigma^m$, and G is a permutation group on $[m]$, then*

$$|\Phi(\Omega, \widehat{G})| = \frac{1}{|G|} \sum_{g \in G} k^{c(g)}.$$

Examples   If $\Sigma = \{0, 1\}$ and G is the symmetric group on $[m]$ then the $m + 1$ orbits consist of, for each $i \in [m + 1]$, those words in Σ^m with the symbol "1" in i positions.

Examples   The set of unlabelled n-vertex graphs (Example 2) can be encoded as orbits in the Pólya-theory setting by using words in Σ^m to encode graphs. In particular, the graph H can be represented by the word corresponding to the upper-diagonal part of H's adjacency matrix.

Further examples can be found in surveys such as [8] and [45].

2.2   Computational questions

We work in the following computational framework which is similar to that of [25]. We specify the input set $\mathcal{I}$, where each input in $\mathcal{I}$ consists of a set Ω and a permutation group G. The inputs are represented in a concise manner, which depends upon $\mathcal{I}$. We study the following computational problems.

1. *exact counting:* given an input $(\Omega, G) \in \mathcal{I}$, output $|\Phi(\Omega, G)|$.

2. *approximate counting:* Given an input $(\Omega, G) \in \mathcal{I}$ and an accuracy parameter $\epsilon \in (0, 1)$, output an integer random variable Y satisfying

$$\Pr\left(e^{-\epsilon} \leq \frac{Y}{|\Phi(\Omega, G)|} \leq e^{\epsilon}\right) \geq \frac{3}{4}.$$

3. *almost-uniform sampling:* Given an input $(\Omega, G) \in \mathcal{I}$, and an accuracy parameter $\epsilon \in (0, 1]$, output a random variable α. Typically, α will be a member of Ω, and will be viewed as a representative of its orbit, but it will be technically useful to allow sampling algorithms to sometimes produce other outputs.

We measure the accuracy of a sampling algorithm by constructing a distribution D based on the output distribution of the algorithm. The domain of D is taken to be $\Phi(\Omega, G) \cup \{\perp\}$, where $\perp$ is an "error" symbol, which records the fact that the output does not represent an orbit. For each orbit $\Delta \in \Phi(\Omega, G)$, the probability of Δ in D is defined to be $\Pr(\alpha \in \Delta)$. Therefore, the probability of $\perp$ in D is equal to $1 - \sum_{\Delta \in \Phi(\Omega, G)} \Pr(\alpha \in \Delta)$. The algorithm is an *almost-uniform sampler* if and only if $d_{tv}(D, \mathcal{U}_{\Omega,G}) \leq \epsilon$.

For each *particular* input set $\mathcal{I}$, we get a particular exact counting problem, approximate counting problem, and almost-uniform sampling problem. We will usually discuss the representation of the inputs when the particular problem is discussed. Typically, we will represent inputs in a concise manner.

In the Pólya-theory setting, we will adopt the following notation from the introduction. The alphabet Σ is fixed. For any permutation group G, we will let $m(G)$ denote the degree of G. For any set $\mathcal{G}$ of permutation groups, $\mathcal{I}(\mathcal{G})$ denotes the input set corresponding to $\mathcal{G}$. In particular,

$$\mathcal{I}(\mathcal{G}) = \{(\Sigma^{m(G)}, \widehat{G}) \mid G \in \mathcal{G}\}.$$

We use the notation $\mathcal{P}$ to denote the set of *all* permutation groups, so $\mathcal{I}(\mathcal{P})$ is the input set corresponding to all inputs. The input $(\Sigma^m, \widehat{G})$ will be presented as a set of $O(m)$ generators for G.[2]

For convenience (in applying complexity-theoretic definitions), we will assume that all inputs to computational problems are encoded as words over the binary alphabet $\{0, 1\}$. This typically does not present any problems. For example a generator of a degree-m permutation group can be encoded as a binary word of length $O(m \log m)$. Note that the input size is typically much smaller than the size of Ω or G. In the Pólya-theory setting, the size of Ω is k^m and the size of $\widehat{G}$ can be as large as $m!$, but the size of the input is bounded from above by a polynomial in m. We are interested in knowing for which input sets $\mathcal{I}$ the computational problems are tractable, in a sense which will be made clear as we go along.

---

[2]The construction of small generating sets is beyond the scope of this article, but Chapter 1 of [7] describes several such constructions, due to Schreier, Sims, Jerrum, McIver and Neumann.

3   Exact counting

In Sections 3.2 and 3.3 we will see that for many natural input sets $\mathcal{I}$ the exact orbit-counting problem is #P-hard. Thus, it is as difficult as counting the number of satisfying assignments of a Boolean formula. In order to give details, we need some definitions, which are given in Section 3.1.

3.1   The complexity class #P

Following Valiant [50], we say that a function $f : \{0,1\}^* \to \mathbb{N}$ is in the complexity class FP if it can be computed by a deterministic polynomial-time Turing machine. We say that it is in #P if there is a nondeterministic polynomial-time Turing machine M such that for all $x \in \{0,1\}^*$ the number of accepting computations of M on input x is $f(x)$. A *polynomial-time Turing reduction* from a function $f : \{0,1\}^* \to \mathbb{N}$ to a function $g : \{0,1\}^* \to \mathbb{N}$ is a deterministic polynomial-time *oracle* Turing machine which, whenever it is supplied with an "oracle" for g, can compute f. Thus, the reduction shows how to compute f in polynomial time, assuming that we have an imaginary means for computing g in polynomial time. A counting problem, that is, a function $f : \{0,1\}^* \to \mathbb{N}$ is said to be #P-*hard* if every function in #P is polynomial-time Turing-reducible to f. If, in addition, $f \in$ #P, then it is said to be #P-*complete*. A #P-complete problem is equivalent in computational difficulty to problems such as counting the number of satisfying assignments of a Boolean formula, or evaluating the permanent of a 0,1-matrix, which are widely believed to be intractable. For background information on #P and its completeness class, see, for example, [14] or [44].

3.2   Automating Pólya theory

In this section, we see that the problem of counting orbits in the Pólya-theory setting is equivalent in computational difficulty to solving a #P-complete problem. Let $\Sigma = [k]$ be a fixed alphabet with $k > 1$. Consider the following computational problem.

Name. #PÓLYAORBITS.

Instance. $O(m)$ generators for a group G of permutations of $[m]$.

Output. $|G| \cdot |\Phi(\Sigma^m, \widehat{G})|$.

The size of a permutation group can be computed in polynomial time from an arbitrary set of generators (see [7]). Thus, #PÓLYAORBITS is computationally equivalent to the orbit-counting problem with input set $\mathcal{I}(\mathcal{P})$. The following theorem quantifies the computational difficulty of this problem.

Theorem 3.1 *[17]* #PÓLYAORBITS *is* #P-*complete.*

Pólya's Theorem (Lemma 2.2) tells us that the appropriate output of #PÓLYAORBITS is $\sum_{g \in G} k^{c(g)}$. From this, it is not difficult to show that #PÓLYAORBITS is in #P. Thus, to prove Theorem 3.1, we need only show that it is #P-hard. The #P-hardness follows from Theorem 2 of [17] and an alternative proof has been given by Jerrum in [27]. Nevertheless, in order to generalise the result in Section 4, we will need a hardness proof which does not use interpolation, so we provide such a proof here.

We start with some definitions. A "cut" of a graph is an unordered partition (S, T) of its vertex set. The "cut edges" corresponding to the cut are those edges of the graph which have one endpoint in S and the other in T. Since the partition (S, T) is unordered, a cut of a connected graph is uniquely determined by the set of cut edges. The "size" of the cut is the number of cut edges. Lemma 12 of Jerrum and Sinclair's paper [30] shows that the following problem is #P-complete.

Name. #LARGECUT.

Instance. A positive integer j and a connected non-bipartite graph H in which no cuts are larger than size j.

Output. The number of size-j cuts of H.

Thus, Theorem 3.1 follows from the following lemma.

Lemma 3.2 *There is a polynomial-time Turing-reduction from #LARGECUT to #PÓLYAORBITS.*

Proof Let j and H be an instance of #LARGECUT. Let V denote the vertex set of H and let E denote the edge set of H. Let $r = |V|^2$. Let G be the degree-$2r|E|$ permutation group constructed as follows. For each edge $e \in E$ and each $i \in [r]$, we will have an object $a_{e,i}$ and an object $b_{e,i}$. For each vertex $v \in V$ we will have a permutation g_v. The action of g_v on the set $\bigcup_{e \in E} \bigcup_{i \in [r]} \{a_{e,i}, b_{e,i}\}$ is as follows. For every edge $e \in E$ which is incident on v, and for every $i \in [r]$, g_v transposes $a_{e,i}$ and $b_{e,i}$. Let G be the group generated by the permutations $\bigcup_v g_v$.

Since the permutations in $\{g_v\}$ commute and have order 2, the permutations in G correspond to subsets of V. The points $a_{e,1}$ and $b_{e,1}$ are transposed in a given permutation if and only if exactly one of the endpoints of e is in the corresponding subset of V. Thus, the permutations in G are in one-to-one correspondence with the cuts of H. A cut of size ℓ corresponds to a permutation with $2|E|r - \ell r$ cycles.

Now let h be the permutation which transposes every pair $(a_{e,i}, b_{e,i})$. Since H is not bipartite, $h \notin G$. Let G' be the permutation group generated by $\{g_v\} \cup \{h\}$. Note that the permutations in this set commute and have order 2. Let C denote the coset of G in G' which is not equal to G. As before, the permutations in C are in one-to-one correspondence with the cuts of H. A cut of size ℓ corresponds to a permutation with $2|E|r - (|E| - \ell)r = |E|r + \ell r$

cycles. Let $P(G)$ denote the output of $\#\text{P\'OLYAORBITS}(G)$ and let N_ℓ denote the number of size-ℓ cuts of H. Then

$$P(G') - P(G) = \sum_{g \in C} k^{c(g)} = \sum_{\ell=0}^{j} N_\ell k^{|E|r + \ell r}, \text{ so}$$

$$\frac{P(G') - P(G)}{k^{|E|r + jr}} = N_j + \sum_{\ell=0}^{j-1} N_\ell k^{(\ell - j)r}.$$

Since $\sum_{\ell=0}^{j-1} N_\ell k^{(\ell-j)r}$ is non-negative and (from the definition of N_ℓ) is at most $2^{|V|} k^{-r}$ we have

$$N_j \leq \frac{P(G') - P(G)}{k^{|E|r + jr}} \leq N_j + 2^{|V|} k^{-r}.$$

Since $r = |V|^2$ and $k \geq 2$ and N_j is an integer,

$$N_j = \left\lfloor \frac{P(G') - P(G)}{k^{|E|r + jr}} \right\rfloor.$$

$\square$

3.3   Counting subtrees of a tree

In this section, we will consider an exact orbit-counting problem that has a rather different flavour from the Pólya-theory problem of the previous section. We will see that this problem too is $\#$P-complete. Thus, exactly counting orbits is computationally difficult, even in a seemingly simple setting.

Suppose that T is a tree with vertex set V and edge set E. We will let Ω_T be the set containing all (labelled) *subtrees* of T. That is, every element T' of Ω_T is a graph with vertex set V, some edge set $E' \subseteq E$, and at most one non-trivial connected component. (At most one connected component of T' will contain edges.) Let G_T be the permutation group acting on Ω_T which is induced by vertex permutations. As in Example 2, the orbits correspond to *unlabelled* subtrees of T. Jerrum and I [20] have shown that this orbit-counting problem, the problem $\#\text{SUBTREES}$ below, is $\#$P-complete.

Name. $\#\text{SUBTREES}$.

Instance. A tree T.

Output. The number of distinct (up to isomorphism) subtrees of T. That is, $|\Phi(\Omega_T, G_T)|$.

The proof that $\#\text{SUBTREES}$ is $\#$P-complete is contained in [20]. The complexity of most variants of the problem is still unknown. For example, the status of the following problem is open.

Name. #SUBFORESTS.

Instance. A tree T.

Output. The number of distinct (up to isomorphism) subforests of T.

The constructions used in [20] involve trees with high-degree vertices, so it is also not clear whether the following problem is #P-complete for any constant $\Delta > 2$.

Name. #ΔTREE-SUBTREES.

Instance. A tree T in which every vertex has degree at most Δ.

Output. The number of distinct (up to isomorphism) subtrees of T.

We will conclude this section by briefly considering the following generalisation of #ΔTREE-SUBTREES.

Name. #ΔGRAPH-SUBTREES.

Instance. A graph H in which every vertex has degree at most Δ.

Output. The number of distinct (up to isomorphism) subtrees of H.

Corollary 6 of [20] shows that #ΔGRAPH-SUBTREES is in the complexity class FP$^{\#P}$. Informally, this is the class of functions which are "as easy" as #P. More formally, a function f is in FP$^{\#P}$ if it is polynomial-time Turing-reducible to a problem in #P. Thus, by the following lemma, which is proved in the appendix, #ΔGRAPH-SUBTREES is complete in FP$^{\#P}$ for every fixed $\Delta \geq 5$.

Lemma 3.3 (Goldberg, Jerrum, Kelk) *For any fixed $\Delta \geq 5$, the problem* #ΔGRAPH-SUBTREES *is #P-hard.*

3.4 Orbit-counting problems and the complexity class #P₁

In the previous section, we have seen that the problem of counting unlabelled subtrees of a given tree T is #P-complete. Now suppose that instead of having a particular n-vertex tree T as input, the input is just n and we are interested in counting *all* unlabelled trees with at most n vertices. We will consider the following computational problem.

Name. #TREES.

Instance. A positive integer n, expressed in unary.[3]

Output. The number of distinct (up to isomorphism) n-vertex trees.

---

[3]The reason that the input to #TREES is expressed in unary is that we are interested in knowing whether there is an algorithm for #TREES whose running time is bounded from above by a polynomial in n. Since there are exponentially many unlabelled n-vertex trees, an algorithm whose running time is bounded from above by a polynomial in $\log n$ would not even have enough time to write down the answer.

The problem #TREES can be viewed as an orbit-counting problem. In particular, it is the orbit-counting problem corresponding to input set $\{(\Omega_n, G_n)\}$ in which Ω_n is the set of n-vertex trees, and G_n is the group induced by vertex permutations (see Example 2).

#TREES can be solved in polynomial time using a generating function for the number of orbits. Once the generating function is given, its coefficients can be computed by dynamic programming. Harary and Palmer's book [23] contains a survey on using generating functions to do unlabelled enumeration. Their book gives a full treatment of the enumeration of unlabelled trees, following the work of Otter [43]. In order to illustrate the principles, we repeat a few of the details here. Let $T(x) = \sum_{n=1}^{\infty} T_n x^n$ be the generating function for *rooted* unlabelled trees. That is, T_n is the number of rooted unlabelled trees with n vertices. Pólya's theorem gives an expression[4] for $T(x)$ which can be manipulated to yield the recurrence

$$T_{n+1} = \frac{1}{n} \sum_{k=1}^{n} \left(\sum_{d|k} d\, T_d \right) T_{n-k+1}, \tag{3.1}$$

where the sum is over all divisors d of k. Using this formula, the coefficients $T_1, T_2, \ldots, T_n$ can be computed in polynomial time by dynamic programming. ("Dynamic programming" just means that the coefficients should be computed in the order $T_1, T_2, \ldots$. Note that a recursive algorithm would not complete in polynomial time unless a device such as a "memory function" is used.) Next, let $t(x) = \sum_{n=1}^{\infty} t_n x^n$ be the generating function for (unrooted) unlabelled trees. It can be shown that

$$t(x) = T(x) - \frac{1}{2} \left(T^2(x) - T(x^2) \right),$$

so the coefficients $t_1, t_2, \ldots, t_n$ can also be computed in polynomial time.

Now that we have seen a polynomial-time algorithm for #TREES, let us consider the following related problem from Example 2.

Name. #GRAPHS.

Instance. A positive integer n, expressed in unary.

Output. The number of distinct (up to isomorphism) n-vertex graphs.

There is no known generating function which would enable us to quickly solve #GRAPHS. In fact, there is no known polynomial-time algorithm (of any type) for this problem.

Both #TREES and #GRAPHS are examples of problems from the complexity class $\#P_1$. The definition of this class is similar to the definition of #P. The only difference is that the input alphabet is now unary rather than binary. Valiant [50] has shown that $\#P_1$ does contain complete problems.

---

[4]This expression was also discovered by Cayley.

Notably, Bertoni, Goldwurm and Sabadini have shown that counting strings of a given length in some context-free language is complete [4]. Nevertheless, no natural combinatorial problem is known to be complete for $\#P_1$ and it seems unlikely that a problem such as #GRAPHS would be complete. Thus, at present, we seem to lack methods for quantifying the computational complexity of #GRAPHS and similar problems. This is an intriguing open question in the complexity theory of counting.

4 Approximate counting

Definition A *randomised approximation scheme* for a function $f : \{0,1\}^* \to \mathbb{N}$ is a probabilistic Turing machine that takes as input a pair $(x, \epsilon) \in \{0,1\}^* \times (0,1)$ and produces as output an integer random variable Y satisfying the condition $\Pr(e^{-\epsilon} \leq Y/f(x) \leq e^{\epsilon}) \geq 3/4$. Such an approximation scheme is said to be a *fully polynomial*[5] randomised approximation scheme (or FPRAS) if its running time is bounded from above by a polynomial in $|x|$ and ϵ^{-1}.

Thus, an algorithm for the approximate counting problem of Section 2.2 is an FPRAS if and only if its running time is bounded from above by a polynomial in the size of the description of the the the input (Ω, G), and in ϵ^{-1}.

Clearly, there is an FPRAS for the problem #TREES, since this problem can be solved (exactly) in deterministic polynomial time (see Section 3.4). We will see in Section 5.6 that there is also an FPRAS for #GRAPHS. It is worth observing at this point that there are asymptotic enumerations of unlabelled graphs based on Pólya's theorem, but these do not seem to be strong enough to give an FPRAS. In particular, let U_n denote the number of unlabelled n-vertex graphs. Pólya showed that U_n is asymptotically equal to $2^{\binom{n}{2}}/n!$. Oberschelp [42] gave a more detailed formula for U_n with improved error terms. (See Chapter 9 of [23].) For example, he showed that there is a constant c such that

$$U_n \leq \frac{2^{\binom{n}{2}}}{n!} \left(1 + \frac{cn^2}{2^n} \right). \tag{4.1}$$

Equation 4.1 is sufficiently accurate when the desired error, ϵ, exceeds $cn^2/2^n$. However, it is not immediately clear how to approximate U_n when the error parameter ϵ is smaller. Note that it takes $\Omega(n!)$ time to apply Pólya's theorem directly and this can exceed poly(ϵ^{-1}) even when ϵ is too small for using Equation 4.1.

We will return to the problem #GRAPHS in Section 5.6, where we will describe an FPRAS. It is not known whether there are efficient approximate counting algorithms for the rest of the problems introduced in Section 3. Before

---

[5]The definitions that we use are taken from [12] but they are closely related to Karp and Luby's definitions from [33].

we say more about these problems, we will look briefly at the complexity-theory context.

4.1 The complexity of approximate counting

From a complexity-theoretic point of view, *exactly* solving a #P-complete problem seems to be much more difficult than *approximately* solving it. The best way to illustrate this point is to introduce the notion of the "polynomial hierarchy". We will just state the relevant facts without giving details or definitions. Details can be found in [14] and [44]. The polynomial hierarchy contains an infinite sequence of complexity classes, $\Sigma_0^p, \Sigma_1^p, \ldots$. The class Σ_0^p is the same as the familiar class P and the class Σ_1^p is the same as NP. It is widely believed that all classes in the hierarchy are distinct. In particular, Σ_i^p is believed to be a proper subset of Σ_{i+1}^p. We can now state the relevance of the polynomial hierarchy — Toda [49] has shown that *every* problem in the entire polynomial hierarchy can be solved in polynomial time using an oracle for any #P-complete problem. Thus, informally, a #P-complete problem is "as hard as" the entire polynomial hierarchy. On the other hand, a result of Valiant and Vazirani [51] implies that every function in #P can be approximated (in the FPRAS sense) by a polynomial-time probabilistic Turing machine equipped with an NP oracle.[6] We can therefore conclude that the approximate counting problems in Sections 3.2 and 3.3 are "as easy as" NP, and we are interested in knowing whether they are easier.

Dyer, Greenhill, Jerrum, and I [12] recently studied the following notion of approximation-preserving reduction. Suppose $f, g : \{0,1\}^* \to \mathbb{N}$ are functions whose complexity (of approximation) we want to compare. An *approximation-preserving reduction* from f to g is a probabilistic oracle Turing machine M that takes as input a pair $(x, \epsilon) \in \{0,1\}^* \times (0,1)$, and satisfies the following three conditions: (i) every oracle call made by M is of the form (w, δ), where $w \in \{0,1\}^*$ is an instance of g and $0 < \delta < 1$ is an error bound satisfying $\delta^{-1} \leq \text{poly}(|x|, \epsilon^{-1})$; (ii) the Turing machine M meets the specification for being a randomised approximation scheme for f whenever the oracle meets the specification for being a randomised approximation scheme for g; and (iii) the run-time of M is polynomial in $|x|$ and ϵ^{-1}. If an approximation-preserving reduction from f to g exists we write $f \leq_{\text{AP}} g$, and say that f *is AP-reducible to g*.

In [12], we identify a class of problems which are complete for #P with respect to AP-reducibility. It is unlikely that any of these problems has an FPRAS. In particular, if any such complete problem has an FPRAS then so does every problem in #P. This, in turn, would imply that RP = NP, which is unlikely.

We will not be using the complexity class RP after Section 4, but for

---

[6]This is Corollary 3.6 of [51]. Only a sketch of the proof appears in [51], but a detailed proof appears in Chapter 10 of [21]. For a related result, see [47].

completeness, we provide a definition. A decision problem (that is, a problem with a "yes"/"no" answer) is in RP (see Chapter 11 of [44]) if there is a randomised polynomial-time algorithm which, for every "no" instance, answers "no" and for every "yes" instance, produces an output ("yes" or "no") which, on any given run, has probability at least 1/2 of being "yes". The relationship between RP and the more familiar classes P and NP is given by $P \subseteq RP \subseteq NP$. It is widely conjectured (for example, Chapter 7 of [21]) that $P = RP$, or at least that $RP \neq NP$.

4.2 Approximately automating Pólya theory

It is an intriguing open question whether #PÓLYAORBITS is complete for #P with respect to AP-reducibility. The most that we can say at this point is that a related problem (in which we work in a coset rather than in a group) is complete in this sense. The relationship between the new problem and #PÓLYAORBITS will be more clear if we first give a new definition of #PÓLYAORBITS, which is equivalent to the original definition by Pólya's theorem (Lemma 2.2). Recall that k is the size of the alphabet Σ in which the words are constructed.

Name. #PÓLYAORBITS.

Instance. $O(m)$ generators for a group G of permutations of $[m]$.

Output. $\sum_{g \in G} k^{c(g)}$.

We now describe the related problem, in which we sum permutations over a coset, rather than over the entire group.

Name. #COSETORBITS.

Instance. $O(m)$ generators for a group G' of permutations of $[m]$, $O(m)$ generators for a subgroup G of G' and a permutation $h \in G'$.

Output. $\sum_{g \in Gh} k^{c(g)}$.

Note that #PÓLYAORBITS corresponds to the special case of #COSETORBITS in which the coset Gh is a group. The following lemma implies that #COSETORBITS is unlikely to have an FPRAS, in which case coset decomposition cannot be used to give an FPRAS for #PÓLYAORBITS.

Lemma 4.1 #COSETORBITS *is complete for #P with respect to AP-reducibility.*

Proof Recall the problem #LARGECUT from Section 3.2. Theorem 1 of [12] shows that #LARGECUT is complete for #P with respect to AP-reducibility. Thus, it will suffice to show that #LARGECUT $\leq_{AP}$ #COSETORBITS. Let j and H be an instance of #LARGECUT and let N_j denote the number of size-j cuts of H. Construct G', G and h as in the proof of Lemma 3.2. Now note that the output of #COSETORBITS corresponding to input (G', G, h), which we denote $\#\text{COSETORBITS}(G', G, h)$, is equal to the quantity $P(G') - P(G)$ in the notation of Lemma 3.2. Thus,

$$N_j = \left\lfloor \frac{\#\text{COSETORBITS}(G', G, h)}{k^{|E|r+jr}} \right\rfloor. \tag{4.2}$$

We conclude that a good approximation to $\#\text{COSETORBITS}(G', G, h)$ gives a good approximation to N_j. We will omit the details about how to choose the accuracy parameter δ in the reduction. If it were not for the floor function in (4.2), we could simply set $\delta = \epsilon$, since division by a constant preserves relative error. The discontinuous floor function could spoil the approximation when its argument is small. However, this is a technical problem and not a real difficulty. For a solution, see the proof of Theorem 3 of [12].   □

We have now shown that the approximation problem corresponding to $\#\text{COSETORBITS}$ is intractable, subject to the standard complexity-theoretic assumption that RP $\neq$ NP. It seems plausible that the approximation problem corresponding to $\#\text{PÓLYAORBITS}$ is also intractable, perhaps in the sense that it is also complete for $\#$P with respect to AP-reducibility.[7] In Section 5.5 we will return to this problem and we will describe some special cases of $\#\text{PÓLYAORBITS}$ for which fully polynomial randomised approximation schemes are known. We close this section by mentioning a surprising fact. Although it is currently unknown whether $\#\text{PÓLYAORBITS}$ has an FPRAS for any fixed integer $k > 1$, Jerrum and I (Theorem 4 of [17] or Theorem 6 of [27]) have shown that if k is allowed to be any fixed rational *that is not an integer* then there is no FPRAS for $\#\text{PÓLYAORBITS}$ unless RP $=$ NP. Our proof for the case in which k is not an integer sheds no light on the intriguing integer case.

5 Almost-uniform sampling

Definition An algorithm for the almost-uniform sampling problem in Section 2.2 is said to be a *fully polynomial* almost-uniform sampler if its running time is bounded from above by a polynomial in the size of the description of the input (Ω, G) and in $\log(\epsilon^{-1})$.

The notion of "fully polynomial almost-uniform sampling" is due to Jerrum, Valiant and Vazirani [31]. The particular definition that we use is based on the one in [11]. Since the running time of a fully polynomial almost-uniform sampler is bounded from above by a polynomial in the *logarithm* of ϵ^{-1} (rather than just by a polynomial in ϵ^{-1}), the output distribution D (see Section 2.2) can be made very close to the uniform distribution $\mathcal{U}_{\Omega,G}$ at modest computational expense. For example, if ϵ is taken to be $e^{-|(\Omega, G)|}$, where $|(\Omega, G)|$ denotes

---

[7]Note that the reduction in Lemma 3.2 is not approximation preserving. In particular, approximations for $P(G')$ and $P(G)$ do not give an accurate approximation for $P(G') - P(G)$. For example, $e^{\epsilon}P(G') - e^{-\epsilon}P(G)$ can be much larger than $e^{\epsilon}(P(G') - P(G))$.

the size of the input (Ω, G), then the variation distance between the two distributions is exponentially small in $|(\Omega, G)|$, even though the running time is only polynomial in $|(\Omega, G)|$. [8]

5.1 Almost-uniform sampling and approximate counting

Jerrum, Valiant and Vazirani [31] have shown that there is a close connection between almost-uniform sampling and approximate counting. In particular, for "self-reducible" combinatorial structures [46], a fully-polynomial almost-uniform sampler exists if and only if an FPRAS exists. We will not give a formal definition of "self-reducible" but intuitively it means that outputs corresponding to a given input can be expressed in terms of outputs corresponding to "smaller" inputs. That is, the family of combinatorial structures has an inductive definition. The techniques from [31] have been used to get similar results for some combinatorial structures that do not seem to be self-reducible (see [11]). Furthermore, Dyer and Greenhill [11] have extended the result to the (related, but larger) class of "self-partitionable" structures. Self-reducibility and self-partitioning do not seem to apply (in general) to orbit-counting problems and there is no known *general* connection between the (approximate) orbit-counting problem and the orbit sampling problem. We shall revisit this point briefly in Section 5.5. The reader is also referred to Jerrum's papers [25] and [27].

One situation in which "counting" technology can be used for sampling orbits is when generating functions for enumerating orbits can be efficiently evaluated. For example, Nijenhuis and Wilf [41] used Equation 3.1 (see Section 3.4) to obtain a polynomial-time algorithm for sampling rooted unlabelled trees. Their algorithm is given in Figure 1 (see also [48]). Note that this is an *exactly* uniform sampling algorithm — its output distribution is exactly the uniform distribution on orbits.

Nijenhuis and Wilf's approach was extended by Wilf [52], who gave a fully-polynomial almost-uniform sampler for the problem #TREES from Section 3.4. Once again, the output distribution of Wilf's algorithm is *exactly* uniform on orbits. Wilf's algorithm is also based on finding a recurrence for the coefficients of the relevant generating function. This approach to sampling has been systematised by Flajolet, Zimmerman and Van Cutsem [13]. In their systematic approach, one specifies a set of structures using a formal grammar involving

---

[8]Note that a similarly demanding definition would not make sense in the context of an FPRAS. If we changed the definition of FPRAS (at the start of Section 4), demanding instead that the running time be bounded from above by a polynomial in $|x|$ and $\log(\epsilon^{-1})$, then finding an FPRAS for a problem would be as difficult as finding an exact algorithm. In particular, for many counting problems f, the quantity $f(x)$ is only exponentially large (as a function of $|x|$). Such problems could be solved exactly in polynomial time by running an FPRAS with $\epsilon \leq 1/(2f(x))$. The close connection between almost-uniform sampling and approximate counting (Section 5.1) indicates that these definitions (less demanding for FPRAS and more demanding for almost-uniform sampling) are the "right" ones.

1. Choose a pair (d, k) such that $k \in [1, n-1]$ and d divides k. The probability that the particular pair (d, k) is chosen should be $\frac{d T_{n-k} T_d}{(n-1) T_n}$.

2. Recursively choose T' uniformly at random (u.a.r.) from $\mathcal{R}_{n-k}$.

3. Recursively choose T'' u.a.r. from $\mathcal{R}_d$.

4. Make k/d copies of T'' and attach the root of each copy to the root of T'.

5. Let the root of T' be the root of the new n-vertex tree and output the new tree.

Figure 1: Let $\mathcal{R}_n$ be set of rooted unlabelled n-vertex trees. Suppose $n > 2$. The tree output by this algorithm of Nijenhuis and Wilf is equally likely to be any element of $\mathcal{R}_n$. The reason for this is given in Equation 3.1 — every n-vertex output comes up $n - 1$ times in the following process. Choose k and d. Choose a d-vertex tree and an $n - k$-vertex tree. Connect these as described in the algorithm. Count the resulting n-vertex tree d times. The quantities T_i are computed using dynamic programming as in Section 3.4.

set, sequence and cycle constructions. Generating functions can be derived automatically from the specification, so uniform sampling can be done automatically using dynamic programming. The combinatorial structures studied in [13] are labelled structures, but the authors observe that similar principles can sometimes be used for sampling "unlabelled structures" (orbits). Jerrum and I have used their approach to sample some tree-like unlabelled structures in Section 4 of [19].

5.2 The orbit-sampling process

We will now describe a general Markov-chain approach for sampling orbits. The approach was proposed[9] by Jerrum [25]. It is essentially a random walk on the bipartite graph which corresponds to the orbit-counting lemma (Lemma 2.1). In particular, consider the bipartite graph in which the left-hand vertex set is a finite set Ω and the right-hand vertex set is a permutation group G acting on Ω. There is an edge between element $\alpha \in \Omega$ and permutation $g \in G$ if and only if $\alpha^g = \alpha$. The Markov chain $M(\Omega, G)$, which we refer to as the "orbit-sampling process", is essentially a random walk on this

---

[9]Jerrum's description of the Markov chain was in terms of the Pólya-theory setting of Section 2.1. However, as Cameron has observed [7], the chain is applicable in the general orbit-counting setting.

graph. In particular, the state space of $M(\Omega, G)$ is the set Ω. The transition probabilities from a state $\alpha \in \Omega$ are specified by the following two-step experiment:

1. Sample g uniformly at random (u.a.r.) from G_α.

2. Sample α' u.a.r. from fix(g).

The new state is α'. The chain $M(\Omega, G)$ is *ergodic* since every state α can be reached from every other in a single transition, by selecting the identity permutation in Step 1. (For Markov-chain definitions, see Chapter 6 of [22]). Let $\pi : \Omega \to [0,1]$ denote the stationary distribution of $M(\Omega, G)$. It is now straightforward to verify that $\pi(\alpha)$ is proportional to the degree of α in the bipartite graph. That is, $\pi(\alpha) = |G_\alpha|/|\Upsilon(\Omega, G)|$. We have thus established the following lemma from [25].

Lemma 5.1 *Let π be the stationary distribution of the Markov chain $M(\Omega, G)$. Then*

$$\pi(\alpha) = \frac{|G_\alpha|}{|\Upsilon(\Omega, G)|} = \frac{|G|}{|\alpha^G||\Upsilon(\Omega, G)|} = \frac{1}{|\alpha^G||\Phi(\Omega, G)|} \tag{5.1}$$

for all $\alpha \in \Omega$. in particular, π assigns equal probability to each orbit α^G.

The second equality in Equation 5.1 follows from Lagrange's Theorem which implies that $|G_\alpha| \times |\alpha^G| = |G|$ and the third follows from Lemma 2.1.

Since the stationary distribution of $M(\Omega, G)$ is uniform on orbits, a reasonable approach to the orbit-sampling problem is to simulate $M(\Omega, G)$ for a sufficient number of transitions (to get "close" to the stationary distribution) and then output the result. Two issues arise at this point.

1. Can the steps of $M(\Omega, G)$ be simulated efficiently?

2. How many transitions have to be simulated before the chain is close to stationarity? In particular, how many transitions have to be simulated before almost-uniform sampling is achieved? (The definition of "almost-uniform sampling" is in Section 2.2.)

Both of these questions depend upon the specific input set $\mathcal{I}$ and the specific representation of the inputs in $\mathcal{I}$.

Let π be the stationary distribution of the Markov chain $M(\Omega, G)$. Let π_t be the distribution of $M(\Omega, G)$ after t transitions, when started in state α_0.

Definition The *mixing time* of $M(\Omega, G)$, given initial state α_0, is a function $\tau_{\alpha_0} : (0,1] \to \mathbb{N}$, from tolerances ϵ to simulation times, defined as follows: for each $\epsilon \in (0,1]$, let $\tau_{\alpha_0}(\epsilon)$ be the smallest t such that $d_{tv}(\pi_{t'}, \pi) \leq \epsilon$ for all $t' \geq t$. We define $\tau(\epsilon)$ to be the maximum of $\tau_{\alpha_0}(\epsilon)$ over all initial states $\alpha_0 \in \Omega$. $M(\Omega, G)$ is said to be *rapidly mixing* if and only if $\tau(\epsilon)$ is at most a polynomial in the size of the input (Ω, G) and in $\log(\epsilon^{-1})$.

Note that if $M(\Omega, G)$ is rapidly mixing, and each transition can be implemented in polynomial time, then $M(\Omega, G)$ is a fully-polynomial almost uniform sampler for orbits.

5.3   The orbit-sampling process and Pólya theory

Let $\mathcal{I}(\mathcal{G})$ be an input set in the Pólya-theory setting. Recall that each input $(\Sigma^m, \widehat{G})$ is represented as a set of $O(m)$ generators for G. Thus, the size of the input is bounded from above by a polynomial in m.

In this framework, Step 2 of each transition is computationally easy: to sample α' u.a.r. from fix($\hat{g}$), one just considers each of the $c(g)$ cycles of g and chooses one of the k alphabet symbols u.a.r. (see Section 2.1). However, Step 1 is apparently difficult. It is equivalent under randomised polynomial-time reductions to the *Setwise Stabiliser* problem, which includes *Graph Isomorphism* as a special case. There are, nevertheless, significant sets $\mathcal{G}$ of groups G for which a polynomial-time implementation exists. Luks has shown that p-groups—groups in which every element has order a power of p for some prime p—is an example of such a set [36]. For the remainder of this section, we will restrict our attention to input sets corresponding to sets $\mathcal{G}$ of permutation groups for which each transition can be implemented in polynomial time.

5.3.1   Negative results

Jerrum [25] asked whether the orbit-sampling process is rapidly mixing for the input set $\mathcal{I}(\mathcal{P})$. Subsequently [18], he and I showed that this is not the case. In particular, we constructed an infinite set $\mathcal{G}$ of permutation groups such that when the inputs $(\Sigma^m, \widehat{G})$ are chosen from $\mathcal{I}(\mathcal{G})$, the mixing time $\tau(1/3)$ of $M(\Sigma^m, \widehat{G})$ is exponential in m.

We will describe the construction (but not the proofs) here. Let k be the size of the fixed alphabet Σ. Let $\lambda = 1/k^2$. We will construct one group for each[10] pair $(l, n(l))$ where l and $n(l)$ are natural numbers satisfying

$$\left| \left(1 - \frac{(1 + 2\lambda)^l - (1 - \lambda)^l}{(1 + 2\lambda)^l + 2(1 - \lambda)^l} \right) - \frac{4 \ln 2}{n(l)} \right| \leq \frac{3}{n(l)^2}.$$

To construct the group $G_{l,n(l)}$, we let $H_{l,n(l)}$ denote the graph which is obtained from the complete graph on $n(l)$ vertices by subdividing each edge, inserting $l - 1$ intermediate vertices of degree two. Thus, $H_{l,n(l)}$ is formed by applying the "l-stretch" operation of Jaeger, Vertigan and Welsh [24] to the complete graph $K_{n(l)}$. Let $V_{l,n(l)}$ and $E_{l,n(l)}$ denote the vertex and edge sets of $H_{l,n(l)}$ (respectively) and let $m_{l,n(l)}$ be $3 |E_{l,n(l)}|$. We will construct a degree-$m_{l,n(l)}$ permutation group $G_{l,n(l)}$.

$G_{l,n(l)}$ acts on the set $K = \bigcup_{e \in E_{l,n(l)}} K_e$, which is the disjoint union of three-element sets K_e. Arbitrarily orient the edges of $H_{l,n(l)}$, so that each edge $e \in E_{l,n(l)}$ has a defined start-vertex e^- and end-vertex e^+. For $e \in E_{l,n(l)}$ and

---

[10]In [18] we prove that there are infinitely many such pairs.

$v \in V_{l,n(l)}$, let h_e be some fixed permutation that induces a 3-cycle on K_e and leaves everything else fixed and let g_v be the generator

$$g_v := \prod_{e:e^+=v} h_e \prod_{e:e^-=v} h_e^{-1}.$$

Finally, let $G_{l,n(l)}$ be $\langle g_v : v \in V_{l,n(l)} \rangle$, the group generated by $\{g_v\}$. Observe that the generators of the group commute and have order three, so each permutation $g \in G_{l,n(l)}$ can be expressed as a product

$$\prod_{v \in V_{l,n(l)}} g_v{}^{\sigma(v)},$$

where $\sigma : V \to \{0,1,2\}$. Thus, for every pair $(l, n(l))$, the group $G_{l,n(l)}$ is Abelian and every permutation $g \in G_{l,n(l)}$ (other than the identity) has order 3.

Let $\mathcal{G} = \{G_{l,n(l)}\}$. In [18], we showed that for any $\delta > 0$, the mixing time of the orbit-sampling process with input set $\mathcal{I}(\mathcal{G})$ satisfies $\tau(1/3) = \Omega(\exp(m(G)^{1/(4+\delta)}))$. Thus, the orbit-sampling process mixes slowly for an infinite set of Abelian 3-groups.

We will not describe the slow-mixing proof here, but the high-level picture is as follows: We can identify two types of permutation $g \in G$ such that, when the chain is in the stationary distribution, the permutation g selected in Step 1 is quite likely to have type 1 and also quite likely to have type 2. On the other hand, it takes the chain a long time to move from a permutation of one type to a permutation of the other type, and this implies slow mixing.

5.3.2 Positive results Despite the slow-mixing result of the previous section, Jerrum [25] has identified two sets of permutation groups for which the orbit-sampling chain is rapidly mixing:

1. $\mathcal{G}$ is the set of symmetric groups, as in Example 2.1;

2. $\mathcal{G}$ is the set of all cyclic groups.

Jerrum showed that the orbit-sampling process is rapidly mixing in both cases, so this process provides a fully-polynomial almost-uniform sampler in these cases.

To illustrate the ideas, we will consider the second case. Let G be a degree-m cyclic group and consider the Markov chain $M(\Sigma^m, \widehat{G})$.

As before, let π be the stationary distribution of $M(\Sigma^m, \widehat{G})$, and let π_t be the distribution after t transitions, starting from state α_0. A (Markovian) *coupling* for $M(\Sigma^m, \widehat{G})$ is a stochastic process (α_t, β_t) on $\Sigma^m \times \Sigma^m$ such that each of (α_t) and (β_t), considered marginally, is a faithful copy of $M(\Sigma^m, \widehat{G})$. In order to prove that $M(\Sigma^m, \widehat{G})$ is rapidly mixing, we want to construct a coupling in which the moves of (α_t) and (β_t) are correlated, so that (α_t) and (β_t) *coalesce* rapidly, ensuring that $\alpha_t = \beta_t$ for all sufficiently large t. The coupling lemma (see, for example, Aldous [1]) says that if β_0 is chosen from π then

$$d_{tv}(\pi_t, \pi) \leq \Pr[\alpha_t \neq \beta_t].$$

Let **1** denote the identity permutation. Let g_i denote the permutation chosen in Step 1 of the i'th transition of $M(\Sigma^m, \widehat{G})$. Jerrum showed that there is a constant ϵ and a polynomial $p(m)$ such that for every permutation $g \in G$ $\Pr(g_{p(m)} = \mathbf{1} \mid g_1 = g) \geq \epsilon$.

Given this fact, the mixing time can be bounded via a straightforward coupling: Let the two copies run independently until they reach a transition during which they both select the identity during Step 1. After that, run the copies together, keeping the second copy in the same state as the first. The probability that coupling has not occurred by time τ is $\exp(-\Omega(\tau/p(m)))$, so the chain is rapidly mixing.

5.4   Open questions regarding the orbit-sampling process

As we observed in the previous section, when the set Ω consists of words in the Pólya-theory framework and the group G is cyclic, the orbit-sampling process visits the identity permutation often, and this implies that it mixes rapidly. Similarly, when the group is the symmetric group, the process visits the word $\alpha = 00 \cdots 0$ often, and it mixes rapidly. I am not aware of any other rapid-mixing results for the orbit-sampling process. It would be interesting to identify a non-trivial input set for which the chain is rapidly mixing, but for some other reason. As a test case, we might ask whether it is rapidly mixing when orbits correspond to unlabelled 2-regular graphs. However, note that unlabelled 2-regular graphs can easily be sampled directly using the connection to integer partitions. See [35].

Cameron illustrated the orbit-sampling process in his textbook [7] by describing the case in which orbits represent unlabelled graphs (Example 2). In this case, Step 1 of the process corresponds to *Graph Isomorphism,* which we do not know how to solve in polynomial time. Nevertheless, as Cameron observes, there are good heuristics for graph isomorphism (for example, McKay's nauty [37]), so implementing the transitions may not represent a serious practical difficulty. It is worth recording the fact that we do not know whether the orbit-sampling process is rapidly mixing for unlabelled graphs. Probably it is. Since the identity permutation is visited often, a proof along the lines of the one sketched in Section 5.3.2 may work. However, as far as I know, nobody has proved this. In particular, even though it is clear that the identity permutation is visited often in the stationary distribution, it is not known whether there are some "bad" starting points from which it takes a long time to reach the identity. It would also be good to know whether the process is rapidly-mixing when orbits correspond to (unlabelled) bounded-degree graphs. In this case, the transitions of the process can be efficiently implemented.

5.5 Approximate counting revisited

There is no known general connection between the problem of approximately counting orbits and the orbit sampling problem (see Section 5.1). This is true even if we restrict attention to the Pólya-theory framework of Section 2.1. Nevertheless, in the Pólya-theory setting, the orbit-sampling *process* can be used for approximate counting.

Recall that

$$\Upsilon(\Omega, G) = \{(\alpha, g) \mid \alpha \in \Omega \text{ and } g \in G \text{ and } \alpha \in \text{fix}(g) \}.$$

Definition A *fully-polynomial almost-uniform* Υ*-sampler* for an input set $\mathcal{I}$ is an algorithm which takes an input $(\Omega, G) \in \mathcal{I}$ and an accuracy parameter $\epsilon \in (0, 1]$ and outputs a random variable. Typically, the output is a member of $\Upsilon(\Omega, G)$. In particular, the variation distance between the output distribution of the algorithm and the uniform distribution on $\Upsilon(\Omega, G)$ should be at most ϵ. Furthermore, the running time of the algorithm should be bounded from above by a polynomial in the size of the description of the input and in $\log(\epsilon^{-1})$.

If we run the orbit-sampling process, and observe the pair (α', g) at the end of each transition, then the stationary distribution of the process is uniform on $\Upsilon(\Omega, G)$ (see Lemma 5.1). Thus, the process is a fully-polynomial almost-uniform Υ-sampler for an input set $\mathcal{I}$ if and only if it is rapidly mixing for $\mathcal{I}$.

Now let $\mathcal{I}(\mathcal{G})$ be an input set in the Pólya-theory setting. The following lemma is due to Jerrum.

Lemma 5.2 *[25] If there is a fully-polynomial almost-uniform Υ-sampler for $\mathcal{I}(\mathcal{G})$ then there is an FPRAS for the corresponding orbit-counting problem.*

Together with Jerrum's rapid-mixing results from Section 5.3.2, Lemma 5.2 implies that the problem #PÓLYAORBITS has an FPRAS if the group G is required to be cyclic or to be a symmetric group. We will not include the proof of the lemma but it will be useful to outline the key ideas, which are frequently used in the "Markov Chain Monte Carlo" area. Note that the proof in [25] uses slightly different definitions, but a general treatment, with definitions similar to ours can be found in [11]. First, since $|\widehat{G}|$ can be computed exactly in polynomial time, Lemma 2.1 implies that it suffices to approximate $|\Upsilon(\Sigma^m, \widehat{G})|$. In this approximation, the self-reducibility in the group structure can be exploited. In particular, it suffices to estimate m ratios of the form

$$\frac{|\Upsilon(\Sigma^m, \widehat{G}_{i-1})|}{|\Upsilon(\Sigma^m, \widehat{G}_i)|} \tag{5.2}$$

where

$$G_j = \{g \in G \mid \ell^g = \ell \text{ for all } \ell < j\}$$

and $i \in \{1, \ldots, m\}$. $|\Upsilon(\Sigma^m, \widehat{G}_m)|$ can be calculated exactly (it is k^m) and this can be multiplied by all of the ratios to yield $|\Upsilon(\Sigma^m, \widehat{G}_0)|$, which is the desired quantity. The ratio in Equation 5.2 can be estimated by sampling from $\Upsilon(\Sigma^m, \widehat{G}_{i-1})$ and checking how many of the samples are in $\Upsilon(\Sigma^m, \widehat{G}_i)$.

Lemma 5.2 tells us that approximate counting is as easy as almost-uniformly sampling from $\Upsilon(\Sigma^m, \widehat{G})$ but it is not known whether the converse is true. In particular, $\Upsilon(\Sigma^m, \widehat{G})$ does not seem to be "self-partitionable" in the sense of [11]. It is easy to see that the set $\Upsilon(\Sigma^m, \widehat{G})$ can be described inductively by breaking $\widehat{G}$ into cosets. However, the problem is that the natural "parts" are cosets rather than groups, and we already know from Lemma 4.1 that approximately counting is difficult over cosets.

In particular, a natural method for sampling from $\Upsilon(\Sigma^m, \widehat{G}_i)$ would be to use counting estimates to determine the relative weight of each coset of $\Upsilon(\Sigma^m, \widehat{G}_{i+1})$, then select a coset (with the appropriate probability) and recursively sample from the coset. But this approach is unlikely to lead to an efficient algorithm because of Lemma 4.1.

5.6   Other orbit-sampling methods

5.6.1   Wormald's method
As in Example 2, let Ω_n be the set of all n-vertex graphs and let G_n be the permutation group acting on Ω_n which is induced by vertex permutations. The orbits of Ω_n under G_n correspond to unlabelled n-vertex graphs. Let $\mathcal{I}$ be the input set $\{(\Omega_n, G_n)\}$. The input (Ω_n, G_n) will be represented by the positive integer n, encoded in unary, as in the problem #GRAPHS. It is unknown whether the orbit-sampling process is rapidly mixing for $\mathcal{I}$. Nevertheless, there is a fully-polynomial almost-uniform Υ-sampler for $\mathcal{I}$. Thus, by Lemma 5.2, there is also an FPRAS for $\mathcal{I}$.[11] The Υ-sampler is due to Wormald [55] and uses the "rejection sampling" method, which is a frequently-used and powerful tool for sampling.

In order to simplify the description of Wormald's algorithm, we introduce the following notation: For every permutation g of a set Ω, let

$$\Upsilon(\Omega, g) = \{(\alpha, g) \mid \alpha \in \Omega \text{ and } \alpha \in \text{fix}(g) \}.$$

Thus, $\Upsilon(\Omega, G) = \bigcup_{g \in G} \Upsilon(\Omega, g)$.

First, suppose that we could estimate $|\Upsilon(\Omega_n, g)|$ and $|\Upsilon(\Omega_n, G_n)|$. Then we could sample from $\Upsilon(\Omega_n, G_n)$ using the following algorithm of Dixon and Wilf [9]:[12]

---

[11] In order to apply Lemma 5.2, we are implicitly using the fact that $\mathcal{I}$ can be encoded in the Pólya-theory setting. See Example 2.1.

[12] Dixon and Wilf's algorithm is more sophisticated than the one that we describe here. In particular, they show that the probabilities in Step 2 are identical for permutations in the same conjugacy class, and by breaking G_n into conjugacy classes, they show how to implement Step 2 in polynomial time *on average* provided the value of $|\Upsilon(\Omega_n, G_n)|$ is known. Details can be found in [9].

1. Input n.

2. Choose $g \in G_n$ with probability $\dfrac{|\Upsilon(\Omega_n,g)|}{|\Upsilon(\Omega_n,G_n)|}$.

3. Choose (α, g) u.a.r. from $\Upsilon(\Omega_n, g)$.

Step 3 of the algorithm is easily implemented; it corresponds to Step 2 of the orbit-sampling process. The main problem is that we do not know how to estimate $|\Upsilon(\Omega_n, G_n)|$. Wormald [55] uses *rejection sampling* to avoid doing this estimation. The basic idea of rejection sampling is as follows. It may be too difficult to sample from a given desired distribution. So what the user does instead is to sample from some other (more tractable) distribution. Imagine the desired distribution as being "scaled down" so that it fits underneath the more tractable distribution. To draw a sample from the desired distribution, the user first draws a sample from the more tractable distribution. The user then uses the sample to determine the probability with which the more tractable distribution over-represents this sample (relative to the "scaled down" desired distribution). With this probability, the sample is rejected (and the value $\bot$ is output instead). Otherwise, the sample is output. The method is useful when it is easy to determine the probability with which a given sample should be rejected (so rejection is fast) and, furthermore, the overall rejection probability is low (so the variation distance between the output distribution of the algorithm and the desired distribution is small).

We will now describe Wormald's algorithm. To simplify the description, we will first omit the accuracy parameter, ϵ, from the input. After we have described the algorithm, we will bound the variation distance between the output distribution of the algorithm and the uniform distribution on $\Upsilon(\Omega_n, G_n)$. We will then say how to modify the algorithm to reduce the variation distance to any desired quantity ϵ. The outline of the algorithm is as follows, where $\mathbf{1}$ denotes the identity permutation (This is a slight abuse of notation, since we use the single symbol $\mathbf{1}$, but when the input is n, we mean the identity permutation on Ω_n.) We will use the symbol $p_{g,n}$ to represent the probability with which permutation g is chosen (so $\sum_g p_{g,n} = 1$). Appropriate choices for $p_{g,n}$ will be discussed below.

1. Input n

2. Choose g with probability $p_{g,n}$.

3. Choose (α, g) u.a.r. from $\Upsilon(\Omega_n, g)$.

4. With probability $\dfrac{p_{1,n}}{|\Upsilon(\Omega_n,1)|} \dfrac{|\Upsilon(\Omega_n,g)|}{p_{g,n}}$ output (α, g); otherwise output $\bot$.

Clearly, we will need to choose the probabilities $p_{g,n}$ in such a way that Criterion 1 (below) is satisfied (so that Step 4 can be implemented). We will also choose the probabilities in such a way that Criteria 2 and 3 are satisfied, so that the algorithm runs in polynomial time.

Criterion 1:   The probabilities $p_{g,n}$ must be chosen so that

$$\frac{p_{1,n}}{|\Upsilon(\Omega_n, 1)|} \frac{|\Upsilon(\Omega_n, g)|}{p_{g,n}} \leq 1.$$

Criterion 2:   The probabilities $p_{g,n}$ must be chosen so that Step 2 can be implemented in polynomial time[13].

Criterion 3:   The probabilities $p_{g,n}$ must be chosen so that Step 4 can be implemented in polynomial time.

It is easy to check that the probability that any given pair (α, g) from $\Upsilon(\Omega_n, G_n)$ is output is $p_{1,n} / |\Upsilon(\Omega_n, 1)|$. With the remaining probability, which we denote π, the algorithm outputs $\perp$. It is now straightforward to verify that the total variation distance between the output distribution of the algorithm and the uniform distribution on $\Upsilon(\Omega_n, G_n)$ is π. Note that the rejection probability is 0 whenever $g = 1$. Thus, $\pi \leq 1 - p_{1,n}$.

If we wish to have an upper bound ϵ on the total variation distance, then we simply run the algorithm for $\lceil \log(\epsilon)/\log(1 - p_{1,n}) \rceil$ iterations. If the output is *always* $\perp$ (for every iteration) then we output $\perp$. Otherwise, we output the first member of $\Upsilon(\Omega_n, G_n)$ which is output by an iteration. We get a fully-polynomial almost-uniform Υ-sampler as long as the total number of times that we run the algorithm is bounded from above by a polynomial in n and $\log(\epsilon^{-1})$. Since

$$\log(\epsilon)/\log(1 - p_{1,n}) = \log(\epsilon^{-1})/\log((1 - p_{1,n})^{-1}),$$

this follows from Criterion 4.

Criterion 4:   The probabilities $p_{g,n}$ must be chosen so that, for some positive constant c and every n, we have $p_{1,n} \geq n^{-c}$.

Wormald [55] showed how to choose the probabilities $p_{g,n}$ so that these criteria are met. Thus, he gave a fully-polynomial almost-uniform Υ-sampler for unlabelled graphs.

---

[13]Note that choosing $p_{g,n} = \frac{|\Upsilon(\Omega_n,g)|}{|\Upsilon(\Omega_n,G_n)|}$ would make Wormald's algorithm equivalent to Dixon and Wilf's. Thus, it would satisfy all criteria except Criterion 2.

5.6.2 Extending Wormald's method Wormald's method can easily be expressed in the general orbit-sampling framework. As before, $p_{g,G}$ denotes the probability with which permutation g is chosen, so $\sum_g p_{g,G} = 1$.

1. Input (Ω, G)

2. Choose $g \in G$ with probability $p_{g,G}$.

3. Choose (α, g) u.a.r. from $\Upsilon(\Omega, g)$.

4. With probability $\dfrac{p_{1,G}}{|\Upsilon(\Omega,1)|} \dfrac{|\Upsilon(\Omega,g)|}{p_{g,G}}$ output (α, g); otherwise output $\perp$.

The analogue of Criterion 4 states that there is a positive constant c such that for every possible input (Ω, G), we must have $p_{1,G} \geq m(G)^{-c}$. On the other hand, the analogue of Criterion 1 implies

$$\frac{|\Upsilon(\Omega, 1)|}{p_{1,G}} \geq \frac{|\Upsilon(\Omega, g)|}{p_{g,G}}, \tag{5.3}$$

which implies

$$p_{1,G} \leq \frac{|\Upsilon(\Omega, 1)|}{|\Upsilon(\Omega, G)|}.$$

Thus, we cannot simultaneously satisfy the two criteria unless there is a positive constant c such that for every possible input (Ω, G),

$$m(G)^{-c} \leq \frac{|\Upsilon(\Omega, 1)|}{|\Upsilon(\Omega, G)|}. \tag{5.4}$$

In other words, we cannot use Wormald's method unless the the part of $\Upsilon(\Omega, G)$ which corresponds to the identity permutation accounts for at least a polynomial fraction of $\Upsilon(\Omega, G)$.

Several natural sets of permutation groups satisfy Equation 5.4. For example, Wormald has shown [55] how to use the method to efficiently sample unlabelled r-regular graphs for $r \geq 3$.

5.6.3 Other possibilities Not much is known about how to sample orbits when the input set does not satisfy Equation 5.4. We have just seen that Wormald's method relies on the identity permutation having a large weight (in the sense that Equation 5.4 must be satisfied). One of the two positive results in Section 5.3.2 (the result showing that the orbit-sampling process is rapidly mixing for cyclic groups) also relies on this fact.

Jerrum and I [19] considered the following orbit-sampling problem, which we chose specifically because Equation 5.4 does not hold.

Examples Let Δ be any fixed constant. For any multigraph H with degree at most Δ, the *degree sequence* of H is a sequence $\mathbf{n} = n_0, \ldots, n_\Delta$, where n_i denotes the number of vertices of H with degree i. Let $\Omega_\mathbf{n}$ be the set of all n-vertex connected multigraphs with degree sequence $\mathbf{n}$. Let $G_\mathbf{n}$ be the permutation group acting on $\Omega_\mathbf{n}$ which is induced by vertex permutations. As in Example 2, the orbits correspond to isomorphism classes. That is, the orbits of $\Omega_\mathbf{n}$ under $G_\mathbf{n}$ correspond to the unlabelled connected multigraphs with degree sequence $\mathbf{n}$.

In [19], we gave a fully polynomial almost-uniform sampler for this orbit-sampling problem (sampling unlabelled connected multigraphs with a given (bounded) degree sequence). Unfortunately, our solution does not contain any new methods; it is really a combination of the methods that have already been described here. Discovering new methods for sampling orbits, particularly methods which do not require Equation 5.4 remains an interesting challenge.

Our algorithm for sampling unlabelled connected multigraphs is based on the following idea. Every unlabelled connected multigraph H is associated with a unique "core"[14] which has no vertices of degree 1 or 2. To randomly generate a multigraph H, the algorithm first generates the core of H and then extends the core by adding trees and chains of trees to obtain H.

The algorithm for generating the core is described using the configuration model of Bender and Canfield [3], Bollobás [6] and Wormald [53]. A configuration (for a given degree sequence) is a labelled combinatorial structure which can be viewed as a refinement of a multigraph with the degree sequence. For any given degree sequence, the orbits of all configurations (with respect to the appropriate permutation group) correspond to the unlabelled multigraphs with the degree sequence. Since the degree sequence of the core has no vertices of degree 1 or 2, a typical core does not have many symmetries and Equation 5.4 is satisfied. (This follows from an extension of Bollobás's analysis of unlabelled regular graphs [5].) Thus, the algorithm uses Wormald's method to generate the core. (If the core is not connected, it is rejected. The fact that this does not happen too often follows from another result of Wormald [54].)

After generating the core of the random multigraph, the algorithm extends the core by adding trees and chains of trees. This part of the algorithm is based on the generating-function approach illustrated in Section 5.1.

It is an open problem to sample unlabelled multigraphs given a *general* degree sequence (in which degrees need not be bounded from above by a constant). Our method is not applicable when the degrees are unbounded. In fact, the problem with unbounded degrees seems to be difficult even in the labelled case (see [29, 39, 10]).

---

[14]For other uses of the "core" idea, see Zhan [56].

6 A related problem: Listing orbits

Consider the following computational problem, which fits into the framework of Section 2.2.

Definition *The orbit-listing problem:* Given an input $(\Omega, G) \in \mathcal{I}$, output exactly one member of each orbit in $\Phi(\Omega, G)$.

There is a vast literature on the problem of listing orbits. The reader is referred particularly to McKay's paper [38] which introduces a new method and also explains the connection between various methods which are used in practice. Further work along these lines can be found in [34]. In this survey we will restrict our attention to *polynomial delay* listing, which is not mentioned in these works.

The notion of "polynomial delay" is due to Johnson, Yannakakis and Papadimitriou [32]. A listing algorithm has polynomial delay if and only if the delay (in time-steps) between each pair of consecutive outputs is bounded from above by a polynomial (in the input size).

When the permutation group is trivial (so the orbits are in one-to-one correspondence with the elements of Ω), listing can be shown to be strictly less difficult than sampling [16], in the sense that the existence of a fully-polynomial almost-uniform sampling algorithm for a given input set implies the existence of a (randomised) polynomial-delay listing algorithm (but not vice-versa). It is not known whether such a result holds for arbitrary input sets, but the idea has been used for at least one non-trivial orbit sampling problem. In particular, Dixon and Wilf [9] suggested using a sampling algorithm for unlabelled graphs in order to list them. Using this idea, one can combine Wormald's unlabelled-graph sampling algorithm from Section 5.6.1 with Babai and Kučera's canonical labelling algorithm [2] to obtain a (randomised) polynomial-delay algorithm for listing unlabelled graphs [15, 16]. The duplicate-elimination contained in this algorithm requires each of the (exponentially-many) orbits to be stored. However, it turns out that there is also a deterministic polynomial-space polynomial-delay algorithm for listing unlabelled graphs [15, 16].

It is not known whether there is a polynomial-delay listing algorithm for listing orbits in the general Pólya-theory framework (that is, for input set $\mathcal{I}(\mathcal{P})$). It would be interesting to know more about this question, and about its connection to the corresponding orbit-sampling problem.

Acknowledgements

This work was supported by the EPSRC Research Grant GR/M96940 and by the ESPRIT Projects RAND-APX and ALCOM-FT. I am grateful to Mark Jerrum for many useful discussions and collaborations in this area and also for helpful comments on an earlier draft of this article. I am also grateful to the referee for helpful comments.

7 Appendix: Proof of Lemma 3.3

The proof is by reduction from the problem #CUBICHAM, which was shown to be #P-complete by Jerrum [26].

Name. #CUBICHAM.

Instance. A graph H in which every vertex has degree at most 3.

Output. The number of Hamiltonian paths in H.

Proof Let F_n be the graph with vertex set $\{c_{i,j} \mid i \in [n], j \in [2n+2]\}$ and edge set

$$\{(c_{i,0}, c_{i',0}) \mid i' = i+1 \pmod{n}\} \cup \bigcup_i \{(c_{i,j}, c_{i,j'}) \mid j' = j+1 \pmod{2n+2}\}.$$

Thus, F_n consists of a "central" length-n cycle $c_{0,0}, \dots, c_{n-1,0}$ and, off of each vertex $c_{i,0}$ in the cycle, there is a length-$(2n+2)$ cycle $c_{i,0}, \dots, c_{i,2n+1}$, which we refer to as a "petal". For every $j \in [n+1]$, let $F_n'[j]$ be the graph obtained from F_n by deleting edges $(c_{0,j}, c_{0,j+1})$ and $(c_{0,j+1}, c_{0,j+2})$. Thus, $F_n'[j]$ is obtained from F_n by removing two adjacent edges from a petal. The shortest path from the central cycle to the two deleted edges is the length-j path from $c_{0,0}$ to $c_{0,j}$. Let F_n' be the union of $n+1$ disjoint graphs, the jth of which is isomorphic to $F_n'[j]$.

Let H be an instance of #CUBICHAM with vertex set $V = \{v_{0,0}, \dots, v_{n-1,0}\}$. Let H' be the graph with vertex set $V \cup \{v_{i,j} \mid i \in [n], j \in \{1, \dots, 2n+1\}\}$ which is constructed from H by adding the edges in

$$\bigcup_i \{(v_{i,j}, v_{i,j'}) \mid j' = j+1 \pmod{2n+2} \text{ and } j \neq i+1\}.$$

Roughly, H' is formed from H by attaching petals, but the $i+1$st edge is deleted from the ith petal.

For any graph Γ, let $N(\Gamma)$ denote the number of distinct (up to isomorphism) subtrees of Γ. We claim that

$$N(H' \cup F_n) - N(H' \cup F_n') = N(F_n) - N(F_n') - \text{\#CUBICHAM}(H),$$

which completes the proof.

To see why the claim is true, note that to form a subtree of F_n, one must delete an edge $(c_{i,0}, c_{i',0})$. Also, for each $i \in [n]$, one must delete an edge $(c_{i,j}, c_{i,j'})$. If one stops at this point, then the subtree is not represented in $N(F_n')$, but if any further edges are deleted, then the subtree is represented in $N(F_n')$. Now we want to know how many subtrees in $N(F_n) - N(F_n')$ are subtrees of H' and this turns out to be the number of Hamiltonian paths in H. $\square$

References

[1] D. Aldous, Random walks on finite groups and rapidly mixing Markov chains, *Séminaire de Probabilités XVII 1981/1982*, (ed. A. Dold and B. Eckmann), *Lecture Notes in Mathematics* **986**, Springer, Berlin, 1983, 243-297.

[2] L. Babai and L. Kučera, Canonical labeling of graphs in linear average time, in *Proceedings 20th IEEE Symposium on Foundations of Computer Science* 1979, 39-46.

[3] E. A. Bender and E. R. Canfield, The asymptotic number of labelled graphs with given degree sequences, *J. Combin. Theory Ser. A* **24** (1978), 296-307.

[4] A. Bertoni, M. Goldwurm and N. Sabadini, The complexity of computing the number of strings of given length in context-free languages, Rapporto Interno n. 26/88, Dipartimento di Scienze dell'Informazione, Università degli Studi di Milano, Milano, 1988.

[5] B. Bollobás, The asymptotic number of unlabelled regular graphs, *J. London Math. Soc.* **26** (1982), 201-206.

[6] B. Bollobás, Almost all regular graphs are Hamiltonian, *European J. Combin.* **4** (1983), 97-106.

[7] P. Cameron, *Permutation Groups*, *London Mathematical Society Student Texts* **45**, Cambridge University Press, Cambridge, 1999.

[8] N. G. De Bruijn, Pólya's theory of counting, *Applied Combinatorial Mathematics*, (ed. E.F. Beckenbach), John Wiley and Sons, New York, 1964.

[9] J.D. Dixon and H.S. Wilf, The random selection of unlabeled graphs, *J. Algorithms* **4** (1983), 205-213.

[10] M. Dyer and C. Greenhill, Polynomial-time counting and sampling of two-rowed contingency tables, *Theoret. Comput. Sci.* **246** (2000), 265-278.

[11] M. Dyer and C. Greenhill, Random walks on combinatorial objects, *Surveys in Combinatorics*, (ed. J.D. Lamb and D.A. Preece), *London Math. Soc. Lecture Note Ser.* **267**, Cambridge University Press, Cambridge, 1999, 101-136.

[12] M. Dyer, C. Greenhill, L.A. Goldberg and M. Jerrum, On the relative complexity of approximate counting problems, in *Proceedings of APPROX Lecture Notes in Computer Science* **1913**, Springer, Berlin, 2000, 108-119.

[13] P. Flajolet, P. Zimmerman and B. Van Cutsem, A calculus for the random generation of labelled combinatorial structures, *Theoret. Comput. Sci.* **132** (1994), 1–35.

[14] M.R. Garey and D.S. Johnson, *Computers and Intractability: A Guide to the Theory of NP-Completeness*, Freeman, San Francisco, 1979.

[15] L. A. Goldberg, Efficient algorithms for listing unlabeled graphs, *J. Algorithms* **13** (1992), 128–143.

[16] L.A. Goldberg, *Efficient Algorithms for Listing Combinatorial Structures*, Cambridge University Press, , 1993.

[17] L.A. Goldberg, Automating Pólya theory: the computational complexity of the cycle index polynomial, *Inform. and Comput.* **105(2)** (1993), 268–288.

[18] L.A. Goldberg and M. Jerrum, The "Burnside process" converges slowly, in *Randomization and Approximation Techniques in Computer Science, Proceedings of RANDOM 1998* (ed. M. Luby, J. Rolim and M. Serna), *Lecture Notes in Computer Science* **1518**, Springer, Berlin, 1998, 331–345.

[19] L.A. Goldberg and M. Jerrum, Randomly sampling molecules, *SIAM J. Comput.* **29(3)** (1999), 834–853.

[20] L.A. Goldberg and M. Jerrum, Counting unlabelled subtrees of a tree is #P-complete, *LMS J. Comput. Math.* **3** (2000), 117-124.

[21] O. Goldreich, *Introduction to Complexity Theory, Lecture Notes Series of the Electronic Colloquium on Computational Complexity* 1999. http://www.eccc.uni-trier.de/eccc-local/ECCC-LectureNotes/

[22] G.R. Grimmett and D.R. Stirzaker, *Probability and Random Processes, Second Edition*, Oxford University Press, Oxford, 1992.

[23] F. Harary and E.M. Palmer, *Graphical Enumeration*, Academic Press, New York, 1973.

[24] F. Jaeger, D.L. Vertigan and D.J.A. Welsh, On the computational complexity of the Jones and Tutte polynomials, *Math. Proc. Cambridge Philos. Soc.* **108** (1990), 35–53.

[25] M. Jerrum, Uniform sampling modulo a group of symmetries using Markov chain simulation, *Expanding Graphs, DIMACS Series in Discrete Mathematics and Theoretical Computer Science* (ed. J. Friedman), **10**, American Mathematical Society, Providence, 37–47. 1993.

[26] M. Jerrum, Counting trees in a graph is #P-complete, *Inform. Process. Lett.* **51(3)** (1994), 111–116.

[27] M. Jerrum, Computational Pólya theory, *Surveys in Combinatorics, London Math. Soc. Lecture Note Ser.* **218**, Cambridge University Press, Cambridge, 1995, 103–118.

[28] M. Jerrum, Mathematical foundations of MCMC, *Probabilistic Methods for Algorithmic Discrete Mathematics*, (ed. M. Habib, C. McDiarmid, J. Ramirez-Alfonsin and B. Reed), Springer, Berlin, 1998, 116–165.

[29] M. Jerrum and A. Sinclair, Fast uniform generation of regular graphs, *Theoret. Comput. Sci.* **73** (1990), 91–100.

[30] M. Jerrum and A. Sinclair, Polynomial-time approximation algorithms for the Ising model, *SIAM J. Comput.* **22** (1993), 1087–1116.

[31] M.R. Jerrum, L.G. Valiant and V.V. Vazirani, Random generation of combinatorial structures from a uniform distribution, *Theoret. Comput. Sci.* **43** (1986), 169–188.

[32] D.S. Johnson, M. Yannakakis and C.H. Papadimitriou, On generating all maximal independent sets, *Inform. Process. Lett.* **27** (1988), 119–123.

[33] R.M. Karp and M. Luby, Monte-Carlo algorithms for enumeration and reliability problems, in *Proceedings of the 24th IEEE Symposium on Foundations of Computer Science* 1983, 56–64.

[34] A. Kerber, *Applied Finite Group Actions, 2nd Edition*, Springer, Berlin, 1999.

[35] D.L. Kreher and D.R. Stinson, *Combinatorial Algorithms: Generation, Enumeration and Search*, CRC Press, Boca Raton, 1999.

[36] E.M. Luks, Isomorphism of graphs of bounded valence can be tested in polynomial time, *J. Comput. System Sci.* **25** (1982), 42–65.

[37] B.D. McKay, nauty user's guide (version 1.5), Technical report TR-CS-90-02, Computer Science Department, Australian National University, 1990.

[38] B.D. McKay, Isomorph-free exhaustive generation, *J. Algorithms* **26** (1998), 306–324.

[39] B.D. McKay and N.C. Wormald, Uniform generation of random regular graphs of moderate degree, *J. Algorithms* **11** (1990), 52–67.

[40] P.M. Neumann, A lemma that is not Burnside's, *Math. Sci.* **4** (1979), 133–141.

[41] A. Nijenhuis and H.S. Wilf, *Combinatorial Algorithms, 2nd Edition*, Academic Press, New York, 1978.

[42] W. Oberschelp, Kombinatorische anzahlbestimmungen in relationen, *Math. Ann.* **174** (1967), 53–58.

[43] R. Otter, The number of trees, *Ann. of Math.* **49** (1948), 583–599.

[44] C.H. Papadimitriou, *Computational Complexity*, Addison-Wesley, Reading, Mass., 1994.

[45] G. Pólya and R.C. Read, *Combinatorial Enumeration of Groups, Graphs, and Chemical Compounds*, Springer, Berlin, 1987.

[46] C.P. Schnorr, Optimal algorithms for self-reducible problems, in *Proceedings of the 3rd International Colloquium on Automata Theory, Languages and Programming* 1976, 322–337.

[47] L. Stockmeyer, The complexity of approximate counting (preliminary version), in *Proceedings of the 15th ACM Symposium on Theory of Computing* 1983, 118–126.

[48] G. Tinhofer, Generating graphs uniformly at random, *Computing, Supp.* **7** (1990), 235–255.

[49] S. Toda, PP is as hard as the polynomial-time hierarchy, *SIAM J. Comput.* **20** (1991), 865–877.

[50] L.G. Valiant, The complexity of enumeration and reliability problems, *SIAM J. Comput.* **8** (1979), 410–421.

[51] L.G. Valiant and V.V. Vazirani, NP is as easy as detecting unique solutions, *Theoret. Comput. Sci.* **47** (1986), 85–93.

[52] H.S. Wilf, The uniform selection of free trees, *J. Algorithms* **2** (1981), 204–207.

[53] N.C. Wormald, Some problems in the enumeration of labelled graphs, Ph.D. Thesis, Department of Mathematics, University of Newcastle, New South Wales, 1978.

[54] N.C. Wormald, The asymptotic connectivity of labelled regular graphs, *J. Combin. Theory Ser. B* **31** (1981), 156–167.

[55] N.C. Wormald, Generating random unlabelled graphs, *SIAM J. Comput.* **16** (1987), 717–727.

[56] S. Zhan, On Hamiltonian line graphs and connectivity, *Discrete Math.* **89** (1991), 89–95.

Department of Computer Science
University of Warwick
Coventry CV4 7AL
United Kingdom
leslie@dcs.warwick.ac.uk

Graph minors and graphs on surfaces

Bojan Mohar

Abstract

Graph minors and the theory of graphs embedded in surfaces are fundamentally interconnected. Robertson and Seymour used graph minors to prove a generalization of the Kuratowski Theorem to arbitrary surfaces [37], while they also need surface embeddings in their Excluded Minor Theorem [45]. Various recent results related to graph minors and graphs on surfaces are presented.

1 Introduction

A graph H is a *minor* of another graph G if H can be obtained from a subgraph of G by contracting edges.

Graph minors and the theory of graphs embedded in surfaces are fundamentally interconnected. The family of all graphs which are embeddable in a fixed surface S is closed under taking minors. Therefore the graphs embeddable in S can be characterized by specifying the list $\mathbf{Forb_0(S)}$ of *minimal forbidden minors*, that is, minor minimal graphs which do not embed in S. (Similarly, they can be characterized by excluding, as subgraphs, all subdivisions of graphs in the set $\mathbf{Forb(S)}$ which is defined as the set of graphs of minimum degree ≥ 3 which cannot be embedded in S but all of whose proper subgraphs have embeddings in S.) Robertson and Seymour used graph minors to prove that $\mathbf{Forb_0(S)}$ (and hence also $\mathbf{Forb(S)}$) is finite for every surface S [37]. This result is a generalization of the Kuratowski Theorem to arbitrary surfaces. On the other hand, Robertson and Seymour needed surface embeddings in their Excluded Minor Theorem [45] where they determine a general structure of graphs which do not have a fixed graph H as a minor. This interplay between the two theories is visible in many other results, some of which are presented here.

The main purpose of this survey is to present up-to-date information on some of the most appealing results about graph minors and their relation to the study of graphs on surfaces.

Besides a stimulating survey article on minors and embeddings by Thomassen [59], there are numerous existing texts that cover this subject. A good introduction to graph minors is Diestel [14, Chapter 12], while excluded minor theorems are treated in Thomas [57]. Graph minors and tree-width are studied in Reed [28]; for tree-width and algorithms we refer to [5] and [6]. Embeddings of graphs in surfaces are treated in Mohar and Thomassen [26]; minors and embeddings are also covered in Robertson and Vitray [54]. The proof of the Graph Minor Theorem is sketched in Robertson and Seymour [29], and a more recent survey with focus on the related disjoint paths problem is [52].

2   Basic definitions

It is convenient to view minors as substructures. Then, a subgraph $\bar{H}$ of G is said to be an *H-minor* in G if $\bar{H}$ can be written as the union of $r = |V(H)|$ pairwise disjoint trees $T_1, \ldots, T_r$ and $m = |E(H)|$ edges $e_1, \ldots, e_m$ such that for $i = 1, \ldots, m$, the edge e_i joins T_j and T_l if the ith edge of H connects the jth and lth vertex of H. In Figure 1, a graph G with subtrees $T_1, \ldots, T_5$ (represented by thick lines) is exhibited to show that the graph K_5 minus an edge is a minor of G.

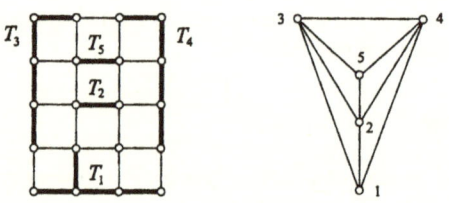

Figure 1: K_5 minus an edge as a minor

A family $\mathcal{F}$ of graphs is *minor closed* if for every graph in $\mathcal{F}$, all its minors are also in $\mathcal{F}$. There are two basic classes of examples of minor closed families. The first class are families related to embeddings in various topological spaces. Such examples include graphs embeddable in a fixed surface, graphs embeddable in $\mathbb{R}^3$ in some specific way, for instance, *linklessly embeddable graphs* [53, 57] (that is, graphs which admit an embedding in $\mathbb{R}^3$ such that no two disjoint cycles of the graph are linked in $\mathbb{R}^3$), *knotlessly embeddable graphs* (every cycle of the graph is embedded as an unknot), etc. The second important class of minor closed families is related to the tree-width. Classes of both types are discussed below.

Every closed surface is either homeomorphic to the orientable surface $\mathbb{S}_g$ of *genus* $g \geq 0$, or to the nonorientable surface $\mathbb{N}_g$ of *nonorientable genus* $g \geq 1$. Surfaces of the same orientability type can be distinguished by their Euler characteristic, and to unify the genus parameters for the surfaces $\mathbb{S}_g$ and $\mathbb{N}_{2g}$, which have the same Euler characteristic, it is convenient to introduce the *Euler genus* which is defined by $\mathbf{eg}(\mathbb{S}_g) = 2g$ and $\mathbf{eg}(\mathbb{N}_g) = g$. An embedding of a graph G in a surface $\mathbb{S}$ is a *2-cell embedding* if every face is homeomorphic to an open disk in the plane. In that case, the number of faces is equal to $|E(G)| - |V(G)| + 2 - \mathbf{eg}(\mathbb{S})$. This relation is known as Euler's formula.

Embeddings of graphs in surfaces, in particular the 2-cell embeddings, can be represented combinatorially. One such combinatorial description, known as the *Heffter-Edmonds-Ringel representation*, can be taken as a definition, and then one can work with combinatorial embeddings without any reference to topology. We refer to Mohar and Thomassen [26] for a thorough combinatorial treatment of surface embeddings. Following [26], we **define** an *embedding* of a connected graph G as a pair $\Pi = (\pi, \lambda)$ where $\pi = \{\pi_v \mid v \in V(G)\}$ is a collection of *local clockwise rotations*, that is, π_v is a cyclic permutation of the edges incident with v ($v \in V(G)$), and $\lambda : E(G) \to \{+1, -1\}$ is a *signature*. The local rotation π_v describes the cyclic clockwise order of edges incident with v on the surface, and the signature $\lambda(uv)$ of the edge uv is positive if and only if the cyclic permutations π_u and π_v both correspond to the clockwise (or both to anticlockwise) cyclic order of edges incident with u and v as seen on the surface when traversing the edge uv. An embedding of the graph G is *orientable* if every cycle of G has an even number of edges with negative signature.

The embedding $\Pi = (\pi, \lambda)$ determines a set of Π-*facial walks*. They are determined by the following process, called the *face traversal procedure*. We start with an arbitrary vertex v and an edge $e = vu$ incident with v. Traverse the edge e from v to u. We continue the walk along the edge $e' = \pi_u(e)$ which follows e in the π-clockwise ordering around u. If $\lambda(e) = -1$, the π-anticlockwise rotation is used instead, that is, $e' = \pi_u^{-1}(e)$. We continue using the π-anticlockwise ordering until the next edge with signature -1 is traversed, and so forth. The walk is completed when the initial edge e is encountered in the same direction from v to u and we are in the same mode (the π-clockwise ordering) which we started with. The other Π-facial walks are determined in the same way by starting with other edges. Two facial walks are considered the same if a cyclic shift of the first one gives rise to the second one or to the reverse of the second walk.

If f is the number of Π-facial walks, then the number

$$\mathrm{eg}(G, \Pi) = 2 - |V(G)| + |E(G)| - f$$

is called the *Euler genus* of the embedding Π. The underlying surface of the embedding Π which is obtained by pasting discs along the facial walks in G has the same Euler genus.

A *tree decomposition* of a graph G is a pair (T, Y), where T is a tree and Y is a family $\{Y_t \mid t \in V(T)\}$ of vertex sets $Y_t \subseteq V(G)$ (called *parts* of the tree decomposition) such that the following two properties hold:

(T1) $\bigcup_{t \in V(T)} Y_t = V(G)$, and every edge of G has both ends in some Y_t;

(T2) if $t, t', t'' \in V(T)$ and t' lies on the path in T between t and t'', then $Y_t \cap Y_{t''} \subseteq Y_{t'}$.

The pair (T, Y) is a *path decomposition* if T is a path. The *width* of the tree decomposition (T, Y) is $\max_{t \in V(T)}(|Y_t| - 1)$.

Figure 2 shows a graph G, a tree decomposition of width 3, and the underlying tree T. Let us observe that the graph G is outerplanar and hence it also has a tree decomposition of width 2.

It was shown in [27] that if a graph G has a tree decomposition of width at most w, then G has a tree decomposition of width at most w that further satisfies the following three properties:

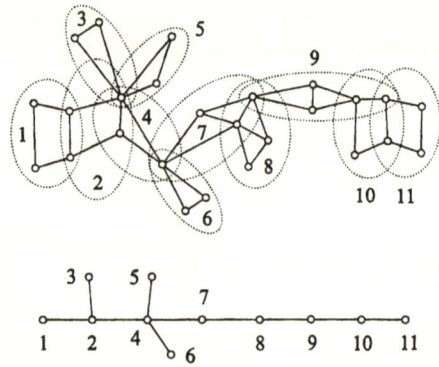

Figure 2: A graph and its tree decomposition of width 3

(T3) for every two vertices t, t' of T and every positive integer k, either there are k disjoint paths in G between Y_t and $Y_{t'}$, or there is a vertex t'' of T on the path between t and t' such that $|Y_{t''}| < k$;

(T4) if t, t' are distinct vertices of T, then $Y_t \neq Y_{t'}$;

(T5) if $t_0 \in V(T)$ and B is a component of $T - t_0$, then $\bigcup_{t \in V(B)} Y_t \setminus Y_{t_0} \neq \emptyset$.

The *tree-width* $\mathbf{tw}(G)$ (*path-width*) of a graph G is the smallest width of a tree decomposition (path decomposition) of G.

Let G_1 and G_2 be vertex disjoint graphs and let k be an integer. Suppose that V_i is a k-clique in G_i, and let G_i' be a subgraph of G_i obtained by deleting some (possibly none) of the edges joining pairs of vertices in V_i, $i = 1, 2$. If a graph G is obtained from $G_1' \cup G_2'$ by pairwise identifying the vertices of V_1 with the vertices of V_2, then we say that G is a *k-sum* of G_1 and G_2, or that G is a *clique sum* of G_1 and G_2 of *order* k.

3   The Excluded Minor Theorem

Robertson and Seymour proved that in any infinite sequence $G_1, G_2, G_3, \ldots$ of graphs there are indices $i < j$ such that G_i is a minor of G_j [30]–[51]. This seminal result, which establishes the well-quasi-ordering[1] of graphs with respect to the minor relation, is known as the **Graph Minor Theorem**. In the proof, one may assume (reductio ad absurdum) that none of the graphs $G_2, G_3, \ldots$ contains G_1 as a minor. Robertson and Seymour then prove that these graphs have a special structure. In particular, if G_1 is a forest, then the

---

[1] A *well-quasi-ordering* of a set X is a reflexive and transitive relation $\preceq$ such that, for every infinite sequence $x_1, x_2, x_3, \ldots$ of elements of X, there are indices i and j such that $i < j$ and $x_i \preceq x_j$.

graphs have bounded path-width [30]. If G_1 is a planar graph, then the graphs have bounded tree-width [34]. It takes a lot of work to reach the **Excluded Minor Theorem** 3.1 [45] which describes the structure of the sequence when a more general graph is an excluded minor. To express this result, an additional definition is needed.

Let G be a graph, $\mathbb{S}$ a surface, and k an integer. We say that G can be *k-nearly embedded* in $\mathbb{S}$ if G has a set A of at most k vertices such that $G - A$ can be written as $G_0 \cup G_1 \cup \cdots \cup G_k$ where the graphs $G_0, G_1, \ldots, G_k$ satisfy the following conditions.

(i) G_0 is embedded in $\mathbb{S}$.

(ii) The graphs $G_1, \ldots, G_k$ are pairwise disjoint.

(iii) For $i = 1, \ldots, k$, let $U_i = \{u_1^{(i)}, u_2^{(i)}, \ldots, u_{r_i}^{(i)}\} := V(G_0) \cap V(G_i)$. Then G_i has a path decomposition $(P_{r_i}, Y^{(i)})$ of width $\leq k$ such that for $t = 1, \ldots, r_i$, $Y_t^{(i)} \cap U_i = \{u_t^{(i)}\}$.

(iv) There are (not necessarily distinct) faces $F_1, \ldots, F_k$ of G_0 in $\mathbb{S}$, and there are pairwise disjoint disks $D_1, \ldots, D_k$ in $\mathbb{S}$, such that for $i = 1, \ldots, k$, $D_i \subset F_i$, $D_i \cap G_0 = U_i$, and the cyclic order of vertices in U_i on the boundary of D_i is $u_1^{(i)}, u_2^{(i)}, \ldots, u_{r_i}^{(i)}$.

Theorem 3.1 (Robertson and Seymour [45]) *For every graph H there exists an integer $k \geq 0$ such that every graph which does not contain H as a minor can be obtained by clique sums of order $\leq k$ from graphs that can be k-nearly embedded in some surface, in which H cannot be embedded.*

The main application of this impressive result is the proof of the Graph Minor Theorem by Robertson and Seymour. As Theorem 3.1 is very general and has not appeared in print till very recently, not many other applications are known. Two such examples, Theorems 3.2 and 3.7 below, have been obtained recently.

Theorem 3.2 (Böhme, Maharry, and Mohar [8, 9]) *For every positive integer k there exists an integer $N = N(k)$ such that every 7-connected graph of order at least N contains $K_{3,k}$ as a minor.*

Theorem 3.2 is sharp in the sense that the 7-connectivity condition cannot be relaxed. There are arbitrarily large 6-connected graphs which can be embedded on the torus. Since $K_{3,7}$ cannot be embedded in the torus, none of these graphs contains $K_{3,7}$ as a minor. The following construction [8] gives arbitrarily large graphs of tree-width $3a - 1$ none of which contain a $K_{a,2a+1}$-minor. Let $m \geq 4$ and $a \geq 3$ be integers, and let $N_{m,a}$ be the graph with vertices $v_{x,y}$ where $1 \leq x \leq m$ and $1 \leq y \leq a$, in which the vertex $v_{x,y}$ is adjacent to another vertex $v_{w,z}$ if and only if $w \in \{x-1, x, x+1\}$ where $x \pm 1$ is considered modulo m.

Theorem 3.3 (Böhme, Maharry, and Mohar [8]) *There is a function* $c :$ $\mathbb{N} \to \mathbb{N}$ *such that for any* $a \geq 3$ *the following holds. For any positive integers* k *and* w *there exists a constant* $N = N(k,w)$ *such that every* $c(a)$-*connected graph of tree-width less than* w *and of order at least* N *contains* $K_{a,k}$ *as a minor.*

Böhme, Maharry, and Mohar [8] conjectured the following extensions.

Conjecture 3.4 *There is a function* $f : \mathbb{N} \to \mathbb{N}$ *such that any* 9-*connected graph on at least* $f(k)$ *vertices contains a* $K_{4,k}$-*minor.*

Conjecture 3.5 *There are functions* $f : \mathbb{N} \to \mathbb{N}$ *and* $c : \mathbb{N} \to \mathbb{N}$ *such that any* $c(a)$-*connected graph on at least* $f(k)$ *vertices contains a* $K_{a,k}$-*minor.*

In [8] it is remarked that the sequence of graphs $K_{a,k}$, where a is fixed and k tends to infinity, is essentially the only family of graphs for which a result like Theorem 3.2 or 3.3 holds. More precisely, there is the following result.

Proposition 3.6 *Let* c *and* $w \geq c$ *be positive integers, and let* H_k $(k \geq 1)$ *be a sequence of graphs such that* $\lim_{k \to \infty} |V(H_k)| = \infty$. *Suppose that for any positive integer* k *there exists an integer* $N(k)$ *such that every* c-*connected graph of tree-width* $\leq w$ *and of order at least* $N(k)$ *contains* H_k *as a minor. Then* H_k *is a minor of* $K_{c,N(k)}$ *for* $k \geq 1$.

Proof Clearly, the graph $K_{c,N(k)}$ is c-connected and has tree-width $c \leq w$. By the assumption on the family H_k, $K_{c,N(k)}$ contains H_k as a minor. □

Böhme, Mohar, and Reed [10] showed that Theorem 3.2 can be strengthened by modifying the connectivity assumptions. Recall that a connected graph G is t-*tough* if for every separating vertex set S, the subgraph $G - S$ of G has at most $|S|/t$ connected components.

If d and k are positive integers, then P_k^d denotes the dth power of the path on k vertices, that is, distinct vertices v_i and v_j of P_k^d are adjacent if and only if $|j - i| \leq d$.

Theorem 3.7 (Böhme, Mohar, and Reed [10]) *For any positive integers* d *and* k *there exist numbers* $t = t(d)$ *and* $N = N(k,d)$ *such that every* t-*tough graph of order at least* N *contains* P_k^d *as a minor.*

4 Excluded minors for a fixed surface

One of the highlights in the Robertson-Seymour theory on graph minors is the proof of the finiteness (for each fixed surface $\mathbb{S}$) of the set **Forb$_0$($\mathbb{S}$)** of the minimal forbidden minors for $\mathbb{S}$.

Theorem 4.1 (Robertson and Seymour [37]) *For each surface* S, *the set* $\mathbf{Forb_0(S)}$ *of minimal forbidden minors is finite.*

Unfortunately, the complete list of graphs in $\mathbf{Forb_0(S)}$ is known only for the 2-sphere, where $\mathbf{Forb_0(S_0)} = \{K_5, K_{3,3}\}$, and for the projective plane $\mathbb{N}_1$, where there are precisely 35 minimal forbidden minors [18, 1].

The original proof of Theorem 4.1 by Robertson and Seymour is non-constructive in the sense that it does not provide a bound on the number or the size of graphs in $\mathbf{Forb_0(S)}$. A constructive proof for the case of nonorientable surfaces was obtained by Archdeacon and Huneke [4], while the first constructive proof for orientable surfaces appeared just recently (Mohar [25]). An independent constructive proof based on graph minors was also obtained by Seymour [55]. Seymour's bound on the size of graphs in $\mathbf{Forb_0(S)}$ is $2^{2^{(3g+9)^9}}$, where g is the Euler genus of S. This number is enormous already for the torus and the Klein bottle ($g = 2$). Even today, it remains a challenge to verify the following conjecture.

Conjecture 4.2 *Every minimal forbidden minor for the torus has less than 30 vertices.*

In the late 90's, Thomassen observed the possibility of obtaining a short proof of Theorem 4.1. He found a very short proof of the following result.

Theorem 4.3 (Thomassen [60]) *Let* $G \in \mathbf{Forb(S_g)}$. *Then* G *contains no* $k \times k$ *grid as a minor, where* $k = \lceil 3300 g^{3/2} \rceil$.

Theorem 4.3 implies Theorem 4.1 when combined with two other important results in the Robertson-Seymour theory, that graphs of large tree-width contain large grid minors [34], and that graphs of bounded tree-width are well-quasi-ordered [33]. For the former of these two results, a short proof with constructive bounds was obtained by Diestel, Gorbunov, Jensen, and Thomassen.

Theorem 4.4 (Diestel, Gorbunov, Jensen, Thomassen [15]) *Let* r, m *be positive integers, and let* G *be a graph of tree-width at least* $r^{4m^2(r+2)}$. *Then* G *contains either* K_m *or the* $r \times r$ *grid as a minor.*

The second result, the well-quasi-ordering of graphs of bounded tree-width, was proved by Robertson and Seymour in [33]. The proof is lengthy and technical as it provides general machinery for the graph minor theory. A shorter direct proof of this result was recently obtained by Geelen, Gerards and Whittle [17]. In the sequel we give a new, much simpler proof of this result restricted to graphs in $\mathbf{Forb_0(S)}$.

Theorem 4.5 *Let* g *and* w *be positive integers and let* S *be a surface of Euler genus* g. *Then there is an integer* N *such that every graph in* $\mathbf{Forb_0(S)}$ *with tree-width* $< w$ *has at most* N *vertices.*

Theorem 4.5 combined with Theorems 4.3 and 4.4 clearly implies Theorem 4.1. Theorem 4.3 is stated for orientable surfaces only but it is not difficult to extend its proof to include the nonorientable case as well.

Proof Suppose that $S \subseteq V(G)$. An S-bridge in G is a subgraph of G which is either an edge with both ends in S or a connected component C of $G - S$ together with all edges joining C with S. We start the proof by establishing some facts about bridges of embedded graphs.

Suppose that x, y is a separating pair of vertices of a graph G. An $\{x, y\}$-bridge B is said to be *nonplanar* if $B + xy$ is a nonplanar graph.

(1) If $G \in \mathbf{Forb}_0(\mathbb{S})$, then every $\{x, y\}$-bridge containing at least two edges is nonplanar.

This is easy to argue since the replacement of a nontrivial planar $\{x, y\}$-bridge by the edge xy would give a proper minor of G but would not decrease the genus of the graph.

Suppose that S is a vertex separating set of a connected graph G which is Π-embedded in $\mathbb{S}$. Let $W = v_1 e_1 v_2 e_2 \ldots v_k e_k v_1$ be a Π-facial walk. A triple $e_{i-1} v_i e_i$ in W (including the triple $e_k v_1 e_1$) is called a *mixed angle* if the edges e_{i-1} and e_i belong to distinct S-bridges in G. Let R be the multigraph embedded in $\mathbb{S}$ obtained by joining vertices of consecutive mixed angles in the Π-facial walks. Then $G \cup R$ has an embedding $\tilde{\Pi}$ in $\mathbb{S}$ which extends the embedding Π. Consider the induced embedding Π^R of R in $\mathbb{S}$. Let us observe that this embedding is not always 2-cell.

(2) The faces of Π^R in $\mathbb{S}$ can be partitioned into two classes, $\mathcal{F}_A$ and $\mathcal{F}_B$, such that every edge of R is incident with a face in $\mathcal{F}_A$ and a face in $\mathcal{F}_B$. The faces in $\mathcal{F}_A$ are 2-cells and correspond to the faces of G with mixed angles. The faces in $\mathcal{F}_B$ and the S-bridges in G which are $\tilde{\Pi}$-embedded in these faces are in bijective correspondence.

The existence of the partition $\mathcal{F}_A \cup \mathcal{F}_B$ is obvious. Let $F \in \mathcal{F}_B$. The boundary of F in $\mathbb{S}$ is composed of one or more closed walks in R. Let e be an edge on one of them, joining vertices v_i and v_j ($i < j$) of the Π-facial walk W. Since $e_{i-1} v_i e_i$ and $e_{j-1} v_j e_j$ are consecutive mixed angles on W, all edges $e_i, e_{i+1}, \ldots, e_{j-1}$ belong to the same S-bridge B. Consider the local clockwise rotation of $\tilde{\Pi}$ around v_j. We may assume that e is followed by e_{j-1}. Then e_{j-1} is followed by some other edges of B (possibly none) until a mixed angle in some face is reached, in which case an edge e' of R would follow the edges of B. Clearly, e' follows e on the boundary of F. By using the same argument at e', etc., we see that the edges of G entering the face F at the considered component of the boundary of F all belong to the same S-bridge B. If the face F has another boundary component, it must be incident with the same bridge; otherwise the embedding of G would not be 2-cell. Clearly, every S-bridge lies in a single face of R. This completes the proof of (2).

(3) Let G be a connected graph and $S \subseteq V(G)$ a separating set such that no vertex of S is a cutvertex and for any two vertices $x, y \in S$, every $\{x, y\}$-bridge containing at least two edges is nonplanar. If G is embedded in $\mathbb{S}$, and $s = |S|$ then

$$|E(R)| \leq 6g + s^2 + 5s - 12. \qquad (4.1)$$

Let $q = |E(R)|$. Since S contains no cutvertices, no facial walk of R has length 1. If a facial walk corresponding to a 2-cell face in $\mathcal{F}_B$ has length 2, then the corresponding S-bridge in that face is planar, hence just an edge joining two vertices of S. The number of such faces is $\leq \binom{s}{2}$. By (2), the sum of the lengths of faces in $\mathcal{F}_B$ is q. This implies that $2\binom{s}{2} + 3(|\mathcal{F}_B| - \binom{s}{2}) \leq q$, hence $3|\mathcal{F}_B| \leq q + \binom{s}{2}$. Similarly, the sum of the lengths of faces in $\mathcal{F}_A$ is q. Therefore, $|\mathcal{F}_A| \leq q/2$. Now, Euler's formula implies that

$$2 - g \leq s - q + |\mathcal{F}_A| + |\mathcal{F}_B| \leq s - \frac{q}{6} + \frac{1}{3}\binom{s}{2},$$

which yields (4.1).

After these preliminary results, we are ready for the proof of Theorem 4.5. Suppose that $G \in \mathbf{Forb}_0(\mathbb{S})$ and that $\mathrm{tw}(G) < w$. By the additivity of the genus (and using induction on g), we may assume that G is 2-connected. Let (T, Y) be a tree decomposition of G of width $< w$ such that (T4)–(T5) hold. Let $S = Y_t$ be a vertex separating set in G. By contracting an edge in one of the S-bridges, a graph embeddable in $\mathbb{S}$ is obtained. Claims (1)–(3) and the upper bound on $|\mathcal{F}_B|$ in the proof of (3) show that there are $\leq d := 2g + 2w + \binom{w}{2} - 4$ S-bridges in G. (T2) and (T5) imply that every vertex of the tree T has degree $\leq d$. By (T1), $|V(T)| \geq \frac{|V(G)|}{w}$. So, assuming G may have as many vertices as we like, T contains a path which is as long as we like. Applying Menger's theorem and the pigeonhole principle to the longest path in T and its subpaths one or more (but at most w) times, one can conclude that there exists an integer $s \leq w$ and there exist separating sets $S_0, \ldots, S_r$ (where r is as large as we want) such that the following hold:

(i) $|S_i| = s$, $i = 0, \ldots, r$;

(ii) there exist disjoint paths $P_1, \ldots, P_s$ from S_0 to S_r which intersect $S_0, S_1, \ldots, S_r$ in that order;

(iii) the path P_1 is *everywhere nontrivial* [8], that is, P_1 has an edge e_i strictly between its intersection with S_{i-1} and S_i, $i = 1, \ldots, r$.

For $i = 1, \ldots, r$, let G_i be the graph obtained from G by contracting the edge e_i of P_1. Let Π_i be an embedding of G_i in $\mathbb{S}$, and let R_i be the corresponding graph on vertices of the mixed angles in Π_i with respect to the separator S_i of G_i. Since every vertex of S_i is incident with at least two S_i-bridges, $V(R_i) = S_i =: \{u_1^i, \ldots, u_s^i\}$, where $u_l^i \in V(P_l)$, $l = 1, \ldots, s$.

For $i = 1, \ldots, r - 1$, let $B^{(i)}$ be the S_i-bridge in G_i which contains the segment of P_1 from S_0 to S_i. Note that $B^{(i)}$ is obtained from the S_i-bridge $B_0^{(i)}$ in G_i containing the same segment of P_1 by contracting the edge e_i.

Let Π_i^R be the embedding of R_i in $\mathbb{S}$. We say that (R_i, Π_i^R) is *strongly homeomorphic* to (R_j, Π_j^R) if there is a homeomorphism $\mathbb{S} \to \mathbb{S}$ whose restriction to R_i induces an isomorphism of the Π_i^R-embedded graph R_i onto the Π_j^R-embedded graph R_j such that $u_l^i \mapsto u_l^j$, $l = 1, \ldots, s$, and such that the face of R_i corresponding to the bridge $B^{(i)}$ is mapped onto the face of R_j corresponding to $B^{(j)}$.

Claim (3) combined with the surface classification theorem implies that the number of strong homeomorphism types of pairs (R_i, Π_i^R) is bounded in terms of g and w. As r can be arbitrarily large, there are indices i and $j > i$ such that (R_i, Π_i^R) and (R_j, Π_j^R) are strongly homeomorphic.

Take the embedding Π_i and delete the S_i-bridge $B^{(i)}$. Let F denote the resulting face in $\mathbb{S}$. Since (R_i, Π_i^R) and (R_j, Π_j^R) are strongly homeomorphic, the S_j-bridge $B^{(j)}$ can be embedded in F so that any vertex u_l^j of $B^{(j)}$ is identified with u_l^i ($l = 1, \ldots, s$) on the boundary of F. This gives rise to an embedding in $\mathbb{S}$ of the graph G' which is obtained from $G_i \setminus B^{(i)}$ by adding a disjoint copy of $B^{(j)}$ and identifying each $u_l^i \in V(G_i \setminus B^{(i)})$ with the vertex $u_l^j \in V(B^{(j)})$, $l = 1, \ldots, s$. Although $B^{(j)}$ is a bridge in G_j but not a bridge in G, it contains as a minor a copy of the S_i-bridge $B_0^{(i)}$ of G. In order to get $B_0^{(i)}$ as a minor, we contract all edges of the paths P_l ($l = 1, \ldots, s$) between S_i and S_j in the copy of $B^{(j)}$ in G'. Now it is clear that the graph G' contains G as a minor. Since G' is embedded in $\mathbb{S}$, also its minor G admits an embedding in $\mathbb{S}$. This contradiction completes the proof. $\qquad\square$

The above proof crystallized as a side result in the search of an efficient algorithm for determining the genus of graphs of bounded tree-width. It turned out that some of the main ingredients in this proof can also be found in the aforementioned work of Seymour [55].

It is well-known that testing planarity [20], constructing embeddings in the sphere $\mathbb{S}_0$ [12], or finding subgraphs that are subdivisions of Kuratowski graphs [62] can be performed by algorithms whose worst case running time is linear. Although the construction of minimum genus embeddings is **NP**-hard (by Thomassen [58]), Filotti, Miller, and Reif [16] proved that for every fixed surface $\mathbb{S}$, there is a polynomial time algorithm for embedding graphs in $\mathbb{S}$. For every fixed surface $\mathbb{S}$, Robertson and Seymour's theory gives an $O(n^3)$ algorithm for testing embeddability in $\mathbb{S}$ using graph minors [37, 52]. Robertson and Seymour recently improved their $O(n^3)$ algorithms to $O(n^2 \log n)$ [42, 50, 51]. An embeddability testing algorithm can be extended to an algorithm which also constructs an embedding in polynomial time (with estimated complexity $O(n^6)$; see Archdeacon [2]). Mohar [25] (and the papers cited therein) improved these results by showing the following result.

Theorem 4.6 (Mohar [25]) *Let* S *be a fixed surface. There is a linear time algorithm that, for an arbitrary graph* G, *either*

(a) *finds an embedding of* G *in* S, *or*

(b) *finds a subgraph* $K \subseteq G$ *which is a subdivision of some graph in* **Forb**(S).

A simpler linear time algorithm for embedding graphs in the projective plane is described by Mohar [23], while a simpler algorithm for the torus was developed recently by Juvan and Mohar [21].

5 Surface minors and the face-width

Given a Π-embedded graph G, every minor H of G can be considered as being obtained by deleting edges and contracting edges on the surface, so that the embedding of G determines an embedding Π' of H. In that case we say that the pair (H, Π') is a *surface minor* of (G, Π). If the embeddings Π and Π' are clear from the context, then we also say that H is a surface minor of G.

The grid graphs $P_k \square P_k$ can serve as a generic class for planar graphs in the following sense:

Theorem 5.1 (Robertson and Seymour [34]) *Let* G_0 *be a plane graph. Then there is an integer* k *such that* G_0 *is a surface minor of the* $k \times k$ *grid* $P_k \square P_k$.

Proof There is a plane graph G_1 with maximum degree 3 such that G_0 is a surface minor of G_1. It is well-known that every planar graph, hence also G_1 has a straight line embedding in the plane. Now, every edge can be modified so that it becomes a polygonal arc whose segments are all vertical or horizontal. Then it is easy to see that, for some large k, the $k \times k$ grid contains a subdivision of G_1. This completes the proof. $\square$

The proof of Theorem 5.1 does not give an explicit bound on the size of the grid. However, it is not difficult to show that the $O(n) \times O(n)$ grid suffices where n is the number of vertices of G_0; see Di Battista, Eades, Tamassia, and Tollis [7] for references.

Let G be a Π-embedded graph. If $\mathbf{eg}(G, \Pi) \geq 1$, the *face-width* $\mathbf{fw}(G, \Pi)$ of Π is the smallest integer r such that G has a Π-noncontractible cycle which is the union of r paths each of which is contained in a single Π-facial walk. If $g(G, \Pi) = 0$, we let $\mathbf{fw}(G, \Pi) = \infty$.

Theorem 5.1 has the following analogue for general surfaces.

Theorem 5.2 (Robertson and Seymour [36]) *Let* G_0 *be a graph that is* Π_0-*embedded in a surface* $S \neq S_0$. *Then there is a constant* k *such that for any graph* G *which is* Π-*embedded in* S *with face-width at least* k, (G_0, Π_0) *is a surface minor of* (G, Π).

Theorem 5.2 does not give explicit bounds on the face-width k that guarantees the presence of (G_0, Π_0) as a surface minor. Quantitative versions for many special cases are known; see [26]. Let us consider some of them.

D. Barnette and X. Zha (private communication) proposed the following conjectures.

Conjecture 5.3 (Barnette, 1982) *Every triangulation of a surface of genus $g \geq 2$ contains a noncontractible surface separating cycle.*

Ellingham and Zha (private communication) proved Conjecture 5.3 for triangulations of the double torus.

Conjecture 5.4 (Zha, 1991) *Every graph embedded in a surface of genus $g \geq 2$ with face-width at least 3 contains a noncontractible surface separating cycle.*

It follows from Theorem 5.2 that large face-width forces the existence of noncontractible surface separating cycles (where "large" may depend on the surface). Zha and Zhao [63] and Brunet, Mohar, and Richter [11] proved that face-width 6 (even 5 for nonorientable surfaces) is sufficient.

If Conjecture 5.3 is true, also the following may hold as suggested in Mohar and Thomassen [26].

Conjecture 5.5 *Let T be a triangulation of an orientable surface of genus g, and let h be an integer such that $1 \leq h < g$. Then T contains a surface separating cycle C such that the two surfaces separated by C have genera h and $g - h$.*

It is even possible that Conjecture 5.5 extends to all embeddings of face-width at least 3.

Suppose that the embedding of the graph G_0 in Theorem 5.2 is a minimum genus embedding. If G_0 is a surface minor of another embedded graph G (in the same surface), then also the embedding of G is a minimum genus embedding. Therefore, a consequence of Theorem 5.2 is that large face-width of an embedding implies that this is a minimum genus embedding.

Suppose now that G_0 is uniquely embeddable in $\mathbb{S}$ and that its embedding has face-width at least three. (Such graphs are easy to find.) If G is a 3-connected graph embedded in $\mathbb{S}$ such that G_0 is a surface minor of G, then also the embedding of G in $\mathbb{S}$ is unique. Consequently, sufficiently large face-width of a 3-connected graph implies uniqueness of the embedding. Both of theses results are treated in Seymour and Thomas [56] and Mohar [24] who proved that face-width of order $O(g \log g)$ $(g = \mathbf{eg}(G, \Pi))$ is sufficient, and this is essentially best possible (Archdeacon [3]).

There are numerous other results where Theorem 5.2 is used. However, the most surprising seems to be the flow-colouring duality on general surfaces discovered recently by Devos, Goddyn, Mohar, Vertigan, and Zhu [13]. The requirement is that the *edge-width* (which is defined as the length of a shortest noncontractible cycle on the surface) is large enough.

Let G be a 2-connected multigraph. The *circular flow number* $\phi_c(G)$ of G is the minimum real number r such that some orientation of G admits a real-valued flow whose absolute values all lie between 1 and $r - 1$. It is easy to see that $\lceil \phi_c(G) \rceil$ is the usual flow number, that is, the smallest integer k such that G admits a nowhere-zero k-flow.

Let G be a loopless multigraph. The *circular chromatic number* $\chi_c(G)$ is the smallest real number r such that there exists a real-valued function $c : V(G) \to [0, r)$ such that for every edge uv of G, $1 \leq |c(u) - c(v)| \leq r - 1$. We refer to the recent survey article by Zhu [64] for additional details on circular colourings and flows.

Theorem 5.6 (Devos, Goddyn, Mohar, Vertigan, Zhu [13]) *There exists a function $w : \mathbb{R}^+ \times \mathbb{N} \to \mathbb{N}$ such that the following holds. If $\varepsilon > 0$ is a real number and G is a graph embedded in the orientable surface of genus g with edge-width $\geq w(\varepsilon, g)$, then*

$$\chi_c(G) - \varepsilon \leq \phi_c(G^*) \leq \chi_c(G),$$

where G^* is the geometric dual graph of G in **S**.

Proof (sketch). The second inequality can be proved in the same way as the well-known flow-colouring duality result of Tutte [61], and so we sketch only the proof of the first inequality.

Suppose that G is a graph embedded in $\mathbf{S}_g$ and that its dual graph G^* admits a circular r-flow. If the edge-width of G is w, there is a graph $\tilde{G}$ in $\mathbf{S}_g$ which contains G as an induced subgraph such that $\mathrm{fw}(\tilde{G}) = w$. Moreover, $\tilde{G}$ can be chosen in such a way that the circular r-flow of G^* extends to a circular r-flow φ of $\tilde{G}^*$. If w is large enough, then by Theorem 5.2, $\tilde{G}$ contains cycles $C_1, \ldots, C_g$ such that after cutting the surface along these cycles (and pasting discs on the resulting holes), one obtains $g + 1$ surfaces, one homeomorphic to the sphere, all others homeomorphic to the torus such that each C_i corresponds to a face in the sphere and to a face in the ith torus. Moreover, we may assume that the face-width of all the torus embeddings is as large as we may need in the sequel. Let $G_0, G_1, \ldots, G_g$ be the corresponding graphs (where G_0 is the planar one), and let $G_0^*, G_1^*, \ldots, G_g^*$ be their dual graphs.

Fix an $i \in \{1, \ldots, g\}$. Since C_i is a surface separating cycle of $\tilde{G}$, the edges of $\tilde{G}^*$ dual to $E(C_i)$ form a cut in $\tilde{G}^*$. Therefore, their φ-sum is equal to 0. This implies that the restriction φ_i of φ in G_i^* is a circular r-flow in G_i^*.

Similarly, the restriction φ_0 of φ to G_0^* is a circular r-flow. Since G_0 is a plane embedding, the circular flow-colouring duality [64] shows that there is a circular $(r + \varepsilon)$-colouring c_0 of G_0 which is dual to the circular $(r + \varepsilon)$-flow $\frac{r+\varepsilon}{r}\varphi_0$.

As the face-width of G_i is large enough, Theorem 5.2 can be used to show that the toroidal $q \times q$ grid R_q is a surface minor in G_i, where $q = \lceil 2r^2/\varepsilon \rceil$. (As proved by Graaf and Schrijver [19], it is sufficient that the face-width is $\geq \frac{3}{2}q + 3$.) The toroidal grid consists of pairwise disjoint "vertical" cycles $A_1, \ldots, A_q$ and pairwise disjoint "horizontal" cycles $B_1, \ldots, B_q$. Let D_{kl} be the disk between A_k, A_{k+1}, B_l, and B_{l+1} (indices modulo q). By taking a slightly larger grid and omitting its part intersecting C_i, we may assume that C_i is disjoint from the grid.

Let D be the plane graph obtained by cutting G_i along A_1 and B_1. The flow φ gives rise to a circular r-flow in the planar dual of D. By the circular flow-colouring duality in the plane [64], there is a circular r-colouring c of D which is dual to φ.

Denote by α the φ-sum (mod r) of the edges dual to $E(A_1)$ (all considered to be oriented so that they cross A_1 from "left" to "right"). By choosing the direction of A_1, we may assume that $\alpha < r/2$. Similarly, we may assume that $\beta < r/2$, where β is the φ-sum (mod r) corresponding to B_1 (or to any B_l). It is not difficult to see that the following assignment defines a circular $(r + \varepsilon)$-colouring c_i of G_i:

$$c_i(v) := \frac{r + \varepsilon}{r} \left(\left(c(v) - \frac{(k-1)\beta}{q} - \frac{(l-1)\alpha}{q} \right) \bmod r \right)$$

if v is a vertex of D_{kl} which is not in $A_{k+1} \cup B_{l+1}$. Recall that $x \bmod r$ is defined as $x - \lfloor \frac{x}{r} \rfloor r$ and that $0 \leq x \bmod r < r$.

Observe that the colouring c_i is dual to the circular $(r + \varepsilon)$-flow $\varphi' := \frac{r+\varepsilon}{r}\varphi$ on all edges which are not part of the $q \times q$ grid in G_i. In particular, this is satisfied on the edges of C_i. Therefore, we may assume that c_i coincides on C_i with c_0 (by possibly replacing c_i with its cyclic shift). Then, the combination of circular $(r + \varepsilon)$-colourings $c_0, c_1, \ldots, c_g$ gives rise to a circular $(r + \varepsilon)$-colouring of G.   □

References

[1] D. Archdeacon, A Kuratowski theorem for the projective plane, *J. Graph Theory* **5** (1981), 243–246.

[2] D. Archdeacon, The complexity of the graph embedding problem, in *Topics in Combinatorics and Graph Theory*, (ed. R. Bodendiek and R. Henn), Physica-Verlag, Heidelberg, 1990, 59–64.

[3] D. Archdeacon, Densely embedded graphs, *J. Combin. Theory Ser. B* **54** (1992), 13–36.

[4] D. Archdeacon and P. Huneke, A Kuratowski theorem for nonorientable surfaces, *J. Combin. Theory Ser. B* **46** (1989), 173–231.

[5] S. Arnborg, S. Hedetniemi and A. Proskurowski (eds.), Efficient algorithms and partial k-trees, *Discrete Appl. Math.* **54** (1994). i–ii and 97–280.

[6] S. Arnborg and A. Proskurowski, Linear time algorithms for NP-hard problems restricted to partial k-trees, *Discrete Appl. Math.* **23** (1989), 11–24.

[7] G. Di Battista, P. Eades, R. Tamassia and I.G. Tollis, Algorithms for drawing graphs: an annotated bibliography, *Comput. Geom.* **4** (1994), 235–282.

[8] T. Böhme, J. Maharry and B. Mohar, $K_{a,k}$ minors in graphs of bounded tree-width, submitted.

[9] T. Böhme, J. Maharry and B. Mohar, Unavoidable minors in large 7-connected graphs, in preparation.

[10] T. Böhme, B. Mohar and B. Reed, Unavoidable minors in large t-tough graphs, in preparation.

[11] R. Brunet, B. Mohar and R.B. Richter, Separating and nonseparating disjoint homotopic cycles in graph embeddings, *J. Combin. Theory Ser. B* **66** (1996), 201–231.

[12] N. Chiba, T. Nishizeki, S. Abe and T. Ozawa, A linear algorithm for embedding planar graphs using PQ-trees, *J. Comput. System Sci.* **30** (1985), 54–76.

[13] M. Devos, L. Goddyn, B. Mohar, D. Vertigan and X. Zhu, Duality of circular colorings and flows in orientable surfaces, in preparation.

[14] R. Diestel, *Graph Theory*, Second edition, Springer, New York, 2000.

[15] R. Diestel, T.R. Jensen, K.Y. Gorbunov and C. Thomassen, Highly connected sets and the excluded grid theorem, *J. Combin. Theory Ser. B* **75** (1999), 61–73.

[16] I.S. Filotti, G.L. Miller and J. Reif, On determining the genus of a graph in $O(v^{O(g)})$ steps, in *Proc. 11th Ann. ACM STOC*, Atlanta, Georgia (1979), 27–37.

[17] J.F. Geelen, A.M.H. Gerards and G. Whittle, Branch width and well-quasi-ordering in matroids, manuscript, 2000.

[18] H.H. Glover, J.P. Huneke and C.-S. Wang, 103 graphs that are irreducible for the projective plane, *J. Combin. Theory Ser. B* **27** (1979), 332–370.

[19] M. de Graaf and A. Schrijver, Grid minors of graphs on the torus, *J. Combin. Theory Ser. B* **61** (1994), 57–62.

[20] J.E. Hopcroft and R.E. Tarjan, Efficient planarity testing, *J. Assoc. Comput. Mach.* **21** (1974), 549–568.

[21] M. Juvan and B. Mohar, A simplified algorithm for embedding graphs in the torus, preprint.

[22] A. Malnič and B. Mohar, Generating locally cyclic triangulations of surfaces, *J. Combin. Theory Ser. B* **56** (1992), 147–164.

[23] B. Mohar, Projective planarity in linear time, *J. Algorithms* **15** (1993), 482–502.

[24] B. Mohar, Uniqueness and minimality of large face-width embeddings of graphs, *Combinatorica* **15** (1995), 541–556.

[25] B. Mohar, A linear time algorithm for embedding graphs in an arbitrary surface, *SIAM J. Discrete Math.* **12** (1999), 6–26.

[26] B. Mohar and C. Thomassen, *Graphs on Surfaces*, Johns Hopkins University Press, Baltimore, 2001.

[27] B. Oporowski, J. Oxley and R. Thomas, Typical subgraphs of 3- and 4-connected graphs, *J. Combin. Theory Ser. B* **57** (1993), 239–257.

[28] B.A. Reed, Tree width and tangles: a new connectivity measure and some applications, in *Surveys in Combinatorics 1997* (ed. R.A. Bailey), Cambridge University Press, Cambridge, 1997, 87–162.

[29] N. Robertson and P.D. Seymour, Graph minors — a survey, in *Surveys in Combinatorics 1985*, (ed. I. Anderson), Cambridge University Press, Cambridge, 1985, 153–171.

[30] N. Robertson and P.D. Seymour, Graph minors. I. Excluding a forest, *J. Combin. Theory Ser. B* **35** (1983), 39–61.

[31] N. Robertson and P.D. Seymour, Graph minors. II. Algorithmic aspects of tree-width, *J. Algorithms* **7** (1986), 309–322.

[32] N. Robertson and P.D. Seymour, Graph minors. III. Planar tree-width, *J. Combin. Theory Ser. B* **36** (1984), 49–64.

[33] N. Robertson and P.D. Seymour, Graph minors. IV. Tree-width and well-quasi-ordering, *J. Combin. Theory Ser. B* **48** (1990), 227–254.

[34] N. Robertson and P.D. Seymour, Graph minors. V. Excluding a planar graph, *J. Combin. Theory Ser. B* **41** (1986), 92–114.

[35] N. Robertson and P.D. Seymour, Graph minors. VI. Disjoint paths across a disc, *J. Combin. Theory Ser. B* **41** (1986), 115–138.

[36] N. Robertson and P.D. Seymour, Graph minors. VII. Disjoint paths on a surface, *J. Combin. Theory Ser. B* **45** (1988), 212–254.

[37] N. Robertson and P.D. Seymour, Graph minors. VIII. A Kuratowski theorem for general surfaces, *J. Combin. Theory Ser. B* **48** (1990), 255–288.

[38] N. Robertson and P.D. Seymour, Graph minors. IX. Disjoint crossed paths, *J. Combin. Theory Ser. B* **49** (1990), 40–77.

[39] N. Robertson and P.D. Seymour, Graph minors. X. Obstructions to tree-decomposition, *J. Combin. Theory Ser. B* **52** (1991), 153–190.

[40] N. Robertson and P.D. Seymour, Graph minors. XI. Circuits on a surface, *J. Combin. Theory Ser. B* **60** (1994), 72–106.

[41] N. Robertson and P.D. Seymour, Graph minors. XII. Distance on a surface, *J. Combin. Theory Ser. B* **64** (1995), 240–272.

[42] N. Robertson and P.D. Seymour, Graph minors. XIII. The disjoint paths problem, *J. Combin. Theory Ser. B* **63** (1995), 65–110.

[43] N. Robertson and P.D. Seymour, Graph minors. XIV. Extending an embedding, *J. Combin. Theory Ser. B* **65** (1995), 23–50.

[44] N. Robertson and P.D. Seymour, Graph minors. XV. Giant steps, *J. Combin. Theory Ser. B* **68** (1996), 112–148.

[45] N. Robertson and P.D. Seymour, Graph minors. XVI. Excluding a non-planar graph, preprint, 1986.

[46] N. Robertson and P.D. Seymour, Graph minors. XVII. Taming a vortex, *J. Combin. Theory Ser. B* **77** (1999), 162–210.

[47] N. Robertson and P.D. Seymour, Graph minors. XVIII. Tree-decompositions and well-quasi-ordering, preprint, 1988.

[48] N. Robertson and P.D. Seymour, Graph minors. XIX. Well-quasi-ordering on a surface, preprint, 1989.

[49] N. Robertson and P.D. Seymour, Graph minors. XX. Wagner's conjecture, preprint,1988.

[50] N. Robertson and P.D. Seymour, Graph minors. XXI. Graphs with unique linkages, preprint, 1992.

[51] N. Robertson and P.D. Seymour, Graph minors. XXII. Irrelevant vertices in linkage problems, preprint, 1992.

[52] N. Robertson and P.D. Seymour, An outline of a disjoint paths algorithm, in *Paths, Flows, and VLSI-Layout*, (ed. B. Korte, L. Lovász, H. J. Prömel and A. Schrijver), Springer-Verlag, Berlin, 1990, 267–292.

[53] N. Robertson, P.D. Seymour and R. Thomas, A survey of linkless embeddings, in *Graph Structure Theory* (Seattle, WA, 1991), (ed. N. Robertson and P.D. Seymour), Contemp. Math. 147, Amer. Math. Soc., Providence, RI, 1993, 125–136.

[54] N. Robertson and R. P. Vitray, Representativity of surface embeddings, in *Paths, Flows, and VLSI-Layout*, (ed. B. Korte, L. Lovász, H.J. Prömel, and A. Schrijver), Springer-Verlag, Berlin, 1990, 293–328.

[55] P.D. Seymour, A bound on the excluded minors for a surface, submitted.

[56] P.D. Seymour and R. Thomas, Uniqueness of highly representative surface embeddings, *J. Graph Theory* **23** (1996), 337–349.

[57] R. Thomas, Recent excluded minor theorems for graphs, in *Surveys in Combinatorics 1999*, (ed. J.D. Lamb and D.A. Preece), Cambridge University Press, Cambridge, 1999, 201–222.

[58] C. Thomassen, The graph genus problem is NP-complete, *J. Algorithms* **10** (1989), 568–576.

[59] C. Thomassen, Embeddings and minors, *Handbook of Combinatorics, Vol. 1*, Elsevier, Amsterdam, 1995, 301–349.

[60] C. Thomassen, A simpler proof of the excluded minor theorem for higher surfaces, *J. Combin. Theory Ser. B* **70** (1997), 306–311.

[61] W.T. Tutte, A contribution to the theory of chromatic polynomials, *Canadian J. Math.* **6** (1954), 80–91.

[62] S.G. Williamson, Depth-first search and Kuratowski subgraphs, *J. Assoc. Comput. Mach.* **31** (1984), 681–693.

[63] X. Zha and Y. Zhao, On nonnull separating circuits in embedded graphs, in *Graph Structure Theory* (Seattle, WA, 1991), (ed. N. Robertson and P.D. Seymour), Contemp. Math. 147, Amer. Math. Soc., Providence, RI, 1993, 349–362.

[64] X. Zhu, Circular chromatic number, a survey, *Discrete Math.*, to appear.

Department of Mathematics
University of Ljubljana
Ljubljana
Slovenia
bojan.mohar@uni-lj.si

Thresholds for colourability and satisfiability in random graphs and boolean formulae

Michael Molloy

Abstract

We survey the progress on two fundamental problems in random graphs and random boolean formulae: How many edges must be added to a random graph until it is not almost surely k-colourable? How many clauses must be added to a random instance of k-SAT until it is not almost surely satisfiable?

1 Introduction

All the graphs considered in this paper are vertex-labelled and finite. Unless specified otherwise, they are always simple and undirected.

The *random graph process* runs as follows. Start with a collection of n vertices and no edges. For $i = 1$ to $\binom{n}{2}$ select an edge uniformly at random from amongst those edges which are not yet present and add it to the graph. (A random choice is *uniform* if all possible choices are equally likely.) Thus, our graph evolves in a random manner from a graph with no edges, to a "typical" graph with a few edges, and so on until we reach the complete graph. We use $G_{n,M}$ to denote a graph obtained by halting the random graph procedure on n vertices after step M. Equivalently, $G_{n,M}$ is a graph selected uniformly at random from amongst all graphs with n vertices and M edges.

We are are usually concerned with $G_{n,M}$ for large values of n and where M is a function of n. So most of our computations are asymptotic, thus allowing for a certain amount of approximation. For any graph property P, we say that $G_{n,M}$ *almost surely* (a.s.) has P if

$$\lim_{n \to \infty} \mathbf{Pr}(G_{n,M} \text{ has } P) = 1.$$

Another commonly studied random graph model is $G_{n,p}$, the graph on n vertices formed by making $\binom{n}{2}$ independent coin flips, one for each potential edge, each time adding the edge to the graph with probability p. These two models are in many respects equivalent. Roughly speaking, for most properties P, if $M \approx \binom{n}{2}p$ then $G_{n,M}$ a.s has P if and only if $G_{n,p}$ a.s. has P. Section 1.4 of [46] contains a more precise statement of this equivalence, but for now, suffice it to say that for all properties P considered in this paper, and for any constant $c > 0$,

$$\lim_{n \to \infty} \mathbf{Pr}(G_{n,M=cn} \text{ has } P) = \lim_{n \to \infty} \mathbf{Pr}(G_{n,p=2c/n} \text{ has } P).$$

This fact is often very convenient. For example, in this paper we are primarily concerned about the model $G_{n,M}$, but sometimes we will prove theorems about it indirectly by proving the corresponding theorems for $G_{n,p}$.

The field of random graph theory was founded by Erdős and Rényi in two seminal papers [33, 34]. In these papers, besides doing a very thorough analysis of many basic properties of a random graph, they asked the following questions, which were obviously inspired by what many people would consider to be the most natural questions to ask about graphs.

Under the random graph process, these are the questions.

A. How long does it typically take until G is connected?

B. For any fixed graph H, how long does it typically take until G contains H as a subgraph?

C. How long does it typically take until G is non-planar?

D. How long does it typically take until G has a perfect matching?

E. For any constant k, how long does it typically take until G has a K_k-subdivision?

F. How long does it typically take until G has a Hamilton path?

G. What is the distribution of the chromatic number of $G_{n,M}$? Specifically, what can be said when $M = cn$ for $c > \frac{1}{2}$?

Erdős and Rényi answered questions A, B and C in these two papers (see [56] for more on question C), but left the others open.

Question D was answered in 1966 [35]. Question E was settled in 1979 [8]; see also [57]. Question F was answered in 1983 [52]; see also [18, 21]. Also, many other basic graph properties became well understood for the random graph, such as the clique number and independence number [20, 59]. But Question G remained elusive.

In the late 80's, Bollobás [16] and independently, Matula and Kucera[61], determined the asymptotics of $\chi(G_{n,M})$ when $M = cn^2$, for any constant $c > 0$, and Bollobás proof extended to $M > n^{2-\delta}$ for any $\delta < 1/3$. Later Luczak[55] determined the asymptotics of $\chi(G_{n,M})$ when M/n is at least a large constant; see section 6 of this paper. To this date, though, we only have a very limited understanding of $\chi(G_{n,M=cn})$ when $c > \frac{1}{2}$ is a small constant.

Focusing on this elusive range of M, Erdős often asked the following refinement of question G; see, for example, his section in [13].

Question 1.1   (a) *For each $k \geq 3$, is there a constant c_k^* such that for every $\epsilon > 0$, a.s. $\chi(G_{n,M=(c_k^*-\epsilon)n}) \leq k$ and a.s. $\chi(G_{n,M=(c_k^*+\epsilon)n}) > k$?*

(b) *If these constants exist, determine them.*

Most of the work towards this problem has been focused on the search for c_3^*, as this is the first difficult case. Thus far, no-one has proven that c_3^* exists, although as we will see in Section 5, Achlioptas and Friedgut[4] came very close.

At first glance, one might think that the reason Question 1.1 is so difficult is that 3-colourability is NP-hard. However, focusing on computational complexity here can be misleading. Since 3-colourability is NP-hard, there is (probably) no simple characterisation of 3-colourable graphs. However, we cannot rule out the possibility that there is a simple characterisation of 3-colourable graphs which fails for only a very small number of graphs; that is, a simple characterization which is a.s. valid on a random graph.

For example, Hamiltonicity is NP-complete. But it turns out that in the random graph process, a.s. the graph will get a Hamilton cycle as soon as it has minimum degree at least 2; that is, "minimum degree at least 2" is a (surprisingly) simple characterization of Hamiltonian graphs which is a.s. valid. Analysing local properties of random graphs, such as vertex degrees, tends to be much easier than analysing global properties, and so this simple characterization is a tremendous help for determining the threshold for Hamiltonicity. It turns out that a.s. the graph becomes Hamiltonian at around $M = \frac{1}{2}n \ln n + \frac{1}{2}n \ln \ln n + \Theta(n)$ [52]; here, $\Theta(z)$ is used to denote a function t of the same asymptotic order as z.

Similarly, for n even, a.s. the graph has a perfect matching as soon as it has minimum degree at least 1. This a.s. valid characterization is much more helpful in answering Question D than Edmonds' polytime algorithm for finding a perfect matching [31] (and its underlying characterization which is valid for all graphs). Also, the graph is a.s. connected as soon as it has minimum degree at least 1, which is more helpful than the fact that we can test for connectivity very easily, for example, with a breadth-first search. It turns out that a.s. the graph becomes connected and gets a perfect matching at around $M = \frac{1}{2}n \ln n + \Theta(n)$ [33, 35].

So, despite the NP-completeness of 3-colourability, it is quite feasible that there is a fairly simple property which is a.s. equivalent to 3-colourability and that this will lead to an answer for Question 1.1. See Questions 4.1 and 7.1 for two approaches along these lines.

The fact that we do not know whether c_k^* exists, makes it awkward to discuss it formally. So instead, we define

$$c_k^- = \sup\{c : \text{a.s. } G_{n,M=cn} \text{ is } k\text{-colourable}\},$$
$$c_k^+ = \inf\{c : \text{a.s. } G_{n,M=cn} \text{ is not } k\text{-colourable}\}.$$

Most researchers in this field think that for $k \geq 3$, c_k^* exists. If it does, then we would have $c_k^- = c_k^+ = c_k^*$. So it is common practice to define

$$c_k = c_k^-,$$

and to discuss c_k, which certainly does exist, instead of discussing c_k^*.

Besides random graphs, we will also be discussing random boolean formulae. The most commonly studied such model is a random instance of k-SAT. An instance of k-SAT is a boolean formula in Conjunctive Normal Form (CNF), in which every clause has size k. For the reader who is unfamiliar with these terms, we provide the following definitions.

We have several boolean variables, say $v_1, ..., v_n$. A *literal* is a variable v_i along with a sign, and is represented as "v_i" (positive) or "$\bar{v}_i$" (negative). Thus, there are $2n$ possible literals. In CNF, a *clause* is a collection of literals, no two corresponding to the same variable. They are joined by "OR", so the clause is satisfied if and only if at least one of the literals is true. The boolean formula is a collection of clauses, all joined by "AND", so the formula is satisfied if and only if every clause is satisfied. A boolean formula is *satisfiable* if and only if at least one of the 2^n possible truth assignments to its variables causes the formula to be satisfied.

Just as with random graphs, we define the *random k-SAT process*. We start with a collection of n variables and no clauses. For $i = 1$ to $2^k \binom{n}{k}$ we select a clause uniformly at random from amongst those clauses which are not yet present and add it to the formula. $F^k_{n,M}$ is the formula after M iterations, or equivalently, it is a formula chosen uniformly at random from amongst all instances of k-SAT with n variables and M clauses.

Alternatively, we define $F^k_{n,p}$ to be the random formula on n variables, where each of the $2^k \binom{n}{k}$ possible k-clauses is present independently with probability p. As was the case with random graphs, these two models are very similar. For any property P considered in this paper, and for any constant $d > 0$,

$$\lim_{n \to \infty} \mathbf{Pr}(F^k_{n,M=dn} \text{ has } P) = \lim_{n \to \infty} \mathbf{Pr}(F^k_{n,p=d \times k!/(2^k n^{k-1})} \text{ has } P).$$

Note that the expected number of clauses in $F^k_{n,p=d \times k!/(2^k n^{k-1})}$ is $dn + o(n)$.

There are many natural questions to ask about random graphs, but with random boolean formula, there is one obvious question which dwarfs all the other ones in importance: What is the probability that the formula is satisfiable? This question is usually asked in the following form.

Question 1.2 (a) *For each $k \geq 3$, is there a constant d^*_k such that for every $\epsilon > 0$, a.s. $F^k_{n,M=(d^*_k-\epsilon)n}$ is satisfiable and a.s. $F^k_{n,M=(d^*_k-\epsilon)n}$ is not satisfiable?*

(b) *If these constants exist, determine them.*

Again, we do not even know whether these constants exist, although Friedgut [39] has shown something very close; see Section 5. So we define

$$d_k = d^-_k = \sup\{d : \text{ a.s. } F^k_{n,M=dn} \text{ is satisfiable}\}.$$

We define d^+_k in the obvious way, analogous to the definition of c^+_k. Most researchers in this field think that, for $k \geq 3$, d^*_k exists. If it does, then it must be equal to d_k (and to d^+_k). So we discuss d_k instead of d^*_k.

Question 1.1 has received a great deal of attention from researchers in random graph theory. Question 1.2 has received a phenomenal amount of attention from researchers in a wide range of fields including probabilistic combinatorics, theoretical computer science, artificial intelligence and statistical physics. The vast majority of this focus has been on finding d_3^*. Sophisticated simulations and some very sophisticated but non-rigorous analysis using the so-called "replica method" has produced strong evidence for the following conjecture.

Conjecture 1.3 $d_3^* \approx 4.2$.

Here, it is implicitly assumed that d_3^* exists.

Questions 1.1 and 1.2 behave like twin problems. For example, most new techniques that help with one problem turn out to help with the other one. By now, most people who work on Question 1.1 will, at least occasionally, work on Question 1.2. However, the other direction is far from true - probably most people in, for example, statistical physics who work on Question 1.2 are not even aware of Question 1.1. This is unfortunate, since I think that a good understanding of both problems is valuable when attacking either one of them.

The purpose of this paper is to present an up-to-date survey of the status of these two questions, placing them side-by-side in order to emphasize their similarities and how certain techniques have proven fruitful for both.

For random graphs, the focus of this paper will be on $G_{n,M}$ for the range $M = O(n)$. Thus, we omit virtually all results on the chromatic number of $G_{n,M}$ for much higher values of M (which includes some beautiful work).

Throughout this paper, we use $\mathbf{Exp}(X)$ to denote the expected value of X. For an event A, we use $\overline{A}$ to denote the complement of A, that is, the event that A does not occur. We use $\Pr(A|B)$ and $\mathbf{Exp}(X|B)$ to denote the probability of A and the expected value of X, both conditioned on the event B occurring. To avoid clutter, we adopt the convention of omitting floor and ceiling signs. Such instances can be formalized by inserting those signs where appropriate, along with the occasional negligible corrective term.

1.1 A few tools

It is a common misperception amongst combinatorialists that if one wishes to work in random graph theory then it is necessary to first become an expert in probability. This is not at all true. An understanding of the basics, such as expected value and variance, and familiarity with a few concentration inequalities will get you well on your way. On the other hand, if you wish to work with "the probabilistic method" then you also need to learn the Lovász Local Lemma.

The first tool that one needs is the following.

Linearity of Expectation

$$\mathbf{Exp}(X_1 + \ldots + X_t) = \mathbf{Exp}(X_1) + \ldots + \mathbf{Exp}(X_t).$$

Proof For any outcome ω of a random experiment, we denote by $X_i(\omega)$ the corresponding value of X_i. Recall that the expected value of X_i is $\sum_\omega \mathbf{Pr}(\omega) \times X_i(\omega)$. Linearity of Expectation follows immediately from this as

$$\sum_\omega \mathbf{Pr}(\omega) \times (X_1(\omega) + ... + X_t(\omega)) = \sum_{i=1}^{t} \left(\sum_\omega \mathbf{Pr}(\omega) \times X_i(\omega) \right).$$

$\square$

We give an explicit illustration of the use of Linearity of Expectation in the proof of Theorem 3.1. Here is our next tool.

Markov's Inequality *For any non-negative random variable X, and any $t > 0$,*

$$\mathbf{Pr}(X \geq t) \leq \frac{\mathbf{Exp}(X)}{t}.$$

Proof **(for discrete variables)** Recall that $\mathbf{Exp}(X)$ can also be expressed as $\mathbf{Exp}(X) = \sum_i i \times \mathbf{Pr}(X = i)$. Therefore, since X is always positive, $\mathbf{Exp}(X) \geq \sum_{i \geq t} i \times \mathbf{Pr}(X = i) \geq t \times \mathbf{Pr}(X \geq t)$. $\square$

The following fact, which follows immediately from Markov's Inequality with $t = 1$, is very useful.

Corollary 1.4 *If $\mathbf{Exp}(X) = o(1)$ then a.s. $X = 0$.*

It is often tempting to think that when $\mathbf{Exp}(X)$ is very large, then a.s. X is large. This is not always true, as the following example indicates.

Consider a lottery with a jackpot of 4^n pounds. The probability of winning the jackpot is 2^{-n}, and there are no other prizes. Your expected winnings are 2^n, but with very high probability, your winnings will be 0.

This type of situation is referred to as a *jackpot phenomenon*. For a more interesting illustration, recall that a.s. the random graph process becomes Hamiltonian when $M \approx \frac{1}{2}n \ln n + \frac{1}{2}\ln\ln n$. By the equivalency of our two models, this implies that a.s. $G_{n,p=\ln n/n}$ is non-Hamiltonian. However, if the reader computes the expected number of Hamilton cycles in $G_{n,p=10/n}$, she will discover a surprising jackpot phenomenon. This will also provide a good exercise in applying the Linearity of Expectation: Let X be the number of Hamilton cycles. Consider the $\frac{1}{2}(n-1)!$ undirected cyclic arrangements of the vertices, $\sigma_1, ..., \sigma_{\frac{1}{2}(n-1)!}$ and for each $1 \leq i \leq \frac{1}{2}(n-1)!$ let $X_i = 1$ if σ_i is a Hamilton cycle and $X_i = 0$ otherwise; here, X_i is an example of an *indicator variable*. Note that $X = \sum_{i=1}^{\frac{1}{2}(n-1)!} X_i$ and apply Linearity of Expectation to compute $\mathbf{Exp}(X)$.

In light of these *jackpot phenomena*, it is often useful to prove that with high probability X is very close to $\mathbf{Exp}(X)$. If this is true, we say that X is highly *concentrated*.

Simple Concentration Bound *Let X be a random variable determined by n independent trials $T_1, ..., T_n$, and satisfying*

changing the outcome of any one trial can affect X by at most c,   (1.1)

then

$$\mathbf{Pr}(|X - \mathbf{Exp}(X)| \geq t) \leq 2e^{-\frac{t^2}{2c^2 n}}.$$

Here, (1.1) rules out the possibility that a single trial can determine a huge jackpot, along with some more subtle variants of this problem. The constant "2" in the exponent can be improved somewhat (it can be moved from the denominator to the numerator), but in this field, the values of the constants in such bounds are rarely important. As stated, the Simple Concentration Bound is a special case of Azuma's Inequality and, with a slight weakening of some constants, of Talagrand's Inequality, two of the most powerful concentration bounds commonly used in combinatorics See [58] for a good survey.

Finally, the following bound, which can be easily verified using Stirling's formula, will often be useful. For $a \geq b \geq 1$,

$$\binom{a}{b} < \left(\frac{ea}{b}\right)^b.$$

2 k=2

The reason that Questions 1.1 and 1.2 both specify $k \geq 3$ is that the case $k = 2$ is very well understood, and the case $k = 1$ is trivial.

Obviously, a random graph with even one edge is not 1-colourable. An instance of 1-SAT is satisfiable if and only if there are no two conflicting 1-clauses, that is, if there is no variable v for which (v) and $(\bar{v})$ are both clauses. It is not hard to show that the first pair of conflicting 1-clauses will occur at around $M = \Theta(n^{1/2})$: this is essentially the birthday problem.

For 2-colourability, Erdős and Rényi [34] proved the following.

Theorem 2.1   (a) *For $0 < c < \frac{1}{2}$, $\mathbf{Pr}(\chi(G_{n,M=cn}) = 2) = \zeta + o(1)$ for some constant $\zeta = \zeta(c)$ with $0 < \zeta < 1$.*

(b) *For $c > \frac{1}{2}$, a.s. $\mathbf{Pr}(\chi(G_{n,M=cn}) > 2)$.*

A graph is 2-colourable if and only if it has no odd cycles. For $0 < c < \frac{1}{2}$, the expected number of odd cycles in $G_{n,M=cn}$ is $C + o(1)$ for a constant $C = C(c) > 0$. Furthermore, it turns out that the probability there are no odd cycles is $\zeta = e^{-C} + o(1)$. To see part (b) of the theorem, we must look at what is the most fundamental result in random graph theory, also from [34].

Theorem 2.2   (a) *For $c < \frac{1}{2}$, a.s. $G_{n,M=cn}$ has the following structure:*

     (i) *the largest component has size $O(\ln n)$; and*

     (ii) *no component has more than one cycle, and all but $O(1)$ components are trees.*

(b) *For $c > \frac{1}{2}$, there are constants $b(c) > a(c) > 0$ such that a.s. $G_{n,M=cn}$ has a component with $a(c)n + o(n)$ vertices and $b(c)n + o(n)$ edges (hence it has many cycles). This is referred to as the giant component. All other components are unicyclic and have size at most $O(\ln n)$.*

It is not hard to show that once the giant component appears, a.s. at least one of its many cycles will be odd.

The case for 2-SAT is similar, the main difference being that the change is much sharper. Independently, Chvátal and Reed [27], Fernandez de la Vega [36] and Goerdt [43] each proved the next result.

Theorem 2.3   (a) *For $c < 1$, a.s. $F^2_{n,M=cn}$ is satisfiable.*

(b) *For $c > 1$, a.s. $F^2_{n,M=cn}$ is not satisfiable.*

To see why this should be true, take any instance of 2-SAT, F, and form a graph $G(F)$ whose vertices are the $2n$ literals of F, and whose edges are the pairs of literals which form 2-clauses in F. For $d < 1$, $G(F^2_{n,M=dn})$ will a.s. not have a giant component and its small simple components will be very easy to satisfy. We can form a digraph $D(F)$ on the same vertex set by representing the clause $(\ell_1 \vee \ell_2)$ by the two directed edges $\bar{\ell}_1 \to \ell_2$ and $\bar{\ell}_2 \to \ell_1$. The significance of $D(F)$ is that $(\ell_1 \vee \ell_2)$ is equivalent to the statements "$\bar{\ell}_1$ implies ℓ_2" and "$\bar{\ell}_2$ implies ℓ_1". For $d > 1$, one can show that $D(F^2_{n,M=dn})$ will a.s. have a directed cycle containing both x and $\bar{x}$ for some variable x. Clearly, this means that the formula is unsatisfiable.

A great deal of work has gone into studying $G_{n,M}$ when $M = \frac{1}{2}n + o(n)$, culminating in [45]. $F^2_{n,M=cn}$ when $M = n + o(n)$ is much more complicated. Some recent pioneering work can be found in [19].

3   Upper bounds

3.1 k-colourability

The first upper bound on the threshold for k-colourability, was obtained by Erdős [32] (in a somewhat different form) by focusing on large stable sets in the random graph.

Theorem 3.1 $c_k \leq k \ln k + k$.

Proof We will work in the model $G_{n,p}$.

Since each colour class is a stable set, it is easy to see that if G is k-colourable, then G has an stable set of size at least n/k.

So we let X denote the number of stable sets of size n/k in $G_{n,p=2c/n}$, and compute $\mathbf{Exp}(X)$. Enumerate the subsets of n/k vertices as $I_1, .., I_{\binom{n}{n/k}}$. Define the random variable X_i to be equal to 1 if I_i is a stable set, and 0 otherwise. Also, $X = \sum X_i$, and $\mathbf{Exp}(X_i) = \mathbf{Pr}(X_i = 1) = (1-p)^{\binom{n/k}{2}}$. Therefore,

$$
\begin{aligned}
\mathbf{Exp}(X) &= \sum \mathbf{Exp}(X_i) \\
&= \binom{n}{n/k}(1-p)^{\binom{n/k}{2}} \\
&< \left(\frac{en}{n/k}\right)^{n/k} e^{-p \times \frac{1}{2}(n/k)((n/k)-1)} \\
&= \exp((n/k) \times (\ln(ek) - (c/k) - O(n^{-1}))) \\
&= o(1),
\end{aligned}
$$

when $c > k \ln k + k$. Thus, by Corollary 1.4, a.s. $X = 0$ and so there are no stable sets of size n/k. □

Remark Below, we will not be so explicit in our use of Corollary 1.4 nor in our use of indicator variables when applying the Linearity of Expectation.

For example, Theorem 3.1 yields $c_3 \leq 6.29...$, $c_4 \leq 9.54...$, and $c_5 \leq 13.04...$. By being a little more careful about the inequality in the calculations above, we get a slight improvement to $c_3 \leq 5.72...$, $c_4 \leq 8.99...$, $c_5 \leq 12.51...$. For large values of k, this bound turns out to be asymptotically optimal (see Section 6), but for small values of k, say $k = 3$, it is rather poor, as we shall see.

Luc Devroye (see [26]) showed that we can obtain much better bounds on c_k by counting the expected number of k-colourings directly, rather than focusing on large stable sets.

Theorem 3.2 $c_k \leq \ln k / \ln(k/(k-1))$.

This reduces the bound of Theorem 3.1 by roughly $2k$, yielding, for example, $c_3 \leq 2.71...$, $c_4 \leq 4.81...$, $c_5 \leq 7.21...$.

Proof Let X be the number of k-colourings of $G_{n,M=cn}$. There are k^n partitions of the n vertices into k colours; consider any one of them, with parts $A_1, ..., A_k$. This partition is a valid k-colouring if and only if the graph chooses no edges from amongst the $\sum \binom{|A_i|}{2} \geq k \times \binom{n/k}{2}$ potential edges between two vertices in the same part. This happens with probability at most

$$
\binom{\binom{n}{2} - k \times \binom{n/k}{2}}{cn} / \binom{\binom{n}{2}}{cn} = \left(\frac{k-1}{k}\right)^{cn+o(n)}.
$$

Therefore,

$$
\mathbf{Exp}(X) < k^n \left(\frac{k-1}{k}\right)^{cn+o(n)} = \exp(n(\ln k - c\ln(k/(k-1)) + o(1))) = o(1),
$$

for $c > \ln k / \ln(k/(k-1))$. □

The inequalities in the above calculation are fairly tight, and so with more care, one can establish that for $c > \ln k / \ln(k/(k-1))$, the expected number of k-colourings in $G_{n,M=cn}$ is exponentially high. It is tempting to guess that this implies there will a.s. be at least one k-colouring, and so the threshold c_k is equal to $\ln k / \ln(k/(k-1))$. However, things are not that simple; a fairly easy calculation shows that, in fact, $c_k < \ln k / \ln(k/(k-1))$; that is, there are values of c for which the expected number of k-colourings is exponentially high, but a.s. there are no k-colourings. The reason for this apparent paradox is that the number of k-colourings exhibit a jackpot phenomena - if there is at least one k-colouring, then there will probably be an exponentially high number of k-colourings, and this causes the expected number to be deceptively high.

Theorem 3.3 $c_k < \ln k / \ln(k/(k-1))$.

Proof The equivalence of our two random graph models will be helpful here. Choose a random graph $G = G_{n,p=2c/n}$. Let Z be the number of vertices of degree 0 in G. $\mathbf{Exp}(Z) = n(1-p)^{n-1} \approx e^{-2c}n$. Define $\zeta = \frac{1}{2}e^{-2\ln k/\ln(k/(k-1))} < \frac{1}{2}e^{-2c}$ and let A be the event that $Z \geq \zeta n$. A straightforward application of the Simple Concentration Bound shows that Z is highly concentrated and so $\mathbf{Pr}(A) = 1 - o(1)$. Now consider some $G' = G_{n,M=cn}$, let Z' be the number of degree 0 vertices in G' and let A' be the event that $Z' \geq \zeta n$. By the equivalence of our models, $\mathbf{Pr}(A') = 1 - o(1) > \frac{1}{2}$ for n sufficiently large.

Let X be the number of k-colourings of G'. If $X > 0$ then $X \geq k^{Z'}$, since we are free to colour the Z' degree 0 variables any way we choose. Therefore, for n sufficiently large,

$$
\begin{aligned}
\mathbf{Exp}(X) &\geq \mathbf{Pr}(A') \times \mathbf{Exp}(X|A') \\
&> \frac{1}{2}\mathbf{Exp}(X|A') \\
&= \frac{1}{2}[\mathbf{Pr}(X=0|A') \times 0 + \mathbf{Pr}(X>0|A') \times \mathbf{Exp}(X|X>0 \wedge A')] \\
&\geq \frac{1}{2}\mathbf{Pr}(X>0|A') \times k^{\zeta n}.
\end{aligned}
$$

Choose $c < \ln k / \ln(k/(k-1))$ such that $\mathbf{Exp}(X) \leq \frac{1}{2}k^{\frac{1}{2}\zeta n}$, which we can do since $\mathbf{Exp}(X) = (t(c) + o(1))^n$ where $t(c)$ is continuous in c. Then, by the above inequalities, $\mathbf{Pr}(X>0|A') \leq k^{-\frac{1}{2}\zeta n}$. Therefore,

$$
\mathbf{Pr}(X>0) \leq \mathbf{Pr}(A') \times k^{-\frac{1}{2}\zeta n} + \mathbf{Pr}(\overline{A'}) = o(1).
$$

Thus, a.s. $\chi(G_{n,M=cn}) > k$ despite the fact that the expected number of k-colourings of $G_{n,M=cn}$ is exponentially high! $\square$

In [63], Molloy and Reed further improved the bound on c_3.

Theorem 3.4 $c_3 \leq 2.571...$

Proof Outline The proof consisted of two steps. (1) The analysis in the proof of Theorem 3.2 extends to random graphs on a fixed degree sequence; that is, suppose G is chosen uniformly at random from amongst all graphs on n vertices with a specified degree sequence that has average degree greater than $2\ln 3/\ln(3/2) = 2.709....$ Then the expected number of 3-colourings of G is o(1).

(2) By repeatedly deleting all vertices of degree less than 3 from $G_{n,p=c/n}$, $c \geq 2.571..$, we will a.s. arrive at a (large) subgraph H of average degree greater than $2.709....$ Furthermore, H is uniformly random with respect to its degree sequence. Hence, the expected number of 3-colourings of H is o(1), and so a.s. it has no 3-colouring. Thus, a.s. G has no 3-colouring. □

In the next section, we will discuss a recent technique developed independently by Kirousis et al. [51] and by Dubois and Boufkhad [28] which was first used to obtain an upper bound on d_3. This technique was applied, in increasing degrees of precision, to obtain the following upper bounds on c_3.

Theorems 3.5 (a) $c_3 \leq 2.602$ (Dunne and Zito [30]).

(b) $c_3 \leq 2.522$ (Achlioptas and Molloy [6])

(c) $c_3 \leq 2.495$ (independently due to Kaporis et al. [49] and McDiarmid and Fountoulakis [37]).

To date, the last of these is the best known upper bound on c_k. It is not hard to show that this bound is not tight and that sufficient effort will produce a better bound. We will discuss these bounds and their proofs in the next subsection, after presenting the technique responsible for them.

We close this section by noting an interesting fact, first mentioned in [26]. Performing the calculations in the proof of Theorem 3.2 using the $G_{n,p}$ model, rather than the $G_{n,M}$ model yields a worse bound: each of the k^n partitions of the vertex set is a valid k-colouring with probability at most $(1-p)^{k \times \binom{n/k}{2}}$, since none of the at least $k \times \binom{n/k}{2}$ edges within a part can be chosen. Thus, the expected number of k-colourings is at most $k^n(1-\frac{2c}{n})^{\frac{n^2}{2k}} \approx \exp(-n(\ln k - c/k))$ which is o(1) for $c > k \ln k$. By being more careful, one can check that these calculations are asymptotically tight in that for $c < k \ln k$, the expected number of k-colourings is exponentially high. The main observation we need is that the vast majority of the k^n partitions have all parts of size very near n/k.

Thus, for $\ln k/\ln(k/(k-1)) < c < k \ln k$, the expected number of k-colourings in $G_{n,p=2c/n}$ is exponentially high, while in $G_{n,M=cn}$ it is exponentially low. An intuitive reason for this is as follows: in the $G_{n,p}$ model, there is a chance that the number of edges is much fewer than $p\binom{n}{2}$. The probability of this happening is very small, but if it does happen, then it will typically result in a very large number of k-colourings. Thus, again we have a jackpot phenomenon which produces a significant increase in the expected number.

3.2 Satisfiability

As we will see, most of the techniques used in subsection 3.1 yield analogous bounds on d_k. The analogue of Theorem 3.2 was first shown by Franco and Paull [38].

Theorem 3.6 $d_k \leq \ln 2/\ln(2^k/(2^k - 1))$.

For example, this gives $d_3 \leq 5.19..., d_4 \leq 10.74..., d_5 \leq 21.80....$

Proof The proof is very similar to that of Theorem 3.2. We work in the model $F^k_{n,M=dn}$. There are 2^n possible truth assignments to the n variables. For any one truth assignment, the probability that it satisfies the random formula is $\left(\frac{2^k-1}{2^k}\right)^{dn}$, since each of the dn clauses has a $\frac{2^k-1}{2^k}$ probability of being signed in an acceptable way. Thus, the expected number of truth assignments is $2^n \left(\frac{2^k-1}{2^k}\right)^{dn} = \exp(n(\ln 2 - d\ln(2^k/(2^k - 1))))$, which is $o(1)$ for $d > \ln 2/\ln(2^k/(2^k - 1))$. $\square$

Again, the same reasoning as in the proof of Theorem 3.3 yields that d_k is actually less than this bound, because variables which appear in no clauses help to create a jackpot phenomenon. Kamath et al. [48] carried out this reasoning very carefully, also accounting for other contributions of small degree variables to the jackpots, and showed that the resulting drop in the bound is significant.

Theorem 3.7 $d_3 \leq 4.758$.

More recently, Kirousis et al. [51] and, independently, Dubois and Boufkhad [28] introduced a very elegant new technique to (partially) account for the jackpot phenomenon. Define $X(f)$ to be the number of satisfying solutions to a formula f. Define $X'(f)$ to be equal to 1 if $X(f) > 0$ and 0 if $X(f) = 0$. Thus the expected number of solutions is the weighted average over all possible f of $X(f)$, while the probability of satisfiability is the weighted average of $X'(f)$. So a key to overcoming the jackpot phenomenon would be to compute an expected value which is closer to the average of X' than to the average of X.

They suggested that if f has at least one satisfying assignment, then we focus on the lexicographically maximum such assignment, that is, the maximum satisfying assignment under the ordering where $A_1 > A_2$ if the two assignments agree on variables $v_1, ..., v_r$ and if $A_1(v_{r+1}) = T$ while $A_2(v_{r+1}) = F$. Of course, this maximum is unique. Therefore $X'(f)$ is equal to the number of lexicographically maximum satisfying assignments of f, and so the expected number of such assignments is equal to the probability of satisfiability.

Unfortunately, computing this expected number seems very difficult. So instead, we consider the expected number of assignments which are close to being maximum. We say that a satisfying assignment is i-maximal if one cannot obtain a lexicographically greater satisfying assignment by changing at most i variables, and we define $X_i(f)$ to be the number of i-maximal satisfying assignments of f. Kirousis et al. [51] showed that just by focusing on $\mathbf{Exp}(X_1)$, we can get a good improvement on our upper bound, with a relatively small amount of effort.

Theorem 3.8 $d_3 \le 4.667$.

Proof We will work in the model $F^3_{n,M=dn}$. Consider any truth assignment A with a variables set True and $n - a$ variables set False. As we saw earlier, the probability that A satisfies our random formula is $\left(\frac{7}{8}\right)^{dn}$. We will bound the probability that A is 1-maximal, conditional on the event that it is satisfying.

Note that A is 1-maximal if and only if for every variable v set False, changing v to True will result in a non-satisfying assignment; that is, there must be at least one clause of the form $(\bar{v} \wedge x \wedge y)$ where x and y are two literals which are negated in A; we call such a clause a *blocker* for v. Consider any v which is set False. There are $\binom{n-1}{2}$ possible blockers for v. Conditioning on A being satisfying, simply reduces the choices of our random clauses to the $7\binom{n}{3}$ clauses which A satisfies. Thus, the probability, q, that v has a blocker satisfies

$$q = 1 - \left(\frac{7\binom{n}{3} - \binom{n-1}{2}}{dn}\right) \bigg/ \left(\frac{7\binom{n}{3}}{dn}\right) = 1 - e^{-3d/7} + o(1).$$

Now consider another variable u that is set False. Conditioning on v having at least one blocker only decreases the chances of u having at least one blocker since the number of clauses is fixed, and no clause can block both v and u; see [51] for a more careful elaboration on this point. Therefore, the probability that both v and u have blockers is at most q^2. Similarly, the probability that all a False variables have blockers is at most q^a, and so the expected number of 1-maximal satisfying solutions is at most

$$\sum_{a=0}^{n} \binom{n}{a} \left(\frac{7}{8}\right)^{dn} \left(1 - e^{-3d/7} + o(1)\right)^a$$

$$= \left(\frac{7}{8}\right)^{dn} \sum_{a=0}^{n} \binom{n}{a} \left(1 - e^{-3d/7} + o(1)\right)^a$$

$$= \left(\frac{7}{8}\right)^{dn} \left(2 - e^{-3d/7} + o(1)\right)^n$$

$$= \exp[n(1 - d\ln(8/7) + \ln(2 - e^{-3d/7} + o(1))],$$

which is $o(1)$ for $d \ge 4.667$. Here, the second equality comes from the identity $\sum_{a=0}^{n} \binom{n}{a} x^a = (1 + x)^n$. $\qquad\square$

Carrying out this analysis for general $k > 3$ yields that d_k is at most the unique positive solution of

$$\left(\frac{2^k - 1}{2^k}\right)^d (2 - e^{-kd/(2^k-1)}) = 1.$$

Of course, we gave a little bit away in our analysis when we upper bounded the probability that all False variables have blockers by what the value would be if the events of having blockers were independent. By being more exact there, Kaporis et al. [50] improved this bound slightly to $d_3 \leq 4.64$. This was obtained independently in a similar (but much more complicated) manner by Dubois and Boufkhad [28]. An even better improvement can be obtained by focusing on $\mathbf{Exp}(X_2)$. In this way, Kirousis et al. [51] obtained $d_3 \leq 4.598$. Janson et al. [47], by being a little more precise, improved it slightly.

Theorem 3.9 $d_3 \leq 4.596$.

This is the best bound to appear at the time this survey was written. Dubois et al. [29] reported that they obtained the better bound of $d_3 \leq 4.506$, but their details are not yet available. They outline their approach as follows.

Given a boolean formula, F, the *degree* of a variable x is $(d^+(x), d^-(x))$, the number of appearances in F of x and $\bar{x}$. Recall that in the previous chapter, the first moment analysis of $G_{n,M=cn}$ yielded a much better bound than that of $G_{n,p=2c/n}$. The reason is that by conditioning on the number of edges being equal to the expected number, we remove the jackpots contributed by graphs with a small number of edges. Dubois et al. report that a similar improvement can be obtained by conditioning on the event that for each (i,j), the number of variables with degree (i,j) is approximately equal to what we expect. This removes, for example, the jackpots contributed by formulas with a larger than expected number of variables of degree 0. So they consider a random formula with a fixed degree sequence such that the number of variables with degree (i,j) is equal to the expected number of such variables in $F^3_{n,M=dn}$, and they bound $\mathbf{Exp}(X_1)$ in that formula. The computations are reportedly very complicated.

Having described the technique introduced in [51], we can now discuss Theorems 3.5. Theorem 3.5a was proved by considering the analogue of X_1 for 3-colourings when we rank the colours $R > B = G$. Theorem 3.5b was proved by considering the same variable when we rank the colours $R > B > G$, and follow the proof of Theorem 3.8. Theorem 3.5c was proved by being a little more precise in the calculations, in the same way that the bound in Theorem 3.5c can be improved to 4.64.

We close this section with a question. It is not hard to show that for any fixed i, the bound on d_k obtained from $\mathbf{Exp}(X_i)$ is not optimal. Define $d_k^{(i)}$ to be equal to that bound, that is, the value for d at which $\mathbf{Exp}(X_i)$ shifts from being exponentially low to exponentially high. Kirousis et al. have asked the following.

Question 3.10 *Does $d_k = \lim_{i\to\infty} d_k^{(i)}$?*

Intuitively, this asks whether all the jackpots inherent in our expected value computations can be accounted for by clusters of many satisfying solutions which differ in $O(1)$ variables.

4 Lower bounds

4.1 Colouring

As described in the introduction, the first lower bound on the threshold for 3-colouring is $c_3 \geq \frac{1}{2}$. Most improvements on this lower bound came from focusing on the appearance of what is known as a 3-*core*. Every graph G with $\chi(G) \geq k+1$ must have a $(k+1)$-critical subgraph, that is, a subgraph H such that $\chi(H) = k+1$ and for any proper subgraph $H' \subseteq H$, $\chi(H') \leq k$. It is well-known, and easy to prove (see [69]) that any $(k+1)$-critical graph must have minimum degree at least k. Thus, any graph which cannot be k-coloured must contain a non-empty subgraph with minimum degree at least k. The k-*core* of a graph G, is defined to be the largest such subgraph, if at least one such subgraph exists.

It is easy to see that the k-core is unique, and can be determined by the following simple procedure: repeatedly delete a vertex of degree less than k from the graph until no such vertices remain. Since a deletable vertex cannot become undeletable, the order in which the vertices are deleted is irrelevant, and the procedure will always leave the same subgraph, namely, the k-core.

For $k \geq 3$, we define

$$\lambda_k = \lambda_k^- = \sup\{\lambda : \text{a.s. } G_{n,M=\lambda n} \text{ has no } k\text{-core}\}.$$

We define λ_k^+ in the obvious way, analogous to the definition of c_k^+ and d_k^+.

Recall that for some of the properties discussed in section 1, the threshold for the property coincided with the threshold for a much simpler, necessary but not always sufficient, property. For example, the threshold for a random graph to have a perfect matching is the same as the threshold for it to have no isolated vertices, and the threshold for the appearance of a Hamilton cycle is the same as the threshold for it to have minimum degree at least 2. Inspired by this trend, Bollobás [18] asked if $c_k = \lambda_k$. In fact, he asked the following somewhat stronger question.

Question 4.1 *Consider $k \geq 3$. Run the random graph process until the first k-core appears. Is the resulting graph a.s. not k-colourable?*

Łuczak [53] showed that when the 3-core appears, it must be very large, of size at least $\frac{1}{5000}n$. In fact, he showed that a.s. a random graph in the relevant range of probabilities will not contain any smaller subgraph with even *average* degree at least 3.

Theorem 4.2 *Consider any $0 < c < 3$. A.s. $G_{n,M=cn}$ has no subgraph with fewer than $\frac{1}{5000}n$ vertices and with average degree at least 3.*

Proof We will work in the model $G_{n,p}$. Consider any $4 \le a \le \frac{1}{5000}n$. If a graph G has a subgraph on a vertices with at least $\frac{3}{2}a$ edges, then by deleting edges, we see that G has a subgraph on a vertices with exactly $\lceil \frac{3}{2}a \rceil$ edges. The expected number of such subgraphs in $G_{n,p=2c/n}$ is at most

$$
\binom{n}{a}\binom{\binom{a}{2}}{\lceil \frac{3}{2}a \rceil}p^{\lceil \frac{3}{2}a \rceil} \le \left(\frac{en}{a}\right)^a \left(\frac{e(\frac{a^2}{2})}{\lceil \frac{3}{2}a \rceil}\right)^{\lceil \frac{3}{2}a \rceil} \left(\frac{2c}{n}\right)^{\lceil \frac{3}{2}a \rceil}
$$

$$
\le \left(\frac{8e^5c^3a}{3^3n}\right)^{a/2}
$$

$$
< \left(\frac{3}{4}\right)^{a/2}.
$$

Furthermore, this expected number is of order $\left(O\left(\frac{a}{n}\right)\right)^{a/2}$ which is less than $O\left(n^{-2/3}\right)$ for $a < n^{1/3}$. Therefore, the expected number of such subgraphs is at most

$$
\sum_{a=1}^{n^{1/3}} O(n^{-2/3}) + \sum_{a=n^{1/3}}^{n/1000} \left(\frac{3}{4}\right)^{a/2} = o(1),
$$

as required.   □

A very similar proof will show that for any c and $\epsilon > 0$ there is an $\alpha = \alpha(c, \epsilon) > 0$ such that a.s. $G_{n,M=cn}$ has no subgraph with fewer than αn vertices and with average degree at least $2 + \epsilon$.

Luczak used this fact to obtain the first improvement on the trivial lower bound $c_3 \ge \frac{1}{2}$.

Theorem 4.3 $c_3 \ge \lambda_3 > .50005$.

Proof The size of the giant component in $G_{n,M=cn}$ is well understood (see [15, 13, 46]). For $c \le .50005$, the giant component a.s. has fewer than $\frac{n}{5000}$ vertices, and so, by Theorem 4.2, it a.s. does not contain a 3-core. A.s., the other components are unicyclic and so they do not contain a 3-core either.
□

Chvátal [26] used an intricate analysis to determine the precise value of c where the expected number of subgraphs with minimum degree 3 shifts from $o(1)$ to exponentially high.

Theorem 4.4 *For $\psi = 1.442...$ we have the following:*

(a) *for $c < \psi$, the expected number of nonempty subgraphs in $G_{n,M=cn}$ with minimum degree 3 is $o(1)$;*

(b) *for $c > \psi$, this expected number is at least z^n for some $z = z(c) > 1$.*

As an immediate corollary, he obtains $c_3 \geq \lambda_3 \geq \psi$.

Perhaps not surprisingly, a jackpot phenomenon occurs here, and λ_3 is in fact much higher than ψ. Molloy and Reed [63] showed that $c_3 \geq \lambda_3 \geq 1.675$. They also showed that $\lambda_3 \leq 1.7941$. Then, Pittel, Spencer and Wormald [67] determined the exact value of λ_k. Setting $\pi_i(x) = 1 - e^{-x}\sum_{j=0}^{i-1}\frac{x^j}{j!}$, the probability that a Poisson variable with mean x is at least i, their result is as follows.

Theorem 4.5

$$\lambda_k = \lambda_k^- = \lambda_k^+ = \min_{\gamma>0}\frac{\gamma}{2\pi_{k-1}(\gamma)}.$$

For example, $\lambda_3 = 1.675..., \lambda_4 = 2.57..., \lambda_5 = 3.40....$

To do this, they used a system of differential equations to analyse the behaviour of a procedure which repeatedly deletes a randomly chosen vertex of degree less than k until no such vertices remain.

This result yields that, asymptotically, $\lambda_k = \frac{1}{2}k + o(k)$. It is an easy exercise to show that $\lambda_k \leq k - 1$, since any graph with average degree at least $2(k-1)$ must contain a k-core. Luczak [55] showed that $c_k = \Theta(k \ln k)$ and so for high values of k, the answer to Bollobás' Question 4.1 is negative, that is, $c_k \neq \lambda_k$. Molloy [64] noted that for $k \geq 4$, $c_k \neq \lambda_k$. To do so, he simply applied the following inequality[1], taken from [22] (see also [26]):

$$c_k \leq c_{k+1} - \frac{1}{2}\ln(2c_{k+1} + 1). \tag{4.1}$$

Using the bound $c_3 \geq \lambda_3 \geq 1.675$, this yields $c_4 \geq 2.58... > \lambda_4$, and similarly for each $k > 4$. Inequality (4.1) is proven in [22] in the following way.

Consider any $G = G_{n,p=2c/n}$. We build a stable set I using a simple greedy procedure whereby we run through the vertices in an arbitrary sequence. When we arrive at a vertex v, we add v to I, and remove all neighbours of v from the sequence. It turns out that a.s. $|I| \geq (\ln(2c+1)/(2c))n - o(n)$. The key observation is that the outcome of this procedure only depends on the edges from I to the rest of the graph. Therefore, after building I, we have exposed no information about the edges in $G - I$, and so the graph induced by $G - I$ is distributed exactly like $G_{n',p}$ where $n' = n - |I| = (1 - \ln(2c+1)/(2c))n - o(n)$. If $G - I$ is k-colourable, then G is $(k + 1)$-colourable. Thus, if $c > c_{k+1}$ then $G - I$ is not a.s. k-colourable; that is, $G_{n',p}$ is not a.s. k-colourable. Therefore, $p > 2c_k/n'$ and so $2c_k < n' \times 2c/n = 2c - \ln(2c + 1) + o(1)$. This yields (4.1).

---

[1]There is a confusing typographical error in [64] - the "+1" term is missing from the inequality.

Of course, using (4.1) to show $c_4 > \lambda_4$ requires the bound $c_3 \geq \lambda_3$ as a "base" and so there was no way to adjust this argument to show $c_3 > \lambda_3$. Eventually, Achlioptas and Molloy [1, 5] used a different technique to show $c_3 > \lambda_3$, and obtain what is, to date, the best known lower bound on c_3.

Theorem 4.6 $c_3 \geq 1.923$.

To prove this theorem, they use a set of differential equations to analyse the following procedure: Begin with a list of the same 3 colours on every vertex of $G_{n,p=c/n}$. Repeatedly pick a random vertex v of minimum list-size, choose a random colour for v from its list, and delete that colour from the list of every neighbour of v. They show that this procedure a.s. fails for $c > 1.923...$ while for $c < 1.923...$ it does not a.s. fail. Corollary 5.5 below allows them to strengthen this latter fact: for $c < 1.923..$, $G_{n,p=c/n}$ is a.s. 3-colourable. They also show that, for large k, the analogous procedure succeeds for $c \leq k \ln k - \frac{3}{2}k$ but a.s. fails for $c > (1 + \epsilon)k \ln k$ for any constant $\epsilon > 0$.

4.2 Satisfiability

Thus far, all lower bounds for d_3 have taken the following approach: (1) define an algorithm which looks for a satisfying solution in an instance of 3-SAT; (2) prove that for d less than a particular constant, this algorithm does not a.s. fail on $F^3_{n,M=cn}$. The astute reader will notice that these two facts alone do not provide a bound on d_3; this point is addressed after Theorem 4.7 below.

Of course, for this approach to work, there has to be a balance between the sophistication of the algorithm, and the ease with which this algorithm can be analyzed. Roughly speaking, increases in the lower bounds on d_3 over the years correspond to increases in the abilities of the researchers to analyse more complicated algorithms. So far, all algorithms studied have been fairly similar, they all involve setting variables one-at-a-time in a greedy manner. Achlioptas has recently written an excellent survey on this subject. So, few details are given here; for more, see [3].

All of the algorithms follow a similar strategy. Each time a variable is set, say $x_1 = T$, any clause that contains x_1 is removed from the formula since it is now satisfied. Any clause that contains $\bar{x}_1$ has $\bar{x}_1$ deleted from it, and it shrinks to a smaller clause. This is because there is no hope for this clause to be satisfied by setting $x_1 = F$, and so the $\bar{x}_1$ term is redundant. So neither x_1 nor $\bar{x}_1$ appears anywhere in the formula. This leaves us with a new formula on one fewer variables, with a few less clauses and a few clauses whose sizes have shrunk by one.

Initially, we have only 3-clauses, but as the procedure progresses we will also have 2-clauses and 1-clauses. 1-clauses are dealt with immediately since, for example, if we wish to satisfy a formula containing the clause $(\bar{x}_7)$ then we have no choice but to set $x_7 = F$. So if there are any 1-clauses, then we choose one of them (usually at random) and set its variable appropriately. Where the various strategies differ is in the way they choose the variable to set when there are no 1-clauses.

The first result in this vein was due to Chao and Franco [24].

Theorem 4.7 $d_3 \geq 8/3 = 2.666...$

To be precise, Chao and Franco proved something slightly weaker - that for $d < 8/3$, $F_{n,M=dn}^3$ is not a.s. unsatisfiable. Applying Corollary 5.5 from section 5 allows us to obtain Theorem 4.7 from their weaker result, that is, to show that for the same range of d, $F_{n,M=dn}^3$ is a.s. satisfiable. Corollary 5.5 came many years after Chao and Franco's work.

The results in [24] come from analyzing what they call the Unit Clause algorithm. If there is a unit clause (a 1-clause) then they set that variable. Otherwise, they pick a variable x uniformly at random and choose its setting, T or F, uniformly at random.

They then improved this simple procedure by not picking the setting of x randomly. Instead, they count the number of 3-clauses that x appears in and the number of 3-clauses that $\bar{x}$ appears in, and set x according to which of these two counts is highest. Analyzing this procedure yields the following.

Theorem 4.8 $d_3 \geq 2.9$.

Again, Chao and Franco only proved that for $d < 2.9$, $F_{n,M=dn}^3$ is not a.s. unsatisfiable; we require Corollary 5.5 to obtain this stronger result. A sharper analysis of the same procedure (see [3]) yields $d_3 \geq 3.001$.

The Unit Clause algorithm can obviously be run on k-SAT for any k. For general k, they proved the following.

Theorem 4.9 *For* $c < \frac{1}{2} \left(\frac{k-1}{k-2}\right)^{k-2} \frac{2^k}{k}$, *the Unit Clause algorithm will find a satisfying assignment for* $F_{n,M=cn}^3$ *with probability at least* $\zeta + o(1)$ *for some* $\zeta = \zeta(c) > 0$.

Corollary 5.5 shows that this yields the following.

Corollary 4.10 $d_k \geq \frac{1}{2} \left(\frac{k-1}{k-2}\right)^{k-2} \frac{2^k}{k}$.

It turns out that, in the interesting range of d, there tend to be many 2-clauses throughout most of the procedure. Chvátal and Reed [27] analysed an algorithm which takes advantage of this fact. If there is no 1-clause then, instead of choosing a random variable, we choose a random 2-clause and set one of its variables to satisfy the clause (we choose which of the two variables to set uniformly at random). If there is no 2-clause then we choose a literal to set uniformly at random. This guarantees that we will satisfy at least one of the 2-clauses. They proved that for general k, this algorithm will a.s. produce a satisfying solution to $F_{n,M=dn}^k$ if $c < \frac{1}{8} \left(\frac{k-1}{k-3}\right)^{k-3} \frac{k-1}{k-2} \times \frac{2^k}{k}$. At first glance, this seems worse than Theorem 4.9, but one should note that here the algorithm succeeds a.s., rather than simply with probability bounded away from 0.

Frieze and Suen [42] studied this algorithm more closely for 3-SAT and used it to show the next result.

Theorem 4.11 $d_3 \geq 3.003$.

Frieze and Suen actually obtained this bound without appealing to Corollary 5.5 (which did not exist at the time). To do this, they introduced a complicated backtracking component to the algorithm. Thus, unlike the versions of Theorems 4.7 and 4.8 found in [24], they showed that their algorithm a.s. succeeds, rather than the weaker fact that it does not a.s. fail.

This remained the best bound on d_3 for several years. Recently, Achlioptas [2] suggested a similar procedure where, instead of setting one of the variables in the chosen 2-clause, we set both variables. The main advantage of this approach is the following: Recall that Chao and Franco increased the performance of the Unit Clause algorithm significantly by choosing between the two possible settings of the selected variable. In Chvátal and Reed's variant, the variable is chosen specifically to satisfy a particular 2-clause and so there is no choice for its setting. However, when setting both variables in a 2-clause, Achlioptas could choose between 3 different settings which satisfy that clause. Choosing the best of these 3 settings leads to improved performance, and Achlioptas proved the following.

Theorem 4.12 $d_3 \geq 3.165$.

Besides setting two variables at once, Achlioptas' procedure involved a second innovation: he did not always set 1-clauses as soon as they appeared. Rather, at each time step he made a random choice as to whether to set a 1-clause or a 2-clause. The probability of choosing to set a 1-clause is chosen so that, in the long haul, he is setting 1-clauses at roughly the same pace as they are formed. Certainly this does not make his algorithm run more efficiently since, as we noted earlier, when a 1-clause appears we have no choice but to set it eventually, so we might as well set it right away. Nevertheless, this innovation makes the algorithm dramatically easier to analyse, since it completely dispenses with analyzing the timing of the 1-clause appearances, which can be very difficult. For a more detailed description of the advantages, see [2, 3].

Most recently, Achlioptas and Sorkin [7] determined what they proved to be the most efficient possible algorithm from amongst all algorithms which fit the general framework described in this section. See their paper for a formal definition of the class of algorithms amongst which theirs is the optimum. The analysis of their algorithm yielded the following.

Theorem 4.13 $d_3 \geq 3.26$.

At the time this article was written, this stood as the best bound thus far.

In the setting of random 3-SAT, the analogue of the 3-core comes from applying what is known as the *pure literal rule*. A *pure literal* is one whose negation does not appear in the formula; a *pure variable* is a variable at least one of whose literals is pure. In this procedure, one repeatedly looks for a pure variable v and sets v appropriately. If we consider the degree of a variable v to be an ordered pair containing the number of clauses that v and $\bar{v}$ each appear in, then the pure literal rule will continue until we find what can be thought of as a (1,1)-core, and so we define it to be an *impure core*.

For $k \geq 3$, we define

$$\gamma_k = \gamma_k^- = \sup\{\gamma : \text{a.s. } F_{n,M=\gamma n}^k \text{ has no impure core}\}.$$

We define γ_k^+ in the obvious way, analogous to the definition of c_k^+, d_k^+ and λ_k^+. Clearly, $d_k \geq \gamma_k$.

The first analysis of γ_3 was due to Broder et al. [23].

Theorem 4.14 $1.63 \leq \gamma_k \leq 1.7$.

The paper of Pittel et al. [67] in determining the threshold for the appearance of a 3-core provided a short and elegant, but non-rigorous, heuristic for predicting the value of λ_k. Their long rigorous proof used a different approach. A number of different people independently noted that this heuristic also yields a good prediction for the value of γ_k, as do other non-rigorous approaches; see, for example, [62]. In [65], Molloy and Pittel found a simple technique for making this heuristic rigorous. In [66], Molloy and Wormald use it to provide the first rigorous proof of the value of γ_k along with the thresholds for many similar cores.

Theorem 4.15

$$\gamma_k = \gamma_k^- = \gamma_k^+ = \min_{\beta>0} \frac{2\beta}{k(1 - e^{-\beta})^{k-1}}.$$

Thus, for example, $\gamma_3 = 1.6369...,\gamma_4 = 1.5445...,\gamma_5 = 1.4035...$. See [44] for another application of the same technique.

5 Sharp thresholds

As described in the introduction, we still do not know whether there exist absolute constants $c_3^*, c_4^*, \ldots$ such that for any $\epsilon > 0$, $G_{n,M=(c_k^*-\epsilon)n}$ is a.s. k-colourable and $G_{n,M=(c_k^*+\epsilon)n}$ is a.s. not k-colourable. However, we know that something very close to this is true. Before describing that result more precisely, we will discuss some work which led up to it.

5.1   Concentration on a few points

If the constants c_k^* existed, then for every $c \geq \frac{1}{2}$ which is not equal to one of them, we would know, with high probability, the chromatic number of $G_{n,M=cn}$. Thus, for each $\frac{1}{2} \leq c \neq c_k^*$, the random variable $\chi(G_{n,M=cn})$ a.s. takes on one particular value; that is, it is concentrated on one point. In [68], Shamir and Spencer proved that for every constant c, $\chi(G_{n,M=cn})$ is concentrated on a small range of at most 4 points.

Theorem 5.1 *For every constant c, there is an integer $u(n)$ such that a.s.*
$$u(n) \leq \chi(G_{n,M=cn}) \leq u(n) + 3.$$

Unfortunately, their proof technique does not help us to compute $\lim_{n \to \infty} u(n)$ in terms of c. In fact, for most constants c, we do not even know whether this limit converges, although it is hard to imagine a situation where it does not.

Proof   It is more convenient to work in the $G_{n,p}$ model. We begin with the following fact.

Fact 1 *Define $u = u(n)$ to be the largest integer such that*
$$\mathbf{Pr}(\chi(G_{n,p=2c/n}) \geq u) > \frac{1}{\ln \ln n}.$$
Then a.s. $G_{n,p=2c/n}$ has a partial u-colouring which colours all but at most $\ln^2 n\sqrt{n}$ vertices.

The theorem follows easily from Fact 1. A.s. $G = G_{n,p=2c/n}$ has a partial u-colouring which leaves only $Y \subset V(G)$ uncoloured where $|Y| \leq \ln^2 n\sqrt{n}$. The same calculations as in the proof of Theorem 4.2 show that a.s. G has no subgraph on $o(n)$ vertices with average degree at least 2.99. Therefore, a.s. the vertices of Y do not contain a 3-core and so they can be coloured using at most 3 new colours.

To prove Fact 1, we let X denote the minimum over all partial u-colourings of the number of uncoloured vertices. Thus G is u-colourable if and only if $X = 0$. We will apply the Simple Concentration Bound to prove that X is concentrated. Our vertex set is $v_1, ..., v_n$. For each i, we let T_i be the random trial which chooses all the edges from v_i to $\{v_1, ..., v_{i-1}\}$. Note that the outcome of each T_i can only affect X by at most 1. Therefore,
$$\mathbf{Pr}(|X - \mathbf{Exp}(X)| > \frac{1}{2}\ln^2 n\sqrt{n}) \leq 2e^{(\frac{1}{2}\ln^2 n\sqrt{n})^2/2n} = 2e^{-\frac{1}{8}\ln^2 n}.$$

Since $\mathbf{Pr}(X = 0) \geq \frac{1}{\ln \ln n}$, this implies $\mathbf{Exp}(X) \leq \frac{1}{2}\ln^2 n\sqrt{n}$ and so $\mathbf{Pr}(X > \ln^2 n\sqrt{n}) = o(1)$ as required.   □

Remarks   Actually, Theorem 5.1 was originally stated with "$u(n)+4$" rather than "$u(n) + 3$", and it was more general in that it allowed c to be as high as

$n^{1/6-\epsilon}$ for any constant $\epsilon > 0$. Also, Fact 1, as stated, does not appear in their paper; instead they use an analogous but more complicated result. Frieze (see [54]) noted that this slightly simplified version follows easily from the ideas contained in their original proof.

To a reader who is very familiar with Azuma's Inequality and its ilk, this seems like a fairly simple proof. However, it should be noted that this was the paper which introduced Azuma's Inequality to the combinatorics community and so, at its time, the proof was quite substantial, not to mention influential.

Shortly thereafter, Łuczak [54] showed that a simple modification of their proof reduces the length of the range from 4 to 2.

Theorem 5.2 *For every constant c, there is an integer $u(n)$ such that a.s.* $u \leq \chi(G_{n,M=cn}) \leq u + 1$.

Proof Outline Again, we use the $G_{n,p}$ model. By Theorem 4.3 we can assume that $c > 1.0001$ and so $u \geq 3$. Consider $G = G_{n,p=2c/n}$ and the set Y as in the proof of Theorem 5.1. Łuczak showed that a.s. we can find a superset $Y' \supseteq Y$ such that (i) $|Y'| < 3\ln^3 n\sqrt{n}$ and (ii) $N(Y')$ is a stable set. ($N(Y')$ is the neighbourhood of Y', that is, the set of vertices in $G - Y'$ which are each adjacent to at least one vertex in Y'.) Thus, we can obtain a $(u+1)$-colouring of G as follows. Take the u-colouring of $V(G) - Y' - N(Y')$ obtained by un-colouring some of the vertices in the partial colouring of $V(G) - Y$. Use a new colour on $N(Y')$. Then colour Y' using 3 of the colours already used in the u-colouring of $V(G) - Y' - N(Y')$.

To prove that Y' exists, we build it greedily, repeating the following step until $N(Y')$ is a stable set: Choose some edge uv such that $u, v \in N(Y')$ and add both u and v to Y'. Clearly, when this stops, $N(Y')$ is a stable set. Each time we add a pair $\{u, v\}$, we add 2 vertices and at least 3 edges to the subgraph induced by Y'. Thus, adding more than $\ln^3 n\sqrt{n}$ pairs would produce a subgraph on fewer than $3\ln^3 n\sqrt{n}$ vertices with average degree at least $3 - o(1) > 2.99$. But a.s. no such subgraph exists. Therefore, a.s. we add fewer than $\ln^3 n\sqrt{n}$ pairs and so $|Y'| < 3\ln^3 n\sqrt{n}$, as required. $\square$

Łuczak proved that, as with Theorem 5.1, Theorem 5.2 holds for p as high as $n^{-5/6}$. More recently, Alon and Krivelevich [10] proved that it holds for p as high as $n^{-1/2-\epsilon}$ for any constant $\epsilon > 0$.

5.2 Friedgut's Theorem

In [39], Friedgut developed an approach which enables us to show that certain properties exhibit a threshold which is almost as sharp as was predicted by Erdős in Question 1.1(a). For simplicity, we will present his theorem in a somewhat weakened form. For one thing, we will only be concerned with $G_{n,p}$ for p of order $\Theta(1/n)$. We begin with a few definitions.

We say that a graph property, P, is *monotonically increasing* if (i) P is invariant under graph automorphisms, and (ii) if H is a subgraph of G on the same vertex set, and H has P, then G must have P.

A monotonically increasing property P exhibits a *transition* if for some $0 < c_1 < c_2$, (i) $G_{n,p=c_1/n}$ a.s. does not have P and (ii) $G_{n,p=c_2/n}$ a.s. has P. Insisting on a property being monotonically increasing rather than monotonically decreasing is a convenience which is easily satisfied by negating the property if necessary. For example, we can let P be the property of non-3-colourability.

A monotonically increasing property P which exhibits a transition is said to have a *sharp threshold* if there exists a function $c(n)$ such that for any $\epsilon > 0$, $G_{n,p=(c(n)-\epsilon)/n}$ a.s. has P and $G_{n,p=(c(n)+\epsilon)/n}$ a.s. does not have P. Note that we must have $c_1 - o(1) \le c(n) \le c_2 + o(1)$.

Note that the only difference between this and the stronger property in Question 1.1(a) is that here we allow c to vary with n, just as in Theorems 5.1, 5.2. All natural properties which are known to have sharp thresholds are conjectured to also satisfy the stronger property where for all n, $c(n)$ can be taken to be some constant c. However, we do not have any general tools for proving this fact.

Given two properties P_1, P_2, their *symmetric difference*, $P_1 \triangle P_2$ is the property of satisfying one but not both of P_1, P_2. Given a set of graphs $\mathcal{H} = \{H_1, ..., H_k\}$, we define $Q(\mathcal{H})$ to be the property of having a subgraph which is isomorphic to one of the members of $\mathcal{H}$.

The following comes from Theorem 1.1 of [39].

Friedgut's Theorem *Let P be any monotonically increasing graph property which exhibits a transition but does not have a sharp threshold. Then there exists some $c(n)$ such that for every $\delta > 0$ there is an infinite sequence of values of n for which the following holds: there is a set of unicyclic graphs $\mathcal{H} = \{H_1, ..., H_k\}$ where each H_i has size at most $K = K(\delta, P)$ (that is, the sizes of the H_i cannot grow with n) such that the probability of $G_{n,p=c(n)/n}$ having $P \triangle Q(\mathcal{H})$ is less than δ.*

In other words, when $p = c(n)/n$, P can be arbitrarily closely approximated in probability by the property of having a subgraph from a fixed list of unicyclic graphs.

For example, suppose that P is the property of being non-bipartite. Then P is equivalent to the property of having a subgraph from a list of unicyclic graphs, namely the list of odd cycles, $\{C_3, C_5, ...\}$. But there is a subtlety here: the sizes of the graphs in this list grow with n, since we include all cycles of odd length up to n. However, it is not important that P be equivalent to the property of containing a subgraph from the list; P need only be closely approximated by that property. And by truncating the list, we can obtain lists of graphs with bounded size which approximate P arbitrarily well, so long as we are in the range $c < 1$. More specifically, it turns out that for every $\delta > 0$, there exists $t = t(\delta)$ such that if $\mathcal{H} = \{C_3, C_5, ..., C_{2t+1}\}$, then the probability of $G_{n,p=1/2n}$ having $P \triangle Q(\mathcal{H})$ is less than δ.

Theorem 1.1 of [39] is in fact much more general than the version presented above, in that (i) this is essentially an "if and only if" statement, and (ii) it allows for more general sharp thresholds which occur when M is not linear in n.

It is also more precise about the manner in which K depends on P. For any fixed n, we can consider $\mathbf{Pr}(P)$ as a continuous function in the variable p, the edge-probability. We denote this function by $\mu(p)$. Friedgut observes that if P exhibits a transition, but does not have a sharp threshold, then there exist absolute constants A and $\epsilon > 0$ such that for arbitrarily large values of n, there will be a point $p^* = p^*(n)$ for which (i) $\zeta < \mu(p^*) < 1 - \zeta$, and (ii) $p^* \times \mu'(p^*) < A$. $K(P, \delta)$ is in fact a function only of A and δ.

Friedgut also showed that an analogue to his theorem applies to random instances of k-SAT, and his first application was to almost answer Question 1.2(a).

Theorem 5.3 *For $k \geq 2$, non-satisfiability of $F_{n,M}^k$ has a sharp threshold.*

Theorem 1.2 of [39] is a more technical variation of Theorem 1.1 of [39], and can be easier to apply. Achlioptas and Friedgut [4] used Theorem 1.2 to prove the following.

Theorem 5.4 *For $k \geq 3$, non-k-colourability of $G_{n,M}$ has a sharp threshold.*

Thus, they came very close to answering Question 1.1(a). These two theorems have the following corollaries, which were helpful towards establishing some of the bounds in sections 4.1, 4.2.

Corollary 5.5 (a) *If $G_{n,M=cn}$ is k-colourable with probability at least $\zeta + o(1)$ for some constant $\zeta = \zeta(c) > 0$, then for each $c' < c$, $G_{n,M=c'n}$ is a.s. k-colourable.*

(b) *If $F_{n,M=dn}^k$ is satisfiable with probability at least $\zeta + o(1)$ for some constant $\zeta = \zeta(c) > 0$, then for each $d' < d$, $F_{n,M=d'n}^k$ is a.s. satisfiable.*

Of course, the analogous corollaries also hold for the $G_{n,p}$, $F_{n,p}^k$ models and when we replace "satisfiable" by "not satisfiable", or "k-colourable" by "not k-colourable"

At first glance, it would seem that the sharpness of 3-colourability might follow quite easily from Friedgut's Theorem, since it is well known that non-3-colourability is not equivalent to containing one of a list of small subgraphs. However, to apply Friedgut's Theorem, one needs to show that it cannot even be closely approximated with such a list, and proving this is far from simple. In fact, to date, Friedgut's Theorem has only been successfully applied to a handful of properties [4, 39, 40, 41]. For example, we do not know how to apply his theorem to k-list-colourability.

Question 5.6 *For $k \geq 3$, does non-k-list-colourability have a sharp threshold?*

We do not even know the answer to the following, for any small value of $k \geq 3$.

Question 5.7 *Consider $k \geq 3$. Run the random graph process until the first k-core appears. Is the resulting graph a.s. not k-list-colourable?*

I think that the answer to this question is almost certainly NO for every $k \geq 3$. In fact, the answer to the following question is probably YES.

Question 5.8 *Consider $k \geq 3$. Run the random graph process until the graph is not k-list-colourable. Is the resulting graph a.s. not k-colourable?*

So, I think that the thresholds for k-list-colourability and k-colourability are equal.

We know that the answer to Question 5.7 is NO for large values of k, since Alon and Krivelevich [11] proved that for large k and $c < \frac{1}{100} k \ln k$, $G_{n,M=cn}$ is a.s. k-list colourable.

6 Asymptotics

As we saw in Sections 3.2, 4.2, we know that

$$(\frac{1}{2} + o(1))\frac{2^k}{k} < d_k \leq \ln 2 / \ln(2^k/(2^k - 1)) < \ln 2 \times 2^k.$$

Subsequent efforts have improved the constant term in this lower bound, but no one has yet been able to improve the asymptotic order of either bound. It would be very interesting to get asymptotic upper and lower bounds with the same order of magnitude.

Question 6.1 *What is the asymptotic order of d_k?*

We understand the asymptotics of c_k much better than those of d_k, thanks to the following theorem of Luczak [55].

Theorem 6.2 $c_k = (1 + o(1))k \ln k.$

Recall that every lower bound on d_3 established in section 4.2 was obtained by analyzing an algorithm which will a.s. produce a satisfying assignment for $F^3_{n,M=cn}$. Each of those algorithms, when extended to general k, will a.s. fail for $c > O(2^k/k)$. It is entirely plausible that $d_k = 2^k(\ln 2 + o(1))$ but that there is no efficient algorithm which will find a satisfying solution for $c > O(2^k/k)$.

Question 6.3 *Is there a constant $\epsilon > 0$ and a polytime algorithm P such that for all $k \geq 3$, a.s. P will find a satisfying solution for $F^k_{n,M=\epsilon \times d_k n}$?*

A similar situation is true for colourability. We know of no algorithm which can guarantee a k-colouring of $G_{n,M=cn}$ for $c > (\frac{1}{2} + \epsilon)k \ln k$. In fact, it may be the case that no such algorithm exists.

Question 6.4 *Is there a constant $\epsilon > 0$ and a polytime algorithm P such that for all $k \geq 3$, a.s. P will find a k-colouring of $G_{n,M=(\frac{1}{2}+\epsilon)k \ln k \times n}$?*

We do not even know of any efficient algorithm which will find a stable set large enough to form one of the colour classes in such a colouring.

Question 6.5 *Is there a constant $\epsilon > 0$ and a polytime algorithm P such that for all $k \geq 3$, a.s. P will find a stable set of size n/k in $G_{n,M=(\frac{1}{2}+\epsilon)k \ln k \times n}$?*

An analogous question can be asked for $G_{n,M}$ for much larger values of M. See [12] for some recent work on this problem.

7 A new question

We close this survey with a new question, posed by Achlioptas and Molloy, which is of the same flavour as Bollobás' Question 4.1.

Consider a graph G and any collection of k disjoint stable sets $V_1, ..., V_k$. We say that this collection is *bilinked* if for every pair V_i, V_j, the bipartite graph induced by $V_i \cup V_j$ is connected.

Question 7.1 *Consider a random graph process on n vertices and $k \geq 3$. Does the chromatic number of the graph jump from k to $k + 1$ within $o(n)$ steps of the appearance of the first bilinked collection of k stable sets, other than a triangle?*

The motivation for this question is as follows.

Consider any bilinked collection, set $a_i = |V_i|$, and let B be the subgraph induced by $V_1 \cup ... \cup V_k$. There are at least $a_i + a_j - 1$ edges between V_i, V_j and so the total number of edges in B is at least $(k - 1)(a_1 + ... + a_k) - \binom{k}{2}$. Since $|B| \geq k$, this is at least $\frac{k-1}{2}|B|$ and so B has average degree at least $k - 1$. Furthermore, for $k = 3$ if B is not a triangle then $|B| \geq 4$ and this shows that the average degree in B is at least 2.5. Therefore, a straightforward argument along the lines of Theorem 4.2 shows that a.s. any bilinked collection of at least 3 stable sets in $G_{n,p=c/n}$ which is not a triangle, must contain at least ϵn vertices for some $\epsilon = \epsilon(c) > 0$.

Let B be the first non-triangle bilinked collection to appear. We conjecture that B will not a.s. induce a uniquely k-colourable subgraph, but that B will a.s. contain a uniquely k-colourable subgraph B' of size $|B| - o(n)$. If this is the case, then with high probability, within a small number of steps, an edge will be added between two vertices in the same colour class of B'. The subgraph induced by B' plus that edge will have chromatic number $k + 1$.

To prove the other side of our question, that a.s. the chromatic number does not jump before the appearance of such a bilinked collection, it suffices to answer the following question in the affirmative.

Question 7.2 *Consider the random graph process. Is it true that a.s. by the time the chromatic number jumps to $k + 1$, a uniquely k-colourable subgraph which is not a triangle has appeared?*

It may be noted that the colour classes of any uniquely k-colourable graph must be bilinked, as otherwise we could find a different colouring by switching on a Kempe chain.

It is easy to see that there are many graphs of chromatic number $k+1$ which do not contain any uniquely k-colourable subgraphs. Nevertheless, we think that such graphs do not arise in the random graph process. If one could prove that the answer to Question 7.2 is YES, then Corollary 1 of Aldous [9] would immediately yield an alternate proof of the sharp threshold for k-colourability. We can ask the same question for satisfiability.

Question 7.3 *Consider the random k-SAT process. Is it true that a.s. by the time the formula is unsatisfiable, a uniquely satisfiable subformula has appeared?*

Similarly, if one could prove that the answer to this question is YES, then we would have an alternate proof of the sharp threshold for satisfiability.

If Question 7.1 is answered in the affirmative, then it would reduce the quest for the threshold for k-colourability to finding the threshold for the appearance of a bilinked collection of k stable sets. This latter threshold seems a little less daunting because this structure is somewhat reminiscent of a core. However, none of the now standard techniques for finding the threshold for the appearance of core-like structures seem to apply here, and so even this latter threshold may be very hard to determine.

Acknowledgements

I would like to thank Dimitris Achlioptas, Ehud Friedgut, Tomasz Łuczak, Colin McDiarmid, Joel Spencer and an anonymous referee for much help during the preparation of this paper.

References

[1] D. Achlioptas, Threshold phenomena in random graph coloring and satisfiability, Ph.D. Thesis, University of Toronto, 1999.

[2] D. Achlioptas, Setting two variables at a time yields a new lower bound for random 3-SAT, *32nd Annual Symposium on Foundations of Computer Science (STOC)*, 2000, 28–37.

[3] D. Achlioptas, A survey of lower bounds for random 3-SAT via differential equations, *Theoret. Comput. Sci.*, to appear.

[4] D. Achlioptas and E. Friedgut, A sharp threshold for *k*-colorability, *Random Structures Algorithms* **14** (1999), 63–70.

[5] D. Achlioptas and M. Molloy, Analysis of a list-colouring algorithm on a random graph, *38th Annual Symposium on Foundations of Computer Science (FOCS)*, 1997, 204–212.

[6] D. Achlioptas and M. Molloy, Almost all graphs with 2.522*n* edges are not 3-colourable, *Electron. J. Combin.* **6** (1999), R29.

[7] D. Achlioptas and G. Sorkin, Optimal myopic algorithms for random 3-SAT, *Proceedings of the 41st Annual Symposium on Foundations of Computer Science (FOCS)*, 2000, 590–600.

[8] M. Ajtai, J. Komlós and E. Szemerédi, Topological complete subgraphs in random graphs, *Studia Sci. Math. Hungar.* **14** (1979), 293–297.

[9] D. Aldous, Threshold limits for cover times, *J. Theoret. Probab.* **4** (1991), 197–211.

[10] N. Alon and M. Krivelevich, The concentration of the chromatic number of random graphs, *Combinatorica* **17** (1997), 303–313.

[11] N. Alon and M. Krivelevich, List colouring of random and pseudo-random graphs, *Combinatorica* **19** (1999), 453–472.

[12] N. Alon, M. Krivelevich and B. Sudakov, Finding a large hidden clique in a random graph, *Random Structures Algorithms* **13** (1998), 457–466.

[13] N. Alon and J. Spencer, *The Probabilistic Method*, Wiley, New York, 1992.

[14] K. Azuma, Weighted sums of certain dependent random variables, *Tohoku Math. J.* **19** (1967), 357–367.

[15] B. Bollobás, *Random Graphs*, Academic Press, London, 1985.

[16] B. Bollobás, The chromatic number of random graphs, *Combinatorica* **7** (1988), 49–55.

[17] B. Bollobás, The evolution of random graphs, *Trans. Amer. Math. Soc.* **286** (1984), 257–274.

[18] B. Bollobás, The evolution of sparse graphs, *Graph Theory and Combinatorics*, (ed. B. Bollobás), Academic Press, London, 1984, 35–57.

[19] B. Bollobás, C. Borgs, J. Chayes, J.H. Kim, and D. Wilson, The scaling window of the 2-SAT transition, *Random Structures Algorithms,* to appear.

[20] B. Bollobás and P. Erdős, Cliques in random graphs, *Math. Proc. Cambridge Philos. Soc.* **80** (1976), 419–427.

[21] B. Bollobás, T. Fenner and A. Frieze, Hamilton cycles in random graphs of minimal degree at least k, *A Tribute to Paul Erdős,* (eds. A. Baker, B. Bollobás, A. Hajnal), 1990, 59–95.

[22] B. Bollobás and A. Thomason, Random graphs of small order, *Random Graphs '83,* (eds. M. Karoński and A. Ruciński), *Ann. Discrete Math.* **28**, 1985, 47–97.

[23] A. Broder, A. Frieze and E. Upfal, On the satisfiability and maximum satisfiability of random 3-CNF formulas, *41st Annual ACM-SIAM Symposium on Discrete Algorithms (SODA),* 1993, 322–330.

[24] M. Chao and J. Franco, Probabilistic analysis of two heuristics for the 3-satisfiability problem, *SIAM J. Comput.* **15** (1986), 1106–1118.

[25] M. Chao and J. Franco, Probabilistic analysis of a generalization of the unit-clause literal selection heuristic for the k-satisfiability problem, *Inform. Sci.* **51** (1990), 289–314.

[26] V. Chvátal, Almost all graphs with 1.44n edges are 3-colorable, *Random Structures Algorithms* **2** (1991), 11–28.

[27] V. Chvátal and B. Reed, Mick gets some (the odds are on his side), *33rd Annual Symposium on Foundations of Computer Science,* 1992, 620–627.

[28] O. Dubois and Y. Boufkhad, A general upper bound for the satisfiability threshold of random r-SAT formulas, *J. Algorithms* **24** (1997), 395–420.

[29] O. Dubois, Y. Boufkhad and J. Mandler, Typical random 3-SAT formulae and the satisfiability threshold, *48th Annual ACM-SIAM Symposium on Discrete Algorithms,* 2000, 126–127.

[30] P. Dunne and M. Zito, An improved upper bound on the non 3-colourability threshold, *Inform. Process. Lett.* **65** (1998), 17–23.

[31] J. Edmonds, Paths, trees and flowers, *Canad. J. Math.* **17** (1965), 449–467.

[32] P. Erdős, On circuits and subgraphs of chromatic graphs, *Mathematika* **9** (1962), 170–175.

[33] P. Erdős and A. Rényi, On random graphs I, *Publ. Math. Debrecen* **6** (1959), 290–297.

[34] P. Erdős and A. Rényi, On the evolution of random graphs, *Math. Inst. of the Hungarian Academy of Sciences* **5** (1960), 17–61.

[35] P. Erdős and A. Rényi, On the existence of a factor of degree one of a connected random graph, *Acta Math. Acad. Sci. Hungar.* **17** (1966), 359–368.

[36] W. Fernandez de la Vega, On random 2-SAT, manuscript, 1992.

[37] N. Fountoulakis, M.Sc. dissertation, Oxford University, 2000.

[38] J. Franco and M. Paull, Probabilistic analysis of the Davis–Putnam procedure for solving the satisfiability problem, *Discrete Appl. Math.* **5** (1983), 77–87.

[39] E. Friedgut, Necessary and sufficient conditions for sharp thresholds of graph properties, and the k-SAT problem, *J. Amer. Math. Soc.* **12** (1999), 1017–1054.

[40] E. Friedgut and M. Krivelevich, Sharp thresholds for certain Ramsey properties of random graphs, *Random Structures Algorithms* **17** (2000), 1–19.

[41] E. Friedgut, V. Rödl, A. Rucinski and P. Tetali, in preparation.

[42] A. Frieze and S. Suen, Analysis of two simple heuristics on a random instance of k-SAT, *J. Algorithms* **20** (1996), 312–355.

[43] A. Goerdt, A threshold for unsatisfiability, *J. Comput. System Sci.* **53** (1996), 469–486.

[44] A. Goerdt and M. Molloy, Analysis of edge deletion processes on random regular graphs, *Proceedings of Latin American Theoretical Informatics*, 2000, 38–47.

[45] S. Janson, D. Knuth, T. Łuczak and B. Pittel, The birth of the giant component, *Random Structures Algorithms* **3** (1993), 233–358.

[46] S. Janson, T. Łuczak and A. Ruciński, *Random Graphs*, Wiley, New York, 2000.

[47] S. Janson, Y. Stamatiou and M. Vamvakari, Bounding the unsatisfiability threshold of random 3-SAT, *Random Structures Algorithms* **17** (2000), 103–116.

[48] A. Kamath, R. Motwani, K. Palem and P. Spirakis, Tails bounds for occupancy and the satisfiability threshold conjecture, *Proceedings of the 35th Annual Symposium on Foundations of Computer Science (FOCS)*, 1994, 592–603.

[49] A. Kaporis, L. Kirousis and Y. Stamatiou, A note on the non-colorability threshold of a random graph, *Electron. J. Combin.* **7** (2000), R29.

[50] A. Kaporis, L. Kirousis, Y. Stamatiou, M. Vamvakari and M. Zito, The unsatisfiability threshold revisited, manuscript.

[51] L. M. Kirousis, E. Kranakis, D. Krizanc, and Y. Stamatiou, Approximating the unsatisfiability threshold of random formulas, *Random Structures Algorithms* **12** (1998), 253–269.

[52] J. Komlós and E. Szemerédi, Limit distributions for the existence of Hamilton cycles in a random graph, *Discrete Math.* **43** (1984), 55–63.

[53] T. Łuczak, Size and connectivity of the k-core of a random graph, *Discrete Math.* **91** (1991), 61–68.

[54] T. Łuczak, A note on the sharp concentration of the chromatic number on random graphs, *Combinatorica* **11** (1991), 295–297.

[55] T. Łuczak, The chromatic number of random graphs, *Combinatorica* **11** (1991), 45–54.

[56] T. Łuczak and J. Wierman, The chromatic number of random graphs at the double jump threshold, *Combinatorica* **9** (1989), 39–49.

[57] T. Łuczak, Cycles in a random graph near the critical point, *Random Structures Algorithms* **2** (1991), 421–439.

[58] C. McDiarmid, Concentration, *Probabilistic Methods for Algorithmic Discrete Mathematics*, (eds. M. Habib, C. McDiarmid, J. Ramirez-Alfonsin and B. Reed), Springer, Berlin, 1998, 195–248.

[59] D. Matula, The largest clique size in a random graph, Tech. Rep., Dept. Comp. Sci., Southern Methodist University, 1976.

[60] D. Matula, Expose-and-merge exploration and the chromatic number of a random graph, *Combinatorica* **7** (1987), 275–294.

[61] D. Matula and L. Kucera, An expose-and-merge algorithm and the chromatic number of a random graph, *Random Graphs '87*, (eds. M. Karoński, J. Jaworski and A. Ruciński), 1990, 175–188.

[62] M. Mitzenmacher, Tight thresholds for the pure literal rule, Technical Note 1997-011, Digital Systems Research Center, Palo Alto, CA, 1997.

[63] M. Molloy, The chromatic number of sparse random graphs, M.Math. Thesis, University of Waterloo, 1992.

[64] M. Molloy, A gap between the appearance of a k-core and a k-chromatic graph, *Random Structures Algorithms* **8** (1996), 159–160.

[65] M. Molloy and B. Pittel, Subgraphs with average degree at least 3 in a random graph, in preparation.

[66] M. Molloy and N. Wormald, in preparation.

[67] B. Pittel, J. Spencer and N. Wormald, Sudden emergence of a giant k-core in a random graph, *J. Combin. Theory Ser. B* **67** (1996), 111–151.

[68] E. Shamir and J. Spencer, Sharp concentration of the chromatic number on random graphs $G_{n,p}$, *Combinatorica* **7** (1987), 121–129.

[69] D. West, *Introduction to Graph Theory*, Prentice Hall, Upper Saddle River, NJ, 1996.

Department of Computer Science
University of Toronto
Toronto
Canada M5S 3G4
molloy@cs.toronto.edu

On the interplay between graphs and matroids

James Oxley

Abstract

"If a theorem about graphs can be expressed in terms of *edges* and *circuits* only it probably exemplifies a more general theorem about matroids." This assertion, made by Tutte more than twenty years ago, will be the theme of this paper. In particular, a number of examples will be given of the two-way interaction between graph theory and matroid theory that enriches both subjects.

1 Introduction

This paper aims to be accessible to those with no previous experience of matroids; only some basic familiarity with graph theory and linear algebra will be assumed. In particular, the next section introduces matroids by showing how such objects arise from graphs. It then presents a minimal amount of theory to make the rest of the paper comprehensible. Throughout, the emphasis is on the links between graphs and matroids.

Section 3 begins by showing how 2-connectedness for graphs extends naturally to matroids. It then indicates how the number of edges in a 2-connected loopless graph can be bounded in terms of the circumference and the size of a largest bond. The main result of the section extends this graph result to matroids. The results in this section provide an excellent example of the two-way interaction between graph theory and matroid theory.

In order to increase the accessibility of this paper, the matroid technicalities have been kept to a minimum. Most of those that do arise have been separated from the rest of the paper and appear in two separate sections, 4 and 10, which deal primarily with proofs. The first of these sections outlines the proofs of the main results from Section 3.

Section 5 begins a new topic, that of removing a cycle from a graph while maintaining the connectivity. The seed for the results in this section is a 1974 graph theorem of Mader. Various extensions and non-extensions of this theorem for 2-connected graphs and matroids are described. The topic of removable cycles continues in Section 6 with the focus moving to the 3-connected case. Once again, the symbiosis between graph theory and matroid theory should be apparent throughout this discussion.

Sections 7–10 turn attention to graph minors and their matroid analogues. In particular, motivated by Robertson and Seymour's Graph-Minors Project and a longstanding matroid conjecture of Rota, the theme is the existence of infinite antichains in graphs and matroids. The ideas of branch-width for graphs and matroids are introduced and the effects are considered both of bounding the branch-width above and of bounding it below. The most technical material in this discussion appears in Section 10 where some proofs are outlined. Finally, Section 11 revisits the topic of unavoidable structures, which were considered in the 2-connected case in Section 3. In particular, the substructures that are guaranteed to appear in all sufficiently large 3-connected graphs and matroids are specified.

2 The bare facts about matroids

This section gives a basic introduction to matroid theory beginning with a description of how matroids arise from graphs. The reader who is familiar with matroids may wish to go directly to Section 3. A more detailed treatment of the material in this section may be found in [39], the terminology and notation of which will be followed here.

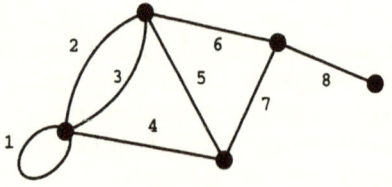

Figure 1

Consider the graph G shown in Figure 1 and let $\mathcal{C}$ be the collection of edge-sets of cycles, simple closed curves, in G. Then

$$\mathcal{C} = \Big\{\{1\}, \{2,3\}, \{2,4,5\}, \{3,4,5\}, \{5,6,7\}, \{2,4,6,7\}, \{3,4,6,7\}\Big\}.$$

Let E be the edge-set of G. Then the pair $(E,\mathcal{C})$ is an example of a matroid. It is called the *cycle matroid* of G and is denoted by $M(G)$.

Another example of a matroid arises from a matrix. Consider the following matrix A over $GF(2)$, the field of two elements.

$$
\begin{array}{c@{\quad}cccccccc}
 & 1 & 2 & 3 & 4 & 5 & 6 & 7 & 8 \\
a & 0 & 1 & 1 & 1 & 0 & 0 & 0 & 0 \\
b & 0 & 1 & 1 & 0 & 1 & 1 & 0 & 0 \\
c & 0 & 0 & 0 & 1 & 1 & 0 & 1 & 0 \\
d & 0 & 0 & 0 & 0 & 0 & 1 & 1 & 1 \\
e & 0 & 0 & 0 & 0 & 0 & 0 & 0 & 1
\end{array}
$$

Let $E = \{1, 2, ..., 8\}$ and let $\mathcal{C}$ be the collection of minimal linearly dependent subsets of E. Again, $(E,\mathcal{C})$ is a matroid. This matroid is called the *vector matroid $M[A]$* of the matrix A. This matrix is the mod 2 vertex-edge incidence matrix of the graph G in Figure 1, and its vector matroid equals the cycle matroid of G. Because this matrix A is over the field $GF(2)$, the associated vector matroid is an example of a *binary matroid*.

In general, a *matroid* consists of a finite set E, called the *ground set*, and a collection $\mathcal{C}$ of non-empty incomparable subsets of E, called *circuits*, that obey the straightforward elimination axiom: *if C_1 and C_2 are distinct circuits and $e \in C_1 \cap C_2$, then $(C_1 \cup C_2) - \{e\}$ contains a circuit.*

The reader will easily verify that this axiom is satisfied by the collection of edge-sets of cycles in a graph and by the collection of minimal linearly dependent sets of columns of a matrix. Indeed, when Whitney [61] introduced matroids in 1935, he sought to provide a unifying abstract treatment of dependence in graph theory and linear algebra. Thus, since their introduction, matroids have been closely tied to graphs. This paper will explore some aspects of this bond.

Another basic example of a matroid is the *uniform matroid* $U_{r,n}$ where r and n are non-negative integers and $r \leq n$. This matroid has ground set $E = \{1, 2, ..., n\}$ and has $\mathcal{C}$ equal to the set of $(r+1)$-element subsets of E.

While the cycle matroid of a graph is most naturally described in terms of its circuits, it is perhaps more natural to describe the vector matroid of a matrix in terms of its *independent sets*. These are the subsets of the ground set that contain no circuit. In the case of the vector matroid of a matrix, a set of column labels is independent if and only if the corresponding set of columns is linearly independent. A matroid is also determined by its *bases*, that is, its maximal independent sets. For the cycle matroid of a connected graph G, the bases are precisely the edge-sets of the spanning trees of G. Thus all the bases of $M(G)$ have the same cardinality, namely $|V(G)| - 1$. This observation generalizes to arbitrary matroids: all bases of a matroid M have the same cardinality, which is called the *rank* $r(M)$ of M. A set X *spans* M if X contains a basis of M.

One of the fundamental operations in graph theory is the construction of a planar dual of a plane graph. This is illustrated in Figure 2 for the graph G in Figure 1. The cycle matroid $M(G^*)$ of G^* has the same ground set E as $M(G)$. Its circuits, the edge-sets of cycles of G^*, are $\{8\}$, $\{6, 7\}$, $\{4, 5, 6\}$, $\{4, 5, 7\}$, $\{2, 3, 4\}$, $\{2, 3, 5, 6\}$ and $\{2, 3, 5, 7\}$. In the original graph G, these sets correspond to minimal edge-cuts or *bonds*, that is, they are the minimal sets of edges of G whose removal increases the number of connected components of the graph. Since the graph G is connected, its bonds are the minimal sets of edges whose removal disconnects the graph. Because the set $\mathcal{C}^*$ of bonds of G equals the set of cycles of G^*, the pair $(E, \mathcal{C}^*)$ is a matroid. In general, for a graph G, planar or not, the set of bonds of G is the set of circuits of a matroid on E, the edge-set of G. This matroid is called the *dual matroid* of $M(G)$ and is denoted by $M^*(G)$.

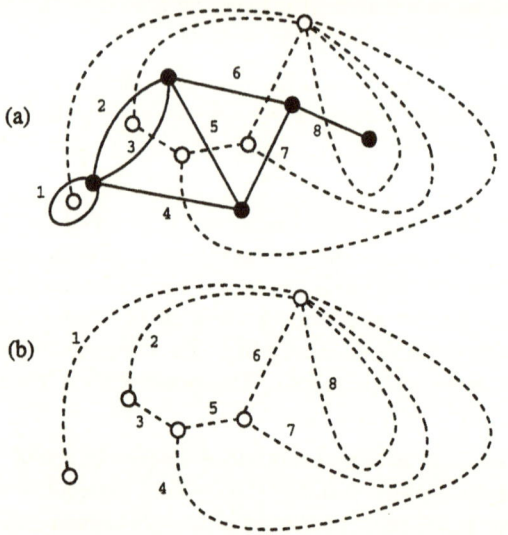

(a)

(b)

Figure 2:   (a) G and G^* superimposed (b) G^*

Every matroid M has a dual M^*, which has the same ground set E as M and has, as its bases, the collection of complements of bases of M. Thus the rank of M^*, which is called the *corank* of M and is denoted by $r^*(M)$, equals $|E| - r(M)$. It is clear that

$$(M^*)^* = M.$$

The circuits of M^* are called the *cocircuits* of M. The cocircuits of M are precisely the complements of the *hyperplanes*, the latter being the maximal subsets of $E(M)$ that do not span M. For a graph G, the cocircuits of $M(G)$ are the bonds of G. It is not difficult to see that a cycle and a bond in a graph cannot have exactly one common edge. This property extends to matroids and is often referred to as *orthogonality*: a circuit and a cocircuit of a matroid cannot have exactly one common element. It follows from the definition that the dual of every uniform matroid is also uniform. In particular,

$$U_{r,n}^* = U_{n-r,n}.$$

The cocircuits of $U_{r,n}$ are all $(n - r + 1)$-element subsets of the ground set.

Certain operations on a matrix do not alter its vector matroid. These include elementary row operations, column permutations, and deletion of a zero row. If M is the vector matroid of an $r \times n$ matrix $[I_r|D]$ over a field $\mathbb{F}$, then M^* is the vector matroid of the matrix $[-D^T|I_{n-r}]$, where the columns of this matrix are labelled in the same order as those of $[I_r|D]$.

The following table summarizes the four different classes of matroids introduced above.

Name	Notation	Ground set E	Set $\mathcal{C}$ of circuits
cycle matroid of graph G	$M(G)$	edge-set of G	edge-sets of cycles of G
bond matroid of graph G	$M^*(G)$	edge-set of G	bonds of G
vector matroid of matrix A	$M[A]$	column labels of A	minimal linearly dependent sets of columns
uniform matroid	$U_{r,n}$	$\{1, 2, \ldots, n\}$	all $(r+1)$-element subsets of $\{1, 2, \ldots, n\}$

A matroid M with ground set $E(M)$ is *graphic* if it is isomorphic to the cycle matroid of some graph, that is, there is a graph G and a bijection $\phi :$ $E(M) \rightarrow E(G)$ such that a subset X of $E(M)$ is a circuit of M if and only if $\phi(X)$ is the edge set of a cycle of G. A *cographic matroid* is one that is isomorphic to the bond matroid of some graph. For a field $\mathbb{F}$, a matroid is $\mathbb{F}$-*representable* if it is isomorphic to the vector matroid of some matrix over $\mathbb{F}$.

Duality is one of the three most basic matroid operations. The other two, deletion and contraction, are, like the first, extensions of natural graph operations.

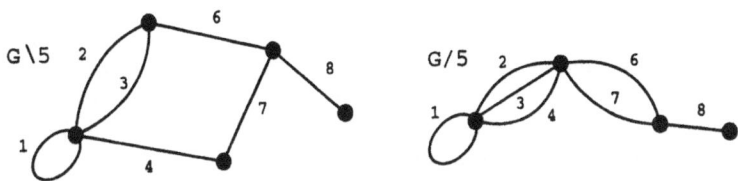

Figure 3

The two graphs in Figure 3 are related to the graph G in Figure 1. The first of these, denoted $G\backslash 5$, is obtained from G by *deleting* the edge 5; the second, denoted $G/5$, is obtained by *contracting* 5, that is, by shrinking the edge 5 to a single vertex, or (equivalently) deleting 5 and then identifying its end-vertices.

To extend these definitions to matroids, we consider how the cycles of $G\backslash e$ and G/e are related to those of G. In general, if T is a subset of the ground set E of a matroid M, then $M\backslash T$, the *deletion* of T from M, is the matroid with ground set $E - T$ whose set of circuits is $\{C \in \mathcal{C}(M) : C \cap T = \emptyset\}$; the *contraction* of T from M, which is denoted M/T, also has $E - T$ as its ground set, but its circuits are the minimal non-empty members of $\{C - T : C \in \mathcal{C}(M)\}$. It is straightforward to check that both $M\backslash T$ and M/T are actually matroids and that if e is an edge of a graph G, then $M(G)\backslash e = M(G\backslash e)$, while $M(G)/e = M(G/e)$. If X is a subset of E, then all the bases of $M\backslash(E - X)$

have the same cardinality, which equals the rank of $M\backslash(E - X)$ and is also called the *rank* $r(X)$ of X.

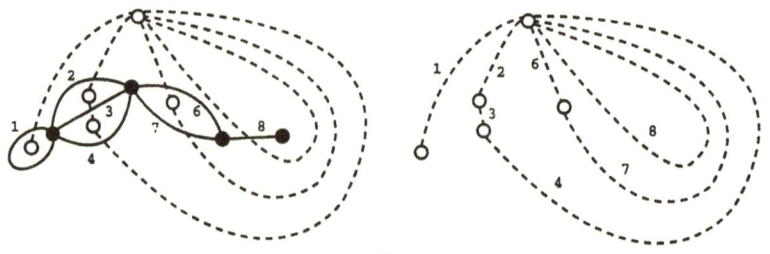

Figure 4

For graphs, deletion and contraction are related through duality. In particular, Figure 4 shows that the graphs $(G/5)^*$ and $G^*\backslash 5$ are equal. This relationship extends to matroids. Specifically, if T is a subset of the ground set of a matroid M, then

$$M^*\backslash T = (M/T)^*, \text{ or (equivalently) } M\backslash T = (M^*/T)^*.$$

A graph H is a *minor* of a graph G if H can be obtained from G by a sequence of deletions, contractions, and deletions of isolated vertices. There is a corresponding notion for matroids with the only difference arising because arbitrary matroids do not have vertices. Thus a *minor* of a matroid M is any matroid that can be obtained from M by a sequence of deletions and contractions. It is not difficult to check that if X and Y are disjoint subsets of the ground set of a matroid M, then

$$M\backslash X/Y = M/Y\backslash X,$$

and every minor of M is determined by the set of elements that are deleted and the set of elements that are contracted. Although it is true that if the graph G_1 is a minor of the graph G_2, then the matroid $M(G_1)$ is a minor of the matroid $M(G_2)$, the converse of this fails. For example, if G_1 is the union of two disjoint copies of K_2 with edge sets $\{1\}$ and $\{2\}$, and G_2 is a copy of K_3 with edge set $\{1, 2, 3\}$, then $M(G_1) \cong U_{2,2}$ and $M(G_2) \cong U_{2,3}$. Certainly $M(G_1)$ is a minor of $M(G_2)$. However, clearly G_1 has four vertices while G_2 has three, so G_1 is not a minor of G_2.

Matroids also arise geometrically as follows. Given a finite set E of points, such as the seven points of the Fano projective plane (see Figure 5(b)), one can define a matroid M on E by first choosing a collection of subsets of E called *lines* such that two distinct lines share at most one common point. The circuits of M are all sets of three collinear points and all sets of four points no three of which are collinear. Such a matroid has rank at most three. When M is derived in this way from the Fano plane, with the lines of the matroid being the seven lines of the projective plane, including $\{4,5,6\}$, we call M the *Fano matroid* F_7. Another important example of such a matroid is the *non-Fano matroid* F_7^- which is represented geometrically in Figure 5(a). It is obtained from F_7 by relaxing $\{4,5,6\}$, which is both a circuit and a hyperplane of F_7. In general, if X is a circuit-hyperplane in a matroid M, then one can obtain a new matroid M' from M by *relaxing* X, that is, by declaring X to be an independent set but leaving the matroid unchanged otherwise. For example, if we apply this operation six times to F_7^-, we obtain the uniform matroid $U_{3,7}$.

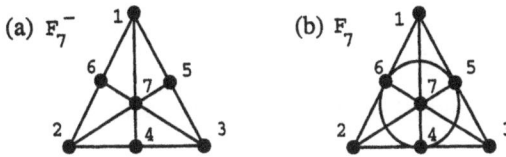

Figure 5

The Fano matroid is isomorphic to the vector matroid of the matrix whose columns consist of all non-zero vectors in the 3-dimensional vector space over $GF(2)$. Hence the Fano matroid is binary where, in general, a matroid M is *binary* if M is isomorphic to the vector matroid of some matrix over the field $GF(2)$. By using the mod-2 vertex-edge incidence matrix, as in the first example, it is straightforward to show that every graphic matroid is binary. The Fano matroid and its dual are examples of non-graphic binary matroids. To see that F_7 is non-graphic, it suffices to observe that there is no simple graph on 4 vertices with exactly 7 edges since the complete graph K_4 has only 6 edges. The matroid $U_{2,4}$ is non-binary because there is no 2×4 matrix A over $GF(2)$ such that $U_{2,4} \cong M[A]$.

A binary matroid is *regular* if it has no minor isomorphic to either F_7 or F_7^*. Such matroids have numerous attractive properties, some of which appear in the next result. A *totally unimodular matrix* is a real matrix all of whose subdeterminants are in $\{0, 1, -1\}$.

Proposition 2.1 *The following statements are equivalent for a matroid M.*

(i) *M is regular.*

(ii) *For all fields $\mathbb{F}$, there is a matrix $A_{\mathbb{F}}$ over $\mathbb{F}$ such that M is isomorphic to the vector matroid of $A_{\mathbb{F}}$.*

(iii) *M is isomorphic to $M[A]$ for some totally unimodular matrix A.*

(iv) *M has no minor isomorphic to any of F_7, F_7^*, or $U_{2,4}$.*

One very special regular matroid R_{10} is the vector matroid of the matrix over $GF(2)$ whose columns are the ten distinct 5-tuples with exactly three ones. In a result that underlies the proof that totally unimodular matrices can be recognized in polynomial time, Seymour [50] established that every regular matroid can be built by sticking together graphic matroids, cographic matroids, and copies of R_{10} using operations that mimic sticking two graphs together at a vertex, along an edge, and across a triangle. This result means that one's best hope of obtaining some matroid extension of a graph result is to be able to extend the result to regular matroids. Some results extend further, to binary matroids, and others further still to, say, all vector matroids. Thus although a graph result may fail to extend to all matroids, there are several natural intermediate classes to which the result may extend.

3 Connectedness, 2-connectedness, and unavoidable stuctures

One area in which the interaction between graphs and matroids has been very fruitful is in the consideration of connectivity. Initially there is some cause for pessimism when we note that the graphs G_1 and G_2 shown in Figure 6 have the same edge sets and have the same sets of edge-sets of cycles. Thus $M(G_1) = M(G_2)$. But G_1 is a connected graph while G_2 is not. We conclude that graph connectedness has no matroid generalization. However, if we look at 2-connectedness, we note that we can describe this property for graphs in terms of edges and circuits only: a loopless graph G is 2-connected if and only if, for every two edges e and f of G, there is a cycle containing $\{e, f\}$. Mimicking this result for graphs, we define a matroid to be *2-connected* if, for every two elements, there is a circuit containing both. This terminology follows Tutte [56]. For many other authors, a matroid with the above property is called *connected*. Thus, for matroids, the terms "connected" and "2-connected" mean precisely the same thing. One very attractive feature of 2-connectedness for matroids, which is a consequence of orthogonality, is that a matroid is 2-connected if and only if its dual is. In a matroid M, a maximal subset X of $E(M)$ for which $M\backslash(E(M) - X)$ is 2-connected is called a *(2-connected) component* of M. Thus, for the graph G_1 in Figure 6, $M(G_1)$ has exactly two components.

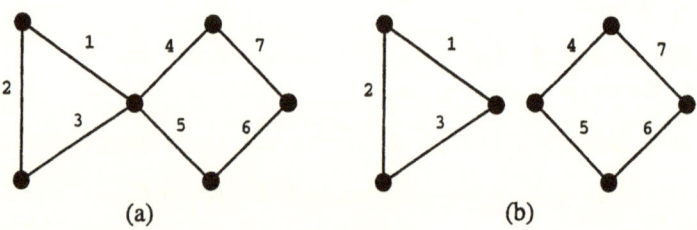

(a)   (b)

Figure 6

Our discussion of areas of interaction between graphs and matroids will begin with the consideration of unavoidable structures in graphs. An easy example of such a result is the observation that a big connected graph has either a long path or a vertex of high degree. This is expressed more formally as follows.

Lemma 3.1 *For each positive integer n, there is an integer $k(n)$ such that every connected graph G with at least $k(n)$ edges contains a path with at least n vertices or a vertex of degree at least n.*

As one might expect, if one imposes a stronger connectivity condition on G, then one can guarantee the presence of a more specific substructure. In particular, if G is 2-connected, then we have the following result.

Proposition 3.2 *Every sufficiently large 2-connected graph contains either a big cycle or a vertex of high degree.*

In a 2-connected loopless graph G, the set of edges incident with a vertex forms a bond of G. Thus we have the following immediate consequence of the last result.

Corollary 3.3 *Every sufficiently large 2-connected loopless graph contains either a big cycle or a big bond.*

Our aim is to extend graph results to matroids so it is natural to ask whether the matroid analogue of the last result holds. Specifically, is it true that every sufficiently large 2-connected matroid contains either a big circuit or a big cocircuit? The question was raised informally by Robin Thomas at the Graph-Minors meeting in Seattle in 1991. It was answered very quickly by Lovász, Schrijver, and Seymour (in [39]) who proved the following result.

Theorem 3.4 *Let M be a 2-connected matroid with at least two elements. If a largest circuit of M has c elements and a largest cocircuit has c^* elements, then M has at most 2^{c+c^*-1} elements.*

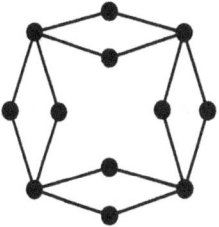

Figure 7: G

An answer to Thomas's question can also be deduced from work of Tuza [59] on set systems. A natural question that arises here is whether the bound in the last theorem can be sharpened for various special classes of matroids such as the class of graphic matroids. For example, for the cycle matroid of the graph G shown in Figure 7, c, which is just the circumference of G, equals 8, while c^* is 4. Thus $2^{c+c^*-1} = 2048$, although G has just 16 edges.

¿From a matroid perspective, one advantage of the bound in Theorem 3.4 is that it is symmetric in c and c^*. Duality is so fundamental in matroid theory that it is desirable to retain this property. As a guide to potential improvements in the bound for graphs, we consider what is already known for graphs. The following is an old result of Erdős and Gallai [14].

Theorem 3.5 *Let G be a simple n-vertex graph. Then*

$$|E(G)| \le \tfrac{1}{2}c(n-1).$$

Motivated by this result and the desire to obtain a bound symmetric in c and c^*, the author asked whether, for graphs, the bound $\tfrac{1}{2}cc^*$ holds. The graph in Figure 7 shows that such a bound would be sharp. Pou-Lin Wu [62] proved this bound.

Theorem 3.6 *Let G be a 2-connected loopless graph with circumference c. If c^* is the size of a largest bond in G, then*

$$|E(G)| \le \tfrac{1}{2}cc^*.$$

The following table compares Wu's bound with the Erdős-Gallai bound.

	Erdős-Gallai	Wu
$|E(G)| \le$	$\tfrac{1}{2}c(n-1)$	$\tfrac{1}{2}cc^*$
NEED	G is simple	G is 2-connected and loopless
SHARP FOR	complete graphs only	cycles etc.

We conclude that Wu's bound, motivated by matroid theory, is often sharper than the Erdős-Gallai bound. Thus, in Theorem 3.6, we have a graph result that was motivated by a matroid result and frequently improves on the previous best graph result.

The natural question that Wu's theorem raises is whether the same bound holds for all 2-connected matroids. Wu was not able to extend his proof to binary matroids. The reader may hope that Seymour's decomposition theorem for regular matroids mentioned above would enable the graph result to be extended to regular matroids but this was not achieved either. Several authors [2, 4, 22, 23, 41, 62, 63] considered this problem and, in particular, Reid [41] proved several attractive results. Eventually, Bonin, McNulty and Reid [2] conjectured that the same bound holds for all matroids. At that stage, there was not even a known bound that was polynomial in c and c^*.

Such a polynomial bound can be derived from the next result of Manoel Lemos and the author [34]. If M is a 2-connected matroid with at least two elements, then every element e of M is in a circuit and a cocircuit. Let c_e be the size of a largest circuit containing e and let c_e^* be the size of a largest cocircuit containing e.

Theorem 3.7 *Let M be a 2-connected matroid with at least two elements. If e is an element of M, then*

$$|E(M)| \leq (c_e - 1)(c_e^* - 1) + 1.$$

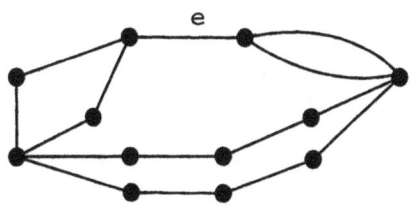

Figure 8

The example given in Figure 8 has $c_e = 8$ and $c_e^* = 3$. Thus $(c_e - 1)(c_e^* - 1) + 1 = 15 = |E(G)|$. Hence the bound in this theorem is sharp. All the matroids that attain this bound were determined in [34]. They are the cycle matroids of certain special series-parallel networks.

The last theorem has an attractive similarity to a bound derived from the *width-length inequality* [30, 16]. In a matroid M, let e be an element that is neither a *loop*, that is, a 1-element circuit, nor a *coloop*, a 1-element cocircuit. Let $l(M)+1$ and $w(M)+1$ be the cardinalities of a smallest circuit containing e and a smallest cocircuit containing e. Lehman [30] showed that if M is regular, then

$$l(M)w(M) + 1 \leq |E(M)|. \tag{3.1}$$

More generally, it follows from a result of Seymour [48] that the last inequality holds for all binary matroids that have no F_7^*-minor using e. In fact, it also holds for F_7^*, although it fails, for example, for $AG(3,2)$. The last matroid is the matroid on the points of the three-dimensional affine space over $GF(2)$ where a set of elements is independent in the matroid if the corresponding set of points is affinely independent. Equivalently, $AG(3,2)$ is the vector matroid of the matrix over $GF(2)$ whose columns are all the 4-tuples whose coordinates sum to one over $GF(2)$.

A straightforward consequence of Theorem 3.7 is the following result for graphs.

Corollary 3.8 *Let u and v be distinct vertices in a 2-connected loopless graph G. Then $|E(G)|$ cannot exceed the product of the length of a longest (u,v)-path and the size of a largest minimal edge-cut separating u from v.*

Guoli Ding, a colleague of the author at Louisiana State University, gave a short proof of Theorem 3.7 by induction. This proof was then modified by Manoel Lemos to prove the following strengthening of the theorem. The proof will be given in the next section.

Theorem 3.9 *Let M be a 2-connected matroid with at least two elements. If $e \in E(M)$, then M has $c_e(M) - 1$ cocircuits each containing e such that the union of these cocircuits is $E(M)$.*

Using the bound Theorem 3.7 and quite a bit more work, Lemos and the author [34] proved Bonin, McNulty and Reid's conjecture. An outline of the proof will be given in the next section.

Theorem 3.10 *Let M be a 2-connected matroid with at least two elements. Then*

$$|E(M)| \leq \tfrac{1}{2}cc^*.$$

For comparison, a lower bound on $|E(M)|$ in terms of c and c^* that holds for all matroids M having non-zero rank and corank is

$$c + c^* - 2 \leq |E(M)|. \tag{3.2}$$

To see this, observe that $c \leq r(M) + 1$ and $c^* \leq r^*(M) + 1$. The bound in (3.2) is sharp. It is attained, for example, by all uniform matroids of non-zero rank and corank.

Neumann-Lara, Rivera-Campo, and Urrutia [37] proved the following extension of Pou-Lin Wu's result for graphs.

Theorem 3.11 *Every 2-connected loopless graph G with circumference c has a collection of c bonds such that every edge of G lies in at least two of them.*

Observe that the last result allows for a bond to be repeated in the collection. It is natural to ask whether this result extends to matroids.

Question 3.12 *Does every 2-connected matroid M have a collection of $c(M)$ cocircuits such that every element is in at least two of them?*

This question is open even for bond matroids of graphs.

Question 3.13 *Let G be a 2-connected loopless graph whose largest bond has c^* edges. Does G have a collection of c^* cycles such that every edge is in at least two of them?*

Pou-Lin Wu [63] completely characterized all the graphs that attain equality in the bound in Theorem 3.10. They turn out to be certain special series-parallel networks that are closely related to the graphs that attain equality in Theorem 3.7. Among arbitrary matroids, the binary affine space $AG(3,2)$ also has exactly $\frac{1}{2}cc^*$ elements. It is the only known example of a non-graphic matroid that attains the bound in Theorem 3.10.

4 Some proofs

In this section, we prove Theorem 3.9 and indicate the main steps in the proof of Theorem 3.10. Tutte [56] proved the following attractive property of 2-connected matroids that is particularly helpful in induction arguments.

Lemma 4.1 *Let e be an element of a 2-connected matroid M. Then $M \backslash e$ or M/e is 2-connected.*

Proof of Theorem 3.9 We argue by induction on $c_e(M)$. If $c_e(M) = 2$, then M is a uniform matroid having rank one, so $E(M)$ is a cocircuit of M and the result follows. Now suppose that the theorem holds for $c_e(M) < n$ and let $c_e(M) = n \geq 3$.

Let C^* be a cocircuit of M that contains e. Clearly $C^* \neq E(M)$. By Lemma 4.1, we may remove the elements of $C^* - e$ from M one at a time by deletion or contraction so as to always maintain a 2-connected matroid. Thus, there is a partition $\{X, Y\}$ of $C^* - e$ such that $M \backslash X/Y$ is 2-connected. Call this minor N. By orthogonality, every circuit of M that contains e must also meet X or Y. It follows that

$$c_e(N) < c_e(M).$$

Since $C^* \neq E(M)$, the matroid N has at least two elements. Thus, by the induction assumption, for some $k \leq c_e(N) - 1$, there are k cocircuits $C_1^*, C_2^*, \ldots, C_k^*$ of N each containing e such that the union of these cocircuits is $E(N)$. For each C_i^*, there is a cocircuit D_i^* of M such that $C_i^* = D_i^* - X$. Hence $C^*, D_1^*, D_2^*, \ldots, D_k^*$ are cocircuits of M each containing e and

$$E(M) = E(N) \cup X \cup Y \subseteq C_1^* \cup C_2^* \cup \cdots \cup C_k^* \cup C^* \subseteq D_1^* \cup D_2^* \cup \cdots \cup D_k^* \cup C^*.$$

Thus we have a family of $k + 1$ cocircuits of M that covers $E(M)$. Since

$$k + 1 \leq (c_e(N) - 1) + 1 = c_e(N) \leq c_e(M) - 1,$$

the result follows by induction.   □

There are two main steps in the proof of Theorem 3.10. These are stated in the next two lemmas. For a circuit C in a matroid M with $|C| \geq 2$, let $c^*(C, M)$ be the size of a largest cocircuit of M that has exactly two elements in common with C. The *components* of a matroid N are the maximal subsets X of $E(N)$ such that $N \backslash (E(N) - X)$ is 2-connected.

Lemma 4.2 *Let M be a 2-connected matroid with at least two elements and let C be a circuit of M. If every component of M/C is a circuit, then*

$$|E(M)| \leq |C| + c(M) \left\lceil \frac{c^*(C, M) - 2}{2} \right\rceil .$$

Lemma 4.3 *Let M be a 2-connected matroid with at least two elements and let C be a circuit of M. If $C = c(M)$, then*

$$|E(M)| \leq c(M) \left\lceil \frac{c^*(C, M)}{2} \right\rceil .$$

The proof of Lemma 4.3, which uses Theorem 3.7, is by induction on the number of components of M/C that are not circuits. If this number is zero, then Lemma 4.3 follows by Lemma 4.2. Using the second of these lemmas, it is not difficult to prove Theorem 3.10.

Proof of Theorem 3.10 We shall first prove the theorem when $c^*(M)$ is even. By Lemma 4.3,

$$|E(M)| \leq c(M) \left\lceil \frac{c^*(C, M)}{2} \right\rceil ,$$

for any circuit C of M such that $|C| = c(M)$. As $c^*(C, M) \leq c^*(M)$, it follows that

$$|E(M)| \leq c(M) \left\lceil \frac{c^*(C, M)}{2} \right\rceil \leq c(M) \left\lceil \frac{c^*(M)}{2} \right\rceil = \frac{c(M)c^*(M)}{2},$$

where the last equality follows since $c^*(M)$ is an even integer.

When $c^*(M)$ is odd, let M' be the matroid obtained from M by inserting an element in parallel with each element of the latter. For graphs, this operation is certainly well-defined. To see that it is well-defined for matroids in general, one needs only to perform the routine check that the collection of circuits of M' obeys the circuit elimination axiom. By orthogonality, $c^*(M') = 2c^*(M)$. Thus $c^*(M')$ is even and hence

$$|E(M')| \leq \frac{c(M')c^*(M')}{2}.$$

As $|E(M')| = 2|E(M)|$ and $c(M') = c(M)$, we get

$$2|E(M)| \leq c(M)c^*(M)$$

and the result follows.   □

5 Removable circuits

Another area in which the interaction between graphs and matroids has been quite fruitful is in the consideration of removable circuits. The following graph result of Mader [35] motivated much of what has been done here.

Theorem 5.1 *Let G be a simple k-connected graph with minimum degree at least $k + 2$. Then G has a cycle C such that the graph obtained from G by deleting all the edges of C is k-connected.*

We shall consider trying to extend this result to matroids. To do this, attention will be restricted to the cases when $k = 2$ and when $k = 3$. The reason for doing this is that, although Tutte defined matroid k-connectedness for all $k \geq 2$, we only really have a good collection of tools for working with k-connected matroids when k is 2 or 3. First consider 2-connected matroids. As noted earlier, in a 2-connected graph without loops, the set of edges meeting at a vertex forms a bond. Thus Mader's theorem immediately implies the following result.

Corollary 5.2 *Let G be a simple 2-connected graph in which every bond has size at least 4. Then G has a cycle C such that $G\backslash C$ is 2-connected.*

Notice that, in the last result, we are deleting the edges but not the vertices of C. Such a cycle C in a 2-connected graph is called *removable*. Analogously, a circuit D of a 2-connected matroid is *removable* if $M\backslash D$ is 2-connected. The last corollary is in an ideal form for generalization to matroids.

Question 5.3 *Let M be a 2-connected matroid such that*

(i) *every circuit has size at least 3; and*

(ii) *every cocircuit has size at least 4.*

Does M have a removable circuit?

It is not difficult to see that the answer to this question is negative. For example, the uniform matroid $U_{3,6}$ of rank 3 on 6 elements has every circuit and every cocircuit of size 4. But every deletion of a circuit is isomorphic to $U_{2,2}$, which is not 2-connected. When a graph result like this fails to extend to matroids in general, it is common to restrict attention to binary matroids, or if it fails even for binary matroids, to restrict attention further to regular matroids. For binary matroids, Question 5.3 appeared as an unsolved problem in [39]. In the original version of [20], Goodyn, van den Heuvel and McGuinness made the following conjecture, which would follow if Question 5.3 had an affirmative answer for the bond matroids of graphs.

Conjecture 5.4 *Let G be a 2-connected graph such that*

(i) *every bond has size at least 3; and*

(ii) *every cycle has size at least 4.*

Then G has a bond C^* such that G/C^* is 2-connected.

The addition of the requirement that G/C^* is loopless would make this conjecture equivalent to the existence of an affirmative answer to Question 5.3 for the bond matroids of graphs.

Manoel Lemos [32] gave the following counterexample to this conjecture and thereby answered Question 5.3. Begin with $K_{5,5}$ having as its two vertex classes $\{1,2,3,4,5\}$ and $\{6,7,8,9,10\}$. For every 3-subset $\{i,j,k\}$ of $\{1,2,3,4,5\}$ and of $\{6,7,8,9,10\}$, add two new vertices v_{ijk} and w_{ijk} each joined to all of i,j,k and nothing else.

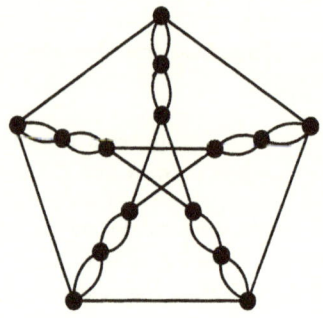

Figure 9

Much of the impetus for the study of removable circuits in graphs seems to have been provided by a question of Arthur Hobbs as to whether every 2-connected Eulerian graph with minimum degree at least four contains a removable cycle. Robertson (in [24]) and, independently, Jackson [24] answered Hobbs's question negatively by producing the modified Petersen graph shown in Figure 9. Clearly this graph is non-simple. Jackson [24] conjectured in 1980 that it is the existence of a Petersen-graph minor that prevents the above graph from having a removable circuit.

Conjecture 5.5 *Let G be a 2-connected graph with minimum degree at least four. If G has no minor isomorphic to the Petersen graph, then G has a removable cycle.*

In 1985, Fleischner and Jackson [15] proved the existence of a removable cycle in a 2-connected *planar* graph with minimum degree at least four. But it was not until 1997 that Jackson's conjecture was proved, by Goddyn, van den Heuvel, and McGuinness [20].

Theorem 5.6 *Let G be a 2-connected graph with minimum degree at least four. If G has no minor isomorphic to the Petersen graph, then G has two edge-disjoint removable cycles.*

We now return to the consideration of bonds in 2-connected graphs that can be contracted to maintain 2-connectedness. The graph in Figure 10 was identified by Goddyn, van den Heuvel, and McGuinness [20] as one having no such bond. This graph is not a counterexample to Conjecture 5.4 since it clearly has numerous bonds of size two. In response to this example, Goddyn, van den Heuvel, and McGuinness [20] conjectured the following analogue of Theorem 5.6.

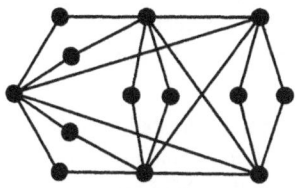

Figure 10

Conjecture 5.7 *Let G be a 2-connected graph such that every cycle has at least four elements. If G has no minor isomorphic to K_5, then G has a bond C^* such that G/C^* is 2-connected.*

This conjecture remains open although McGuinness [36] has proved the following partial result.

Theorem 5.8 *Let G be a 2-connected bipartite graph that is not a multiple edge. If G has no minor isomorphic to K_5, then G has a bond C^* such that G/C^* is 2-connected.*

We know from Robertson and Jackson's example that the requirement that G be simple in Theorem 5.1 is essential. However, Sinclair [51] showed that one could eliminate this condition if one increases the lower bound on the minimum degree.

Theorem 5.9 *Let G be a 2-connected graph with minimum degree at least 5. Then G has a removable cycle.*

This raises the question as to whether, by increasing the lower bound on circuit or cocircuit size in Question 5.3, one can guarantee the existence of a removable circuit in a 2-connected matroid. However, for all $r \geq 2$, the uniform matroid $U_{r,2r}$ is 2-connected and has no removable circuit although all its circuits and cocircuits have exactly $r + 1$ elements. None of these uniform matroids is binary and the question as to whether the analogue of Theorem 5.9 holds for binary matroids remains open. Specifically, Goddyn and Jackson [19] asked the following question.

Question 5.10 *Does there exist an integer t such that every 2-connected binary matroid for which every cocircuit has size at least t has a removable circuit?*

Examples given earlier show that such an integer t must exceed four, so the following is the natural first special case of the last question.

Question 5.11 *If M is a 2-connected binary matroid in which every cocircuit has size at least 5, then does M have a removable circuit?*

Although the answer to Question 5.3 is negative, Goddyn and Jackson [19] were able to prove that 2-connected regular matroids do have removable circuits. More precisely, they proved the following result.

Theorem 5.12 *Let M be a 2-connected binary matroid in which every cocircuit has size at least 5. If M does not minors isomorphic to both F_7 and F_7^*, then M has a circuit C such that $M \backslash C$ is 2-connected and $r(M \backslash C) = r(M)$.*

While the examples given above leave open the possibility that a *binary* 2-connected matroid with no small cocircuits may have a removable circuit, they suggest that, for matroids in general, a change in direction is needed. Mader's original result for a 2-connected graph G includes the hypothesis that all vertex degrees are at least 4. This implies that

$$|E(G)| \geq 2|V(G)|. \tag{5.1}$$

Recall that the rank of a matroid is the size of a largest set that contains no circuit. Thus, the rank of the cycle matroid $M(G)$ of a graph G is the number of edges in a spanning tree of G, that is, $|V(G)| - 1$. Condition (5.1) implies that $M(G)$ satisfies the rank condition,

$$|E| \geq 2r(M) + 2, \tag{5.2}$$

where we recall that the rank of a matroid is the size of a largest set that contains no circuit.

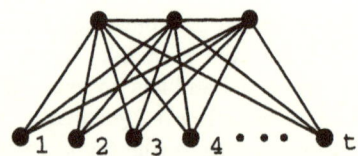

Figure 11

In view of Lemos's example and the fact that (5.2) is implied by Mader's original degree condition, one may guess that a 2-connected simple matroid M satisfying (5.2) need not have a removable circuit. This turns out to be true although Lemos's example cannot be used to establish this since it does not satisfy (5.2). For an integer t exceeding two, consider the graph $K_{3,t}''$, shown in Figure 11, that is obtained from $K_{3,t}$ by adding an edge from one of the degree-t vertices to each of the other degree-t vertices. The cycle matroid M of this graph has no removable circuits since every circuit must meet a vertex of degree three. However,

$$|E| = 3r(M) - 4.$$

This prompts one to ask whether the existence of a removable circuit is guaranteed if one replaces (5.2) by a stronger condition of the same type. The next theorem [32] answers this question.

Theorem 5.13 *Let M be a 2-connected matroid with at least two elements and let C' be a largest circuit of M. If*

$$|E(M)| \geq 3r(M) + 3 - c(M),$$

then M has a removable circuit C that is disjoint from C' such that $r(M \backslash C) = r(M)$. In particular, if C' spans M and

$$|E(M)| \geq 2r(M) + 2,$$

then M has a removable circuit.

Observe that when M has a spanning circuit, condition (5.2) is sufficient to guarantee the existence of a removable circuit. In that case, the bound on $|E(M)|$ is sharp as one can see by taking $M = M(G)$ where G is obtained from an n-cycle with $n \geq 2$ by replacing all but one of the edges by two edges in parallel. To see that the first bound on $|E(M)|$ in the last result is sharp, consider the cycle matroid of $K_{3,t}''$.

The next result follows by applying the last theorem to graphic matroids.

Corollary 5.14 *Let G be a 2-connected loopless graph and C' be a largest cycle in G. If*
$$|E(G)| \geq 3|V(G)| - c(G),$$
then G has a removable cycle C that has no common edges with C'. In particular, if G is Hamiltonian and

$$|E(G)| \geq 2|V(G)|,$$

then G has a removable cycle.

The assertion in the last sentence is easily deduced directly as follows. If C' is a Hamilton cycle in G, then, since $|E(G)| \geq 2|V(G)|$, we have $|E(G \backslash C')| \geq |V(G \backslash C')|$. Hence $G \backslash C'$ certainly has a cycle C. This cycle is clearly removable in G. In the case that G is not Hamiltonian, the assertion of the corollary seems far less obvious. Theorem 5.13 can also be applied to the bond matroid of a 2-connected loopless graph G in which case it gives a necessary condition for G to have a bond C^* such that G/C^* is 2-connected and loopless.

To summarize what occurred above, we began with Mader's theorem for 2-connected graphs and tried to get a matroid analogue of it. While our initial attempts failed, eventually we obtained a rather loose analogue which is not only a new theorem for matroids but also a new theorem for graphs.

6   Removing circuits from 3-connected matroids

Although we have focussed so far on 2-connected matroids, Mader's theorem, with which we began this discussion, actually holds for k-connected graphs for all k. We have noted already that it is for k in $\{2, 3\}$ that Tutte's definition of k-connectedness for matroids has been most successfully analyzed. This leads us to ask whether the last result has an analogue for 3-connected matroids, although we have yet to define 3-connectedness for matroids. As a guide to how to do this, we look again at graphs.

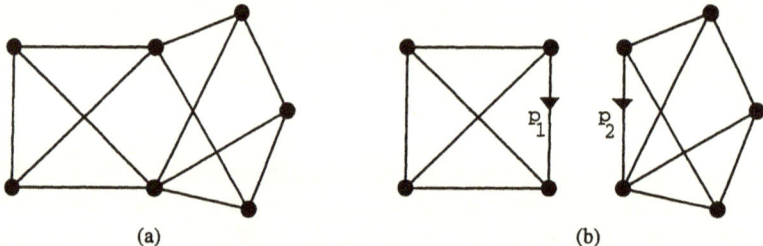

(a)   (b)

Figure 12: (a) G (b) G_1 and G_2

The graph G in Figure 12(a) is 2-connected but not 3-connected. Indeed, G is the 2-sum of the graphs G_1 and G_2 shown in Figure 12(b), that is, G can be obtained from G_1 and G_2 by identifying the directed edges p_1 and p_2 and then deleting the resulting edge. The crucial observation here about G, G_1, and G_2 is that the edge sets of cycles of G can be specified in terms of the cycles of G_1 and the cycles of G_2. Thus, the 2-sum operation for graphs generalizes to matroids as follows: let M_1 and M_2 be 2-connected matroids on disjoint ground sets E_1 and E_2 each of which has at least three elements and suppose that $p_1 \in E_1$ and $p_2 \in E_2$. The 2-*sum* $M_1 \oplus_2 M_2$ of M_1 and M_2 with respect to p_1 and p_2 is the matroid with ground set $(E_1 - p_1) \cup (E_2 - p_2)$ whose circuits are

- all circuits of M_1 avoiding p_1;

- all circuits of M_2 avoiding p_2;

- all sets of the form $(C_1 - \{p_1\}) \cup (C_2 - \{p_2\})$ where C_i is a circuit of M_i containing p_i.

A 2-connected matroid is *3-connected* if it cannot be written as a 2-sum of two other matroids. This definition is able to simultaneously generalize the corresponding graph notion and to incorporate invariance under duality. In particular, if G is a graph without isolated vertices and with at least four vertices, then the cycle matroid $M(G)$ of G is 3-connected if and only if G is simple and 3-connected. An arbitrary matroid is 3-connected if and only if its dual is 3-connected.

The next theorem, which was proved by Lemos and Oxley [33], is an analogue of Theorem 5.13 for 3-connected matroids.

Theorem 6.1 *Let M be a 3-connected matroid and C' be a largest circuit of M. If*

$$|E(M)| \geq \begin{cases} 3r(M) + 1 & \text{when } c(M) = r(M) + 1, \\ 4r(M) + 1 - c(M) & \text{otherwise,} \end{cases}$$

then M has a circuit C that is disjoint from C' such that $M\backslash C$ is 3-connected and $r(M\backslash C) = r(M)$.

As with Theorem 5.13, we can apply the last result to graphs to obtain a new graph result.

Corollary 6.2 *Let G be a simple 3-connected graph and C' be a largest cycle of G. Suppose that*

$$|E(G)| \geq \begin{cases} 3|V(G)| - 2 & \text{if } G \text{ is Hamiltonian,} \\ 4|V(G)| - 3 - c(G) & \text{otherwise.} \end{cases}$$

Then G has a cycle C that has no common edges with C' such that $G\backslash C$ is 3-connected.

7   Minors and infinite antichains

Minors are basic substructures of graphs. Indeed, one of the best-known theorems in graph theory is Kuratowski's Theorem [28, 60] characterizing planar graphs.

Theorem 7.1 *A graph G is planar if and only if it has no minor isomorphic to K_5 or $K_{3,3}$.*

This section begins with a discussion of such excluded-minor theorems for graphs and matroids. It concludes by considering a problem for matroid minors that is motivated by a recent celebrated graph result. Once again, the interaction between graph theory and matroid theory should be evident.

It was noted in Section 2 that, for a graphic matroid M, all single-element deletions and all single-element contractions are also graphic. It follows that every minor of M is graphic, and we say that the class of graphic matroids is *minor-closed*.

In Section 2, we noted that the matroid $U_{2,4}$ is not binary and that every graphic matroid is binary. It follows that $U_{2,4}$ is not graphic, though it is just as easy to show this directly. Every single-element deletion of $U_{2,4}$ is isomorphic to $U_{2,3}$, that is, to $M(K_3)$. Since $U_{2,4}^* \cong U_{2,4}$, we deduce that every single-element contaction of $U_{2,4}$ is isomorphic to $U_{1,3}$, that is, to $M(C_3^*)$ where C_3^* is the graph consisting of two vertices joined by three edges. We conclude that $U_{2,4}$ is an *excluded minor* for the class of graphic matroids because it is not in the class, yet all of its proper minors are in the class. Another excluded minor for the class is the Fano matroid: every single-element deletion of F_7 is isomorphic to $M(K_4)$, while all of its single-element contractions are isomorphic to $M(C_3^2)$, where C_3^2 is the graph obtained from a 3-cycle by replacing every edge by two parallel edges. Since the planar duals of K_4 and C_3^2 are K_4 and $K_{2,3}$, respectively, it follows that every proper minor of F_7^* is also graphic. The following excluded-minor characterization of the class of graphic matroids, which was proved by Tutte [54], is a generalization of Theorem 7.1.

Theorem 7.2 *A matroid is graphic if and only if it has no minor isomorphic to any of the matroids $U_{2,4}$, F_7, F_7^*, $M^*(K_5)$, and $M^*(K_{3,3})$.*

On combining this theorem with its dual and using Kuratowski's Theorem, we obtain the following result.

Theorem 7.3 *A matroid is isomorphic to the cycle matroid of a planar graph if and only if it has no minor isomorphic to any of the matroids $U_{2,4}$, F_7, F_7^*, $M^*(K_5)$, $M(K_5)$, $M^*(K_{3,3})$, and $M(K_{3,3})$.*

Next we shall show that, for every field $\mathbb{F}$, the class of $\mathbb{F}$-representable matroids is minor-closed. To see this, suppose that $M = M[A]$ for some matrix A over $\mathbb{F}$. If e is an element of M, then $M \backslash e$ is represented by the matrix that is obtained from A by deleting the column of A labelled by e. To show that M/e is $\mathbb{F}$-representable, we argue as follows. If the column labelled by e is zero, then $M[A]/e = M[A] \backslash e$, and so the contraction of e is certainly $\mathbb{F}$-representable. If the column labelled by e is non-zero, then we choose a non-zero entry of this column, say the entry in row i. We now add suitable multiples of row i to the other rows of A to produce a matrix A' in which the column labelled by e is the ith unit vector. It is not difficult to check that $M[A'] = M[A]$. The contraction $M[A']/e$ is represented by the matrix that is obtained from A' by deleting row i and column e. We conclude that the class of $\mathbb{F}$-representable matroids is minor-closed.

It was noted in Section 2 that the dual of every $\mathbb{F}$-representable matroid is $\mathbb{F}$-representable. Therefore the dual of every excluded minor for the class of $\mathbb{F}$-representable matroids is also an excluded minor. Now let $\mathbb{F} = GF(2)$. The excluded-minor characterization of the class of binary matroids was proved by Tutte [56]. We have already identified the only excluded minor for this class.

Theorem 7.4 *A matroid is binary if and only if it has no minor isomorphic to $U_{2,4}$.*

The classes of $GF(3)$-representable and $GF(4)$-representable matroids are called, respectively, the classes of *ternary* and *quaternary* matroids. Proving the excluded-minor characterizations of these classes was far more difficult than proving Theorem 7.4. Ralph Reid announced the result for ternary matroids in 1971 but never published a proof. The first published proofs are due to Bixby [1] and Seymour [49].

Theorem 7.5 *A matroid is ternary if and only if it has no minor isomorphic to any of the matroids $U_{2,5}$, $U_{3,5}$, F_7, and F_7^*.*

The next theorem was even more difficult to obtain than the last. It was proved by Geelen, Gerards, and Kapoor [17]. The non-Fano matroid, F_7^-, which is shown in Figure 5(a), is also equal to the matroid that is represented over $GF(3)$ by the matrix whose columns consist of all 3-tuples of zeros and ones except $(0,0,0)^T$. The matroid P_6 is represented geometrically by six points placed in the plane so that exactly one 3-element subset is collinear. The matroids P_8 and $P_8^=$ are represented over $GF(3)$ and $GF(5)$ by the matrices $[I_4|A_1]$ and $[I_4|A_2]$ where A_1 and A_2 are, respectively,

$$\begin{bmatrix} 0 & 1 & 1 & -1 \\ 1 & 0 & 1 & 1 \\ 1 & 1 & 0 & 1 \\ -1 & 1 & 1 & 0 \end{bmatrix} \text{ and } \begin{bmatrix} 1 & 1 & 1 & 1 \\ 1 & 1 & -2 & -1 \\ 1 & -1 & 0 & -1 \\ 1 & 2 & 1 & 0 \end{bmatrix}.$$

All of P_6, P_8, and $P_8^=$ are isomorphic to their duals. Next we describe a very attractive geometric representation for the matroid P_8. Begin with a cube in 3-space, its eight vertices being the elements of the matroid. Then, in the plane of the top face, rotate the face through $45°$ about its centre. The resulting configuration represents P_8, its circuits being all sets of four coplanar points and all sets of five points no four of which are coplanar. We obtain $P_8^=$ from P_8 by relaxing the top and bottom faces, both of which are circuit-hyperplanes of the twisted cube P_8.

Theorem 7.6 *A matroid is quaternary if and only if it has no minor isomorphic to any of the matroids $U_{2,6}$, $U_{4,6}$, P_6, F_7^-, $(F_7^-)^*$, P_8, and $P_8^=$.*

The last result verifies a special case of a conjecture of Rota [47], which was made after Theorems 7.4 and 7.5 had been announced. This conjecture is probably the most important unsolved problem in the study of representable matroids. It remains open for all $q \geq 5$.

Conjecture 7.7 *If q is a prime power, then the set of excluded minors for representability over $GF(q)$ is finite.*

Not only are the lists of excluded minors in the last three matroid theorems finite, but so too are the lists in Theorems 7.2 and 7.3. In each case, there are only finitely many obstructions to the specified matroid property. Wagner conjectured that every minor-closed class of graphs has a finite list of excluded minors; see, for example, [42, p. 155]. We call a set of graphs or a set of matroids an *antichain* if no member of the set is isomorphic to a minor of another member of the set. As the culmination of a long series of difficult papers, Robertson and Seymour [46] proved Wagner's conjecture by establishing the following result.

Theorem 7.8 *There is no infinite antichain of graphs.*

An immediate consequence of this theorem is the following generalization of Kuratowski's Theorem.

Theorem 7.9 *If S is a surface, then there is a set $\{G_1, G_2, \dots, G_n\}$ of graphs such that an arbitrary graph G can be embedded in S if and only if G has none of $G_1, G_2, \dots, G_n$ as a minor.*

Given the intimate links between graphs and matroids, it is natural to ask whether Theorem 7.8 extends to matroids. But, even before that theorem was proved, it was known that infinite antichains of matroids do exist; see, for example, [3, p. 155]. The set $\{C_3, C_4, C_5, \dots\}$ of cycles with at least three vertices is clearly infinite, and no member of this set is isomorphic to a subgraph of another. We can use this set of graphs, or, indeed, any infinite set of simple graphs with the last property, to build an infinite antichain of matroids as follows: embed each C_n in the plane so that no three vertices are collinear, viewing its vertices as points of a rank-3 matroid and its edges as lines of the matroid. Then add one extra point on each line to get a matroid M_n that consists of a ring of n three-point lines. It is not difficult to check that $\{M_3, M_4, M_5, \dots\}$ is an infinite antichain of matroids. For each prime power q, the points of the projective plane $PG(2, q)$ are the points of a rank-3 matroid in which the circuits are all sets of three collinear points and all sets of four points with no three collinear. Another infinite antichain of matroids is $\{PG(2, p) : p$ is prime$\}$.

Let $\mathcal{M}_{a,b,c}$ be the class of matroids such that, by deleting at most a elements and contracting at most b elements, one obtains a matroid in which each component has at most c elements. Thus each of the infinite antichains above is contained in $\mathcal{M}_{0,3,1}$. Ding, Oporowski, and Oxley [10] determined precisely when $\mathcal{M}_{a,b,c}$ contains an infinite antichain.

Theorem 7.10 *The class* $\mathcal{M}_{a,b,c}$ *contains an infinite antichain if and only if none of the following holds:*

(i) $\min\{a,b\} = 0$ *and* $\max\{a,b\} = 1$;

(ii) $\min\{a,b\} = 0$, $\max\{a,b\} = 2$, *and* $c = 2$;

(iii) $\max\{a,b\} \le 2$ *and* $c \le 1$.

By contrast, insisting on representability over a fixed finite field completely changes the result.

Theorem 7.11 *For all prime powers q and all non-negative integers a, b, and c, there is no infinite antichain in* $\mathcal{M}_{a,b,c}$ *in which all the members are $GF(q)$-representable.*

Yet another infinite antichain of matroids was constructed by Lazarson [29] to establish the following result which contrasts strikingly with Conjecture 7.7.

Theorem 7.12 *If $\mathbb{F}$ is a field of characteristic zero, then the set of excluded minors for $\mathbb{F}$-representability is infinite.*

We now know that Theorem 7.8 fails for the class of all matroids. Since each M_i constucted at the top of this page is $\mathbb{R}$-representable, it also fails for the class of $\mathbb{R}$-representable matroids. A problem that is currently attracting much research attention is whether Theorem 7.8 can be generalized to the class of binary matroids. More generally, we have the following.

Question 7.13 *Let q be a prime power. Is there an infinite antichain of $GF(q)$-representable matroids?*

This question and Rota's conjecture appear quite similar. But they are different. Question 7.13 asks about the existence of an infinite set of matroids within the class of $GF(q)$-representable matroids such that no member of the set is a minor of another member of the set. Conjecture 7.7 asserts that, within the set of matroids that are not $GF(q)$-representable, there are only finitely many minor-minimal members. Indeed, it is not clear how settling one of Question 7.13 and Conjecture 7.7 would assist in settling the other. We have remarked already on the importance of Conjecture 7.7. In the context of this paper, describing how graphs and matroids interact, Question 7.13 is the most important open problem for it seeks to obtain a natural matroid extension of what is probably the most difficult theorem ever proved in graph theory.

8 Branch-width and infinite antichains

Recently, Geelen, Gerards, and Whittle [18] have proved an important partial result towards resolving Question 7.13 in the negative. One interesting feature of their result is that it can be used to give an alternative proof of the corresponding result for graphs, which was an important early step [43] in the derivation of Theorem 7.8. To explain this new matroid result, we shall need to define branch-width for matroids. This is an analogue of its namesake for graphs. The latter is closely related to the more widely used notion of tree-width. Robertson and Seymour [45] proved that a family of graphs has bounded branch-width if and only if it has bounded tree-width. Although neither of these notions has yet been defined here, we mention this result for the reader who is already familiar with the concept of tree-width. In order to simplify the somewhat complex discussion to follow, we shall restrict attention to branch-width.

Before defining branch-width for matroids, we return to 2- and 3-connected-ness for matroids and consider what these concepts mean in terms of the rank function. A matroid M is 2-connected if and only if, for every partition $\{X, Y\}$ of $E(M)$, there is a circuit meeting both X and Y, or, equivalently,

$$r(X) + r(Y) - r(M) \geq 1.$$

In other words, M is 2-connected if and only if it has no 1-separation where, for a positive integer k, a partition $\{X, Y\}$ of the ground set of a matroid M is a *k-separation* of M if

$$r(X) + r(Y) - r(M) + 1 \leq k \text{ and}$$

$$\min\{|X|, |Y|\} \geq k.$$

For $n \geq 2$, a matroid M is *n-connected* if there is no k in $\{1, 2, \ldots, n-1\}$ such that M has a k-separation. This definition implies, for example, that an n-connected matroid with at least $2n - 2$ elements has no circuits or cocircuits of size less than n. The Fano and non-Fano matroids, which are shown in Figure 5, are both examples of 3-connected matroids that are not 4-connected.

The quantity $r(X) + r(Y) - r(M) + 1$, which, following Tutte [58], we denote by $\xi(M; X, Y)$, also features prominently in the definition of branch-width. The degree-one vertices of a tree are called *leaves*. A *ternary tree* is a tree in which all vertices except the leaves have degree three. A *branch-decomposition* of a matroid M consists of a ternary tree T with exactly $|E(M)|$ leaves together with a labelling of these leaves so that each is labelled by a different element of $E(M)$. For each edge e of such a branch-decomposition T of M, the graph $T \backslash e$ has exactly two components. Thus the set of leaves of T, and hence $E(M)$, is partitioned into two subsets X and Y, say. The width of the edge e is defined to be $\xi(M; X, Y)$ and the width of T is the maximum of the widths of the edges of T. If $|E(M)| \geq 2$, the *branch-width* $bw(M)$ of M

is the minimum, over all such labelled ternary trees T, of the width of T. If $|E(M)| \leq 1$, then its branch-width is zero.

The next result [6] summarizes some basic properties of branch-width.

Proposition 8.1 *Let M be a matroid. Then the following hold:*

(i) $bw(M^*) = bw(M)$;

(ii) $bw(N) \leq bw(M)$ *for every minor N of M;*

(iii) *if $|E(M)| \geq 3$ and $e \in E(M)$, then both $bw(M \backslash e)$ and $bw(M/e)$ are in $\{bw(M) - 1, bw(M)\}$;*

(iv) *if M has a component with at least two elements, then $bw(M)$ is the maximum of the branch-widths of its components;*

(v) $bw(M) \leq 2$ *if and only if every component of M is isomorphic to a series-parallel network;*

(vi) *if M is n-connected and $n \geq 3$, then $bw(M) \geq n$ if and only if $|E(M)| \geq 3n - 5$.*

The last part of this proposition says, loosely speaking, that if a matroid is highly connected and has a lot of elements, then its branch-width is also high. To try to convey some intuition for what it means for branch-width to be small, we observe that a vector matroid has small branch-width if and only if it can be obtained from small matroids by sticking these together in a tree-like structure across subspaces of the underlying vector space.

Theorem 7.8 and Proposition 8.1(v) imply that there is no infinite antichain of matroids of branch-width at most two. The antichain of rank-3 matroids obtained from $\{C_3, C_4, C_5, \dots\}$ following Theorem 7.9 implies that there are infinite antichains of representable matroids of branch-width at most four. Geelen, Gerards, and Whittle [18] improved this by showing that infinite antichains of representable matroids arise within the class of matroids of branch-width three.

Theorem 8.2 *There is an infinite antichain of matroids each of which has branch-width three and is representable over all infinite fields.*

The main result of Geelen, Gerards, and Whittle [18] is the following theorem which is an important step towards resolving Question 7.13. This theorem extends Theorem 7.11 since, by Proposition 8.1, every member of $\mathcal{M}_{a,b,c}$ has branch-width at most $a + b + c$.

Theorem 8.3 *For all prime powers q and all positive integers n, there is no infinite antichain of $GF(q)$-representable matroids each having branch-width at most n.*

The techniques developed to prove the last result are very interesting and will be discussed in Section 10. In addition, the last theorem can be used to prove the following graph result of Robertson and Seymour [43], a significant early step in the proof of Theorem 7.8.

Theorem 8.4 *For all positive integers n, there is no infinite antichain of graphs all having branch-width at most n.*

To this point, the notion of branch-width has only been defined here for matroids although we have briefly mentioned the corresponding notion for graphs. One may hope that these two notions coincide but this has yet to be proved. Before proceeding further with this discussion, we shall define branch-width for graphs. The definition mimics that of branch-width for matroids but replaces a matroid M and its connectivity function $\xi(M; X, Y)$ by a graph G and its connectivity function $\eta(G; X, Y)$. The latter is defined, when $\{X, Y\}$ is a partition of $E(G)$, to be the number of vertices common to $G[X]$ and $G[Y]$, the subgraphs of G induced by the edges in X and Y. Now

$$\xi(M(G); X, Y) = \eta(G; X, Y) + \omega(G) - \omega(G[X]) - \omega(G[Y]) + 1 \qquad (8.1)$$

where $\omega(H)$ is the number of components of a graph H. Thus if $\beta(G)$ is the branch-width of a connected graph G, then

$$bw(M(G)) \leq \beta(G). \qquad (8.2)$$

The last inequality is strict, for example, when G is the graph that is obtained from K_2 by adding a loop at each vertex. In that case, $bw(M(G)) = 1$ and $\beta(G) = 2$. In addition, the inequality fails when G consists of the disjoint union of k copies of K_2 for some $k \geq 2$. In that case, $\beta(G) = 0$ but $bw(M(G)) = 1$. The examples in which equality fails to hold in (8.2) seem very specialized. Indeed, Geelen, Gerards, and Whittle (private communication, 2000) have proposed the following conjecture, towards which some partial results were previously established by Dharmatilake [5].

Conjecture 8.5 *Let G be a graph that cannot be obtained from a forest by adjoining loops. Then*

$$bw(M(G)) = \beta(G).$$

To obtain Theorem 8.4 from Theorem 8.3, one combines (8.2) with a well-known consequence of a result of Higman [21] that if there is no infinite antichain within a class $\mathcal{G}$ of connected finite graphs, then there is no infinite antichain within the class of finite graphs for which every component is in $\mathcal{G}$.

9 The implications of large branch-width

Theorem 8.3 is a significant step in trying to generalize to matroids one of the major achievements of the Graph-Minors Project, Theorem 7.8. An important part of the proof of the last result involves determining what happens in a graph when the branch-width is large. The $n \times n$ *grid* is the graph with vertex-set $\{(i,j) : i,j \in \{1,2,\dots,n\}\}$ such that (i,j) and (i',j') are adjacent if and only if $|i - i'| + |j - j'| = 1$. The next result follows immediately by combining two theorems of Robertson and Seymour [44, 45].

Theorem 9.1 *For each positive integer n, there is an integer $k(n)$ such that every graph of branch-width at least $k(n)$ has a minor isomorphic to the $n \times n$ grid.*

A matroid generalization of the last result has recently been conjectured by Johnson, Robertson, and Seymour. In order to state it, we shall require some more definitions. The cycle and bond matroids of a graph have already been defined. There is another interesting, but less-well-studied, matroid that arises from a graph G. The cycle matroid of G can be defined as the graph on $E(G)$ for which the circuits are the edge-sets of all subgraphs that are subdivisions of a loop. The *bicircular matroid* of G is the matroid on $E(G)$ in which the circuits are the edge-sets of all subgraphs that are subdivisions of one of the following three graphs: two vertices joined by three edges; two loops at the same vertex; two loops at distinct vertices that are joined by a single edge. The $n \times n$ *griddle* is the bicircular matroid of the $n \times n$ grid. The $n \times n$ *girdle* is the dual of the $n \times n$ *griddle*. Johnson, Robertson, and Seymour's [26] conjecture is as follows.

Conjecture 9.2 *For each positive integer n, there is an integer $k(n)$ such that every matroid of branch-width at least $k(n)$ has a minor isomorphic to one of $U_{n,2n}$, the cycle matroid of the $n \times n$ grid, the $n \times n$ griddle, or the $n \times n$ girdle.*

By using Theorems 7.4 and 7.5, it is straightforward to check that none of $U_{3,6}$, the 3×3 griddle, and the 3×3 girdle is binary or ternary. Thus, for example, the following is the specialization of the last conjecture to binary matroids.

Conjecture 9.3 *For each positive integer n, there is an integer $k(n)$ such that every binary matroid of branch-width at least $k(n)$ has a minor isomorphic to the cycle matroid of the $n \times n$ grid.*

If true, this conjecture would generalize Theorem 9.1 provided Conjecture 8.5 is also true. Some progress has been made towards Conjecture 9.3 by Johnson, Robertson, and Seymour [25].

10 Some proof outlines

In this section, we provide some of the details of the proofs of Theorems 7.9, 8.3, and 8.2. Theorem 7.8 occurs in the twentieth of a sequence of long and difficult papers. An outline of the proof appears in [7, Chapter 12]. By contrast, the proof of Theorem 7.9, Kuratowski's Theorem for arbitrary surfaces, is far more accessible and, next, we indicate the main steps in this proof. By Theorem 8.4, whose proof is relatively short, a counterexample to Theorem 7.9 would contain graphs of arbitrarily high branch-width. By Theorem 9.1, of which a short self-contained proof has been given by Diestel, Jensen, Gorbunov, and Thomassen [8], such graphs contain arbitrarily large grid-minors. Finally, Thomassen [53] has given a short proof that a minor-minimal graph that does not embed on a surface does not contain a large grid minor.

There are three main tools in Robertson and Seymour's proof of Theorem 8.4:

(i) a lemma on rooted trees that generalizes Kruskal's theorem [27] that, under the relation of topological containment, there is no infinite antichain of trees;

(ii) Menger's Theorem; and

(iii) a theorem of Thomas [52] on linked tree-decompositions, that is, tree-decompositions with a certain Menger-like property.

Geelen, Gerards, and Whittle's proof of Theorem 8.3 has the same basic structure. It uses (i) together with a matroid generalization of Menger's Theorem due to Tutte [57], and a new linked-branch-decomposition theorem. We shall concentrate here on the last two of these and refer the reader to [43] or [18] for the details of (i).

Tutte's matroid generalization of Menger's Theorem is a relatively old result whose significance appears not to have been appreciated until relatively recently. In addition to being used by Geelen, Gerards and Whittle, it also plays a role in the proof of Johnson, Robertson, and Seymour's partial result towards Conjecture 9.3. We state the theorem next and then derive Menger's Theorem as a corollary of it.

Theorem 10.1 *Let S and T be disjoint non-empty subsets of a matroid M. Then the minimum value of $\xi(M; X, Y)$ over all subsets X and Y such that $X \supseteq S$ and $Y \supseteq T$ equals the maximum value of $\xi(N; S, T)$ over all minors N of M having ground set $S \cup T$.*

Corollary 10.2 *Let s and t be distinct non-adjacent vertices in a graph G. Then the minimum number of vertices needed to separate s from t equals the maximum number of internally disjoint paths joining s and t.*

Proof Suppose that the minimum number of vertices needed to separate s from t is m. The non-trivial part of the corollary is to establish that G has m internally disjoint paths joining s and t. Clearly this holds if $m \leq 1$. Thus we may assume that $m \geq 2$. There is no loss of generality in assuming that G is connected. Let S and T be the sets of edges incident with s and t, respectively. Then $G[S]$ and $G[T]$ are connected and it is not difficult to check that the minimum value of $\xi(M(G); X, Y)$ can be achieved by some X and Y such that $G[X]$ and $G[Y]$ are connected. For this choice of X and Y, we have, by (8.1), that $\eta(G; X, Y) = \xi(M(G); X, Y)$. As $G[X]$ and $G[Y]$ must have at least m common vertices, we deduce that $\xi(M(G); X, Y) \geq m$. Since it is not difficult to construct X' and Y' with $X' \supseteq S$ and $Y' \supseteq T$ such that $\eta(G; X', Y') = m$, we deduce, by (8.1), that $\xi(M(G); X', Y') \leq m$. Since $\xi(M(G); X, Y) \leq \xi(M(G); X', Y')$, we conclude that $\xi(M(G); X, Y) = m$. Thus, by Theorem 10.1, G has a minor H with edge set $S \cup T$ such that $\xi(M(H); S, T) = m$. Thus, since $H[S]$ and $H[T]$ are both connected, $\omega(H) \leq 2$ and, by (8.1), $m = \eta(H; S, T) + \omega(H) - 1$. If $\omega(H) = 2$, then $\eta(H; S, T) = 0$ and we obtain the contradiction that $m = 1$. Thus $\omega(H) = 1$, so $\eta(H; S, T) = m$, that is, $H[S]$ and $H[T]$ have m common vertices. Since S and T must be the sets of edges incident with s and t in H, it follows that H has m internally disjoint paths joining s and t. Therefore so does G. $\qquad\square$

The third main tool in the proof of Theorem 8.3 is the analogue of Thomas's linked-tree-decomposition result. Geelen, Gerards, and Whittle were able to avoid many of the technicalities of Thomas's proof by using branch-width rather than tree-width. In addition, they took advantage of the similarities between the functions $\xi(M; X, Y)$ and $\eta(G; X, Y)$. Since Y is the complement of X, each of ξ and η can be viewed as a function of the variable X. As such, each is both submodular and symmetric where, in general, a function λ on the set of subsets of a set E is *submodular* if $\lambda(A) + \lambda(B) \geq \lambda(A \cup B) + \lambda(A \cap B)$ for all $A, B \subseteq E$; and λ is *symmetric* if $\lambda(A) = \lambda(E - A)$ for all $A \subseteq E$. Submodular functions have arisen earlier in the paper, though not explicitly, for a basic property of the rank function of a matroid is that it is submodular.

One can define a branch-decomposition of an arbitrary symmetric submodular function λ on a set E in just the same way as it was defined for ξ. A subset X of E is *displayed* by a branch-decomposition T if there is an edge e of T such that the sets of vertex labels that occur (on the leaves of T) in the two components of $T \backslash e$ are X and $E - X$. The width of such an edge is the common value of $\lambda(X)$ and $\lambda(E - X)$. This leads to the definition of the branch-width of λ. Let f and g be two edges in a branch-decomposition T of λ and let F and G be, respectively, the component of $T \backslash f$ avoiding g and the component of $T \backslash g$ avoiding f. Each edge on the shortest path P in T containing f and g displays a subset of E that contains F and avoids G. Thus the widths of the edges of P are upper bounds on $\lambda(F, G)$, the minimum value of $\lambda(X)$ taken over all X such that $F \subseteq X \subseteq E - G$. We say that

F and G are *linked* if $\lambda(F, G)$ actually equals the minimum of the widths of the edges of P. A branch-decomposition is *linked* if all of its edge pairs are linked. Such a branch-decomposition enables one to use Tutte's matroid form of Menger's Theorem, for it means that there will be no small separations between two disjoint displayed sets other than those that can be seen from the branch-decomposition. Geelen, Gerards and Whittle's result is the following.

Theorem 10.3 *An integer-valued symmetric submodular function that has branch-width n has a linked branch-decomposition of width n.*

To conclude this section, we shall prove Theorem 8.2. The proof will be geometric and will involve a fundamental class of matroids called spikes. Geometrically, an n-spike with tip consists of n three-point lines all passing through a common point but otherwise placed as freely as possible in rank-n space. More formally, for $n \geq 3$, a rank-n matroid M is an *n-spike* with *tip p* if it satisfies the following two conditions:

(a) $E(M)$ is the union of n three-element circuits, $L_1, L_2, \ldots, L_n$, all of which contain the element p;

(b) for all k in $\{1, 2, \ldots, n-1\}$, the union of any k of $L_1, L_2, \ldots, L_n$ has rank $k + 1$.

Thus, for example, each of F_7^- and F_7 in Figure 5 is a 3-spike with tip 1. In general, if M is an n-spike with tip p, then

(i) $(L_i \cup L_j) - \{p\}$ is a circuit of M for all distinct i and j;

(ii) apart from $L_1, L_2, \ldots, L_n$ and those sets listed in (i), every non-spanning circuit of M avoids p, is a circuit-hyperplane, and contains a unique element from each of $L_1 - \{p\}, L_2 - \{p\}, \ldots, L_n - \{p\}$;

(iii) M/p can be obtained from an n-element circuit by replacing each element by two elements in parallel; and

(iv) if $\{x, y\} = L_i - \{p\}$ for some i, then each of $M\backslash p/x$ and $(M\backslash p\backslash x)^*$ is an $(n-1)$-spike with tip y.

Spikes play an important role in matroid structure theory and will be discussed further in the next section. Sometimes spikes are considered with their tips deleted (and still called spikes) because such matroids equal their duals. We show next how spikes can be used to prove Theorem 8.2.

Proof of Theorem 8.2 From property (iii) of spikes and Proposition 8.1(v), it follows that, for $n \geq 3$, every n-spike has branch-width three. We complete the proof by identifying a special class of n-spikes.

For $n \geq 4$, let M_n be the n-spike with tip p such that $L_i = \{p, a_i, b_i\}$ for all i and M_n has exactly two circuit-hyperplanes, namely $\{a_1, a_2, \dots, a_n\}$ and $\{b_1, b_2, \dots, b_n\}$. We shall show that $\{M_n : n \geq 5\}$ is an antichain in which all the members are representable over every infinite field.

To see that $\{M_n : n \geq 5\}$ is an antichain, we shall consider the minors of some M_n with $n \geq 6$. To produce a matroid of lower rank, we may assume that some element of M_n is contracted. But, from (iii) above, contracting the tip produces a matroid that does not have any M_k as a minor. By symmetry, all other single-element contractions of M_n are isomorphic to M_n/a_1. The last matroid is obtained from an $(n-1)$-spike by adding b_1 in parallel with the tip. Thus if some M_k is a proper minor of M_n, then it is a minor of the $(n-1)$-spike $M/a_1 \backslash b_1$. But the last matroid has a unique circuit-hyperplane, namely $\{a_2, a_3, \dots, a_n\}$. Thus every proper spike-minor of M_n of rank at least five has at most one circuit-hyperplane, and $\{M_n : n \geq 5\}$ is indeed an antichain.

Next we describe geometrically how to construct a matrix that represents M_n over an arbitrary infinite field $\mathbb{F}$. For each vector in $V(n, \mathbb{F})$ the n-dimensional vector space over $\mathbb{F}$, we shall consider the corresponding point of the projective space $PG(n-1, \mathbb{F})$. This will enable us to argue geometrically. For each i in $\{1, 2, \dots, n-1\}$, let b_i correspond to the ith unit vector in $V(n, \mathbb{F})$. Let a_n, p, and b_n correspond, respectively, to the nth unit vector, the all-ones vector, and the vector with zero as its last entry and every other entry equal to one. Then $\{b_1, b_2, \dots, b_n\}$ is a circuit in the resulting matroid. For each i in $\{1, 2, \dots, n-1\}$, we now need to add a_i to the line spanned by b_i and p so that $\{a_1, a_2, \dots, a_n\}$ and $\{b_1, b_2, \dots, b_n\}$ are the only circuits of the form $\{d_1, d_2, \dots, d_n\}$ with d_i in $\{a_i, b_i\}$ for all i. If we add $a_1, a_2, \dots, a_{n-1}$ one at a time, then, at each stage, there are only finitely many points on the line spanned by p and b_i that, if they were used for a_i would create unwanted circuits. (Equivalently, there are only finitely many 1-dimensional subspaces of $V(n, \mathbb{F})$ in the 2-dimensional subspace spanned by p and b_i such that, if a_i were chosen in one of them, unwanted circuits would be created.) By choosing a_i to avoid these points, which can certainly be done since $\mathbb{F}$ is infinite, we avoid all such unwanted circuits. Since a_n was chosen at the outset, a little extra care must be employed in placing a_{n-2}. Its placement will immediately determine a_{n-1} since we want $\{a_1, a_2, \dots, a_n\}$ to be a circuit. This creates additional restrictions on the placement of a_{n-2} since we need to avoid not only unwanted circuits involving a_{n-2} but also unwanted circuits involving the resulting a_{n-1}. But, once again, the number of restrictions is finite and we can place a_{n-2} to produce the desired matrix representation of M_n. $\square$

11 Unavoidability revisited

Section 3 began with a discussion of unavoidable structures in graphs and developed the thread of an interesting interaction that occurs between graphs and matroids in the 2-connected case. There has been a similarly successful interaction in the 3-connected case, and this is described briefly in this section.

Corollary 3.3, which asserts that every large loopless 2-connected graph has a big cycle or a big bond, can be restated in terms of minors as follows. Recall that C_n denotes an n-edge cycle, and let C_n^* be the graph consisting of two vertices that are joined by n edges.

Corollary 11.1 *For each positive integer n, there is an integer $k(n)$ such that every 2-connected loopless graph with at least $k(n)$ edges has a minor isomorphic to C_n or C_n^*.*

The matroid generalization of this is an immediate consequence of Theorem 3.4. The cycle matroids of C_n and C_n^* are isomorphic to $U_{n-1,n}$ and $U_{1,n}$, respectively.

Corollary 11.2 *For each positive integer n, there is an integer $k(n)$ such that every 2-connected matroid with at least $k(n)$ elements has a minor isomorphic to $U_{n-1,n}$ or $U_{1,n}$.*

It is natural to ask what can be said about unavoidable minors in graphs and matroids when the connectivity is increased. One of the best-known and most important 3-connected graphs is the n-spoked wheel $\mathcal{W}_n$. For $n \geq 3$, this graph is formed from an n-cycle, the *rim*, by adding an extra vertex and joining this to each vertex on the rim. The following result of Tutte [55] means that every simple 3-connected graph can be built from a wheel by adding edges or splitting vertices so that one maintains a 3-connected simple graph throughout.

Theorem 11.3 *Let n be an integer exceeding two. The following are equivalent for a simple 3-connected graph G with $n + 1$ vertices.*

(i) *For all edges e of G, neither $G \backslash e$ nor G/e is both simple and 3-connected.*

(ii) $G \cong \mathcal{W}_n$.

In view of this result, it is probably not surprising to see wheels occur among the set of unavoidable minors of 3-connected graphs. The following result was proved by Oporowski, Oxley, and Thomas [38].

Theorem 11.4 *For each integer n exceeding two, there is an integer $k(n)$ such that every 3-connected simple graph with at least $k(n)$ edges has a minor isomorphic to $\mathcal{W}_n$ or $K_{3,n}$.*

Obtaining a matroid generalization of the last result proved to be quite difficult and followed a somewhat familiar pattern. The first generalization proved was to binary matroids [11]. We denote by J_n and $\mathbf{1}$ the $n \times n$ and $1 \times n$ matrices of all ones. The vector matroid of the matrix $[I_n | J_n - I_n | \mathbf{1}]$ over $GF(2)$ is an n-spike with tip corresponding to the column $\mathbf{1}$. In general, $M[I_n | J_n - I_n | \mathbf{1}]$ is the unique binary spike of rank n. When $n = 3$, this binary spike is the Fano matroid F_7.

Theorem 11.5 *For each integer n exceeding two, there is an integer $k(n)$ such that every 3-connected binary matroid with at least $k(n)$ elements has a minor isomorphic to one of $M(\mathcal{W}_n)$, $M(K_{3,n})$, $M^*(K_{3,n})$, or $M[I_n|J_n - I_n|1]$.*

For 3-connected matroids in general, one would expect the list of unavoidable minors to grow and to include, in particular, the whirls, which we now define. In $M(\mathcal{W}_n)$, the rim is a circuit-hyperplane. By relaxing this, we obtain the *rank-n whirl*, $\mathcal{W}^n$. One of the fundamental results for 3-connected matroids, and another striking example of the successful generalization of a graph result to matroids, is Tutte's Wheels-and-Whirls Theorem [58], which is stated next.

Theorem 11.6 *Let n be an integer exceeding two. The following are equivalent for a 3-connected matroid M of rank n.*

(i) *For all elements e of M, neither $M\backslash e$ nor M/e is 3-connected.*

(ii) *M is isomorphic to $M(\mathcal{W}_n)$ or $\mathcal{W}^n$.*

The next result [12] extends Theorem 11.5 to arbitrary 3-connected matroids. Its proof, which required the development of some new tools, is outlined in [40].

Theorem 11.7 *For every integer n exceeding two, there is an integer $k(n)$ such that every 3-connected matroid with at least $k(n)$ elements has a minor isomorphic to one of $M(\mathcal{W}_n)$, $\mathcal{W}^n$, $M(K_{3,n})$, $M^*(K_{3,n})$, $U_{2,n+2}$, $U_{n,n+2}$, or an n-spike.*

One may wonder whether the story stops with the 3-connected case. For matroids, the last result is as far as the theory has been developed. But, for graphs, Oporowski, Oxley, and Thomas [38] determined the families of unavoidable minors in the 4-connected case. Although their technique could possibly be extended to prove the corresponding result for 5-connected graphs, this has not been done. Moreover, a new technique will be needed to extend the result for k-connected graphs with $k \geq 6$. Let $n \geq 3$. We denote by D_n the graph that is obtained from $\mathcal{W}_n$ by adding a new vertex and joining it to every vertex of the rim. The *zig-zag circular ladder* Z_n is obtained from two vertex-disjoint cycles $u_1 u_2 \ldots u_n$ and $v_1 v_2 \ldots v_n$ by joining each u_i to both v_i and v_{i+1}, where $v_{n+1} = v_1$. The *zig-zag Möbius ladder* V_n is obtained from a cycle $w_1 w_2 \ldots w_{2n+1}$ by joining each w_i to both w_{i+n} and w_{i+n+1}, where all subscripts are interpreted modulo $2n + 1$. Hence Z_3 is the graph of the octahedron and $V_3 \cong K_5$.

Theorem 11.8 *For each integer n exceeding three, there is an integer $k(n)$ such that every 4-connected simple graph with at least $k(n)$ edges has a minor isomorphic to one of D_n, Z_n, V_n, or $K_{4,n}$.*

Oporowski, Oxley, and Thomas also proved slightly stronger versions of Theorems 11.4 and 11.8 by establishing the corresponding results for topological minors.

We close this section with one further interesting development in this area that relates to graphs but not matroids. Corollary 11.1 can be strengthened as follows; see, for example, [38]. Notice that the magnitude condition is now one on vertices rather than edges, and the condition that the graph be loopless has been dropped. Similar modifications can be made to the other graph results appearing earlier in this section.

Theorem 11.9 *For each positive integer n, there is an integer $k(n)$ such that every 2-connected graph with at least $k(n)$ vertices has a minor isomorphic to C_n or $K_{2,n}$.*

Instead of having two possible unavoidable classes, we could seek a result in which there was just a single such class. The following is an immediate consequence of Theorem 11.4. For a graph H, a graph G has an *H-minor* if G has a minor isomorphic to H.

Corollary 11.10 *For each positive integer n, there is an integer $k(n)$ such that every 3-connected graph with at least $k(n)$ vertices has a $K_{1,n}$-minor.*

Note that, while the goal of reducing to a single class of unavoidable minors has been achieved, we have lost the property that the members of that class maintain the connectivity of the starting graphs. The natural extension of the last result to 3-connected matroids would be that every sufficiently large 3-connected matroid M has a big cocircuit. We know that this fails even when M is regular since $M(K_{3,n})$ has no circuits of size exceeding six so its dual has no cocircuits of size exceeding six.

Ding [9] has determined the connectedness needed to ensure that every sufficiently large graph has a big $K_{2,n}$-minor.

Theorem 11.11 *For each positive integer n, there is an integer $k(n)$ such that every 5-connected graph with at least $k(n)$ vertices has a $K_{2,n}$-minor.*

Ding has also conjectured the following natural extension of the last result, but he believes that he may not be the first to have conjectured this.

Conjecture 11.12 *For each positive integer n, there is an integer $k(n)$ such that every 7-connected graph with at least $k(n)$ vertices has a $K_{3,n}$-minor.*

Finally, Ding has significantly extended the last conjecture as follows.

Conjecture 11.13 *There are functions $f(m)$ and $g(m,n)$ defined for all positive integers m and n such that the following hold:*

(i) *$f(m) \to \infty$ as $m \to \infty$;*

(ii) *$g(m,n) \to \infty$ as $n \to \infty$ for all fixed m;*

(iii) *every m-connected graph with at least n vertices has a $K_{f,g}$-minor.*

12   Conclusion

This paper highlights several broad areas in which interesting interactions occur between graphs and matroids. These relate to bounding the size of 2-connected matroids, finding removable circuits in 2- and 3-connected graphs and matroids, finding infinite antichains in matroids and graphs, and finding unavoidable classes of graph and matroid minors. Graph theorems that "can be expressed in terms of edges and circuits only" [56] are always going to suggest new matroid results. One of the aims of this paper has been to show that, by taking a matroid perspective on graphs, one can frequently produce new results not only for matroids but also for graphs.

Acknowledgements

The author thanks Bogdan Oporowski and Geoff Whittle for helpful discussions during the preparation of this paper. The author's work was partially supported by grants from the National Security Agency.

References

[1] R.E. Bixby, On Reid's characterization of the ternary matroids, *J. Combin. Theory Ser. B* **26** (1979), 174–204.

[2] J. Bonin, J. McNulty, and T.J. Reid, The matroid Ramsey number $n(6,6)$, *Combin. Probab. Comput.* **8** (1999), 229–235.

[3] T. Brylawski, Constructions, in *Theory of Matroids* (ed. N. White), *Encyclopedia of Math. and Its Applications* **26**, Cambridge University Press, Cambridge, 1986, 127–223.

[4] T. Denley and T. J. Reid, On the number of elements in matroids with small circuits and small cocircuits, *Combin. Probab. Comput.* **8** (1999), 529–537.

[5] J.S. Dharmatilake, Binary matroids of branch-width 3, Ph. D. thesis, Ohio State University, 1994.

[6] J.S. Dharmatilake, A min-max theorem using matroid separations, in *Matroid Theory* (eds. J.E. Bonin, J.G. Oxley, B. Servatius), *Contemporary Math.* **197**, Amer. Math. Soc., Providence, 1996, 333–342.

[7] R. Diestel, *Graph Theory*, Springer, New York, 1997.

[8] R. Diestel, T.R. Jensen, K.Y. Gorbunov, and C. Thomassen, Highly connected sets and the excluded grid theorem, *J. Combin. Theory Ser. B* **75** (1999), 61–73.

[9] G. Ding, Graphs with no $K_{2,n}$ minor, Colloquium talk, Louisiana State University, 2000.

[10] G. Ding, B. Oporowski, and J. Oxley, On infinite antichains of matroids, *J. Combin. Theory Ser. B* **63** (1995), 21–40.

[11] G. Ding, B. Oporowski, J. Oxley, and D. Vertigan, Unavoidable minors of large 3-connected binary matroids, *J. Combin. Theory Ser. B* **66** (1996), 334–360.

[12] G. Ding, B. Oporowski, J. Oxley, and D. Vertigan, Unavoidable minors of large 3-connected matroids, *J. Combin. Theory Ser. B* **71** (1997), 244–293.

[13] P. Erdős and G. Szekeres, A combinatorial problem in geometry, *Compositio Math.* **2** (1935), 463–470.

[14] P. Erdős and T. Gallai, On maximal paths and circuits of graphs, *Acta Math. Acad. Sci. Hungar.* **10** (1959), 337–356.

[15] H. Fleischner and B. Jackson, Removable cycles in planar graphs, *J. London Math. Soc. (2)* **31** (1985), 193–199.

[16] D.R. Fulkerson, Networks, frames, blocking systems, in *Mathematics of the Decision Sciences, Part 1 Lectures in Applied Math.* **11**, Amer. Math. Soc., Providence, 1968, 303–334.

[17] J.F. Geelen, A.M.H. Gerards, and A. Kapoor, The excluded minors for $GF(4)$-representable matroids, *J. Combin. Theory Ser. B* **79** (2000), 247–299.

[18] J.F. Geelen, A.M.H. Gerards, and G. Whittle, Branch width and well-quasi-ordering in matroids and graphs, Victoria University of Wellington, New Zealand, 2000.

[19] L.A. Goddyn and B. Jackson, Removable circuits in binary matroids, *Combin. Probab. Comput.* **6** (1999), 539–545.

[20] L.A. Goddyn, J. van den Heuvel, and S. McGuinness, Removable circuits in multigraphs, *J. Combin. Theory Ser. B* **71** (1997), 130–143.

[21] G. Higman, Ordering by divisibility in abstract algebras, *Proc. London Math. Soc.* **2** (1952), 326–336.

[22] F. Hurst and T.J. Reid, Some small circuit-cocircuit Ramsey numbers for matroids, *Combin. Probab. Comput.* **4** (1995), 67–80.

[23] F. Hurst and T.J. Reid, Ramsey numbers for cocircuits in matroids, *Ars Combin.* **45** (1997), 181–192.

[24] B. Jackson, Removable cycles in 2-connected graphs of minimum degree at least four, *J. London Math. Soc. (2)* **21** (1980), 385–392.

[25] T. Johnson, N. Robertson, and P. Seymour, Grids in binary matroids of large branch width, Conference talk by T. Johnson, Oberwolfach, 1999.

[26] T. Johnson, N. Robertson, and P. Seymour, Grids in binary matroids, Conference talk by P. Seymour, Oberwolfach, 1999.

[27] J. Kruskal, Well-quasi-ordering, the tree theorem, and Vázsonyi's conjecture, *Trans. Amer. Math. Soc.* **95** (1960), 210–225.

[28] K. Kuratowski, Sur le problème des courbes gauches en topologie, *Fund. Math.* **15** (1930), 271–283.

[29] T. Lazarson, The representation problem for independence functions, *J. London Math. Soc.* **33** (1958), 21–25.

[30] A. Lehman, On the width-length inequality, *Math. Programming* **17** (1979), 403–417.

[31] M. Lemos and J. Oxley, On packing minors into connected matroids, *Discrete Math.* **189** (1998), 283–289.

[32] M. Lemos and J. Oxley, On removable circuits in graphs and matroids, *J. Graph Theory* **30** (1999), 51–66.

[33] M. Lemos and J. Oxley, On size, circumference and circuit removal in 3-connected matroids, *Discrete Math.* **220** (1999), 145–157.

[34] M. Lemos and J. Oxley, A sharp bound on the size of a connected matroid, *Trans. Amer. Math. Soc.*, in press.

[35] W. Mader, Kreuzungfreie a, b-Wege in endlichen Graphe, *Abh. Math. Sem. Univ. Hamburg* **42** (1974), 187–204.

[36] S. McGuinness, Contractible bonds in bipartite graphs, University of Umeå, submitted, 1998.

[37] V. Neumann-Lara, E. Rivera-Campo, and J. Urrutia, A note on covering the edges of a graph with bonds, *Discrete Math.* **197/198** (1999), 633–636.

[38] B. Oporowski, J. Oxley, and R. Thomas, Typical subgraphs of 3- and 4-connected graphs, *J. Combin. Theory Ser. B* **57** (1993), 239–257.

[39] J.G. Oxley, *Matroid Theory*, Oxford University Press, New York, 1992.

[40] J.G. Oxley, Unavoidable minors in graphs and matroids, in *Graph Theory and Combinatorial Biology* (eds. L. Lovász, A. Gyárfás, G. Katona, A. Recski, L. Székely), *Bolyai Soc. Math. Stud.* **7**, János Bolyai Math. Soc., Budapest, 1999, 279–305.

[41] T.J. Reid, Ramsey numbers for matroids, *European J. Combin.* **18** (1997), 589–585.

[42] N. Robertson and P.D. Seymour, Graph minors – a survey, in *Surveys in Combinatorics 1985* (eds. I. Anderson), *London Math. Soc. Lecture Notes* **103**, Cambridge University Press, Cambridge, 1985, pp. 153–171.

[43] N. Robertson and P.D. Seymour, Graph minors. IV. Tree-width and well-quasi-ordering, *J. Combin. Theory Ser. B* **48** (1990), 227–254.

[44] N. Robertson and P.D. Seymour, Graph minors. V. Excluding a planar graph, *J. Combin. Theory Ser. B* **41** (1986), 92–114.

[45] N. Robertson and P.D. Seymour, Graph minors. X. Obstructions to tree-decomposition, *J. Combin. Theory Ser. B* **52** (1991), 153–190.

[46] N. Robertson and P.D. Seymour, Graph minors. XX. Wagner's conjecture, *J. Combin. Theory Ser. B* , in press.

[47] G.-C. Rota, Combinatorial theory, old and new, in *Proc. Internat. Cong. Math. (Nice, 1970), Vol. 3*, Gauthier-Villars, Paris, 1971, 229-233.

[48] P.D. Seymour, The matroids with the max-flow min-cut property, *J. Combin. Theory Ser. B* **23** (1977), 189–222.

[49] P.D. Seymour, Matroid representation over $GF(3)$, *J. Combin. Theory Ser. B* **26** (1979), 159–173.

[50] P.D. Seymour, Decomposition of regular matroids, *J. Combin. Theory Ser. B* **28** (1980), 305–354.

[51] P.A. Sinclair, Strong snarks and the removal of edges from circuits in graphs, Ph. D. thesis, University of London, 1998.

[52] R. Thomas, A Menger-like property of tree-width: The finite case, *J. Combin. Theory Ser. B* **48** (1990), 67–76.

[53] C. Thomassen, A simpler proof of the excluded minor theorem for higher surfaces, *J. Combin. Theory Ser. B* **70** (1997), 306–311.

[54] W.T. Tutte, Matroids and graphs, *Trans. Amer. Math. Soc.* **90** (1959), 527–552.

[55] W.T. Tutte, A theory of 3-connected graphs, *Nederl. Akad. Wetensch. Proc. Ser. A* **64** (1961), 441–445.

[56] W.T. Tutte, Lectures on matroids, *J. Res. Nat. Bur. Standards Sect. B* **69B** (1965), 1–47.

[57] W.T. Tutte, Menger's theorem for matroids, *J. Res. Nat. Bur. Standards Sect. B* **69B** (1965), 49–53.

[58] W.T. Tutte, Connectivity in matroids, *Canad. J. Math.* **18** (1966), 1301–1324.

[59] Z. Tuza, On two intersecting set systems and k-continuous Boolean functions, *Discrete Appl. Math.* **16** (1987), 183–185.

[60] K. Wagner, Über eine Erweiterung eines Satzes von Kuratowski, *Deut. Math.* **2** (1937), 280–285.

[61] H. Whitney, On the abstract properties of linear dependence, *Amer. J. Math.* **57** (1935), 509–533.

[62] P.-L. Wu, An upper bound on the number of edges of a 2-connected graph, *Combin. Probab. Comput.* **6** (1997), 107–113.

[63] P.-L. Wu, Extremal graphs with prescribed circumference and cocircumference, *Discrete Math.* **223** (2000), 299–308.

Department of Mathematics
Louisiana State University
Baton Rouge, Louisiana 70803-4918
USA
oxley@math.lsu.edu

Ovoids, spreads and m-systems of finite classical polar spaces

J.A. Thas

Abstract

A survey of the most important results on partial m-systems and m-systems of finite classical polar spaces will be given. Also, the paper contains several recent results on the topic. Finally, many applications of m-systems to strongly regular graphs, linear projective two-weight codes, maximal arcs, generalized quadrangles and semi-partial geometries are mentioned.

1 Introduction

Let P be a finite polar space of rank $r \geq 2$. An *ovoid* O of P is a pointset of P, which has exactly one point in common with each generator of P, that is, with each maximal totally singular subspace of P. A *spread* S of P is a set of generators, which constitutes a partition of the pointset. It appears that $|O| = |S|$ for any ovoid O and any spread S of any given polar space P; this common number will be denoted by μ_P. Ovoids and spreads have many connections with and applications to projective planes, circle geometries, generalized polygons, strongly regular graphs, partial geometries, semi-partial geometries, codes, designs.

A *partial m-system* of P, with $0 \leq m \leq r - 1$, is any set $\{\pi_1, \pi_2, \ldots, \pi_k\}$ of $k(\neq 0)$ totally singular m-spaces of P such that no generator containing π_i has a point in common with $(\pi_1 \cup \pi_2 \cup \cdots \cup \pi_k) - \pi_i$, with $i = 1, 2, \ldots, k$. For any partial m-system M of P the bound $|M| \leq \mu_P$ holds. If $|M| = \mu_P$, then the partial m-system M of P is called an *m-system* of P. For $m = 0$, the m-system is an ovoid of P; for $m = r - 1$, with r the rank of P, the m-system is a spread of P. The fact that $|M|$, with M any m-system of the polar space P, is independent of m gives an explanation why an ovoid and a spread of a polar space P have the same size.

Here a survey of the most important results on the topic will be given in the case that P is classical; that is, we exclude the non-classical finite generalized quadrangles. The paper also includes several recent results on (partial) m-systems. Finally, many applications of m-systems to strongly regular graphs, linear projective two-weight codes, maximal arcs, sets with the BLT property and SPG reguli are mentioned.

2   Finite classical polar spaces

Let P be a finite *classical polar space* of *rank* r, with $r \geq 2$; see, for example, Hirschfeld and Thas [18]. We use the following notation:

$W_n(q)$: the polar space arising from a symplectic polarity η of $\mathrm{PG}(n,q)$, n odd and $n \geq 3$; here the rank is $r = (n+1)/2$;

$Q(2n,q)$: the polar space arising from a non-singular quadric Q in $\mathrm{PG}(2n,q)$, $n \geq 2$; here the rank is $r = n$;

$Q^+(2n+1,q)$: the polar space arising from a non-singular hyperbolic quadric Q^+ in $\mathrm{PG}(2n+1,q)$, $n \geq 1$; here the rank is $r = n+1$;

$Q^-(2n+1,q)$: the polar space arising from a non-singular elliptic quadric Q^- in $\mathrm{PG}(2n+1,q)$, $n \geq 2$; here the rank is $r = n$;

$H(n,q^2)$: the polar space arising from a non-singular Hermitian variety H in $\mathrm{PG}(n,q^2)$, $n \geq 3$; for n odd the rank is $r = (n+1)/2$, for n even the rank is $r = n/2$.

Remark   Usually, for a non-singular quadric, respectively Hermitian variety, of rank r, with $r \geq 2$, and the corresponding polar space P, the same notation P will be used.

Let $|P|$ denote the number of points of P, and let $\Sigma(P)$ be the set of all *generators*, that is, maximal subspaces (or maximal totally singular subspaces), of P; all elements of $\Sigma(P)$ have dimension $r - 1$.

For a proof of the following theorems, see, for example, Hirschfeld and Thas [18].

Theorem 2.1 *The number of points of the finite classical polar spaces are as follows:*

   (i) $|W_n(q)| = (q^{n+1} - 1)/(q - 1)$;

   (ii) $|Q(2n,q)| = (q^{2n} - 1)/(q - 1)$;

   (iii) $|Q^+(2n+1,q)| = (q^n + 1)(q^{n+1} - 1)/(q - 1)$;

   (iv) $|Q^-(2n+1,q)| = (q^n - 1)(q^{n+1} + 1)/(q - 1)$;

   (v) $|H(n,q^2)| = (q^{n+1} + (-1)^n)(q^n - (-1)^n)/(q^2 - 1)$.

Theorem 2.2 *The number of generators of the finite classical polar spaces are as follows:*

   (i) $|\Sigma(W_{2n+1}(q)| = (q+1)(q^2+1)\ldots(q^{n+1}+1)$;

   (ii) $|\Sigma(Q(2n,q))| = (q+1)(q^2+1)\ldots(q^n+1)$;

   (iii) $|\Sigma(Q^+(2n+1,q))| = 2(q+1)(q^2+1)\ldots(q^n+1)$;

(iv) $|\Sigma(Q^-(2n+1,q))| = (q^2+1)(q^3+1)\ldots(q^{n+1}+1)$;

(v) $|\Sigma(H(2n,q^2))| = (q^3+1)(q^5+1)\ldots(q^{2n+1}+1)$;

(vi) $|\Sigma(H(2n+1,q^2))| = (q+1)(q^3+1)\ldots(q^{2n+1}+1)$.

3   Ovoids and spreads of finite classical polar spaces

Let P be a finite classical polar space of rank $r \geq 2$. An *ovoid* O of P is a pointset of P, which has exactly one point in common with each generator of P. A *spread* S of P is a set of generators, which constitutes a partition of the pointset. The next theorem easily follows from Theorems 1.1 and 1.2.

Theorem 3.1 *Let O be an ovoid and let S be a spread of the finite classical polar space P. Then*

(i) *for* $P = W_{2n+1}(q), |O| = |S| = q^{n+1} + 1$;

(ii) *for* $P = Q(2n,q), |O| = |S| = q^n + 1$;

(iii) *for* $P = Q^+(2n+1,q), |O| = |S| = q^n + 1$;

(iv) *for* $P = Q^-(2n+1,q), |O| = |S| = q^{n+1} + 1$;

(v) *for* $P = H(2n,q^2), |O| = |S| = q^{2n+1} + 1$;

(vi) *for* $P = H(2n+1,q^2), |O| = |S| = q^{2n+1} + 1$.

4   Existence of ovoids and spreads

4.1   Spreads

Open problems   Determine the existence or non-existence of spreads in the following cases:

(a) $Q(6,q)$, for q odd, with $q \equiv 1 \pmod 3$ and not a prime;
   $Q(4n+2,q)$, for $n > 1$ and q odd;

(b) $Q^+(7,q)$, for q odd with $q \equiv 1 \pmod 3$ and not a prime;
   $Q^+(4n+3,q)$, for $n > 1$ and q odd;

(c) $Q^-(2n+1,q)$, for $n > 2$ and q odd;

(d) $H(4,q^2)$, for $q > 2$;
   $H(2n,q^2)$, for $n > 2$.

Table 1: Existence of spreads

Polar space	Existence	References
$W_{2n+1}(q)$	Yes	[1], [22],[25], [32], [40], [49]
$Q(2n,q)$, $n \geq 2$ and q even	Yes	[12], [42],[43], [49]
$Q(6,q)$, q odd with q prime	Yes	[9], [12], [20], [21],[22], [28],
or $q \equiv 0$ or 2 (mod 3)		[34], [42], [43] (*)
$Q(4n,q)$, q odd	No	[38], [46]
$Q^+(4n+1,q)$	No	[18]
$Q^+(4n+3,q)$, q even	Yes	[12], [42],[43]
$Q^+(3,q)$	Yes	
$Q^+(7,q)$, q odd with q prime	Yes	See (*)
or $q \equiv 0$ or 2 (mod 3)		
$Q^-(2n+1,q)$, q even	Yes	[12], [42], [43]
$Q^-(5,q)$	Yes	[31], [44], [49]
$H(2n+1,q^2)$	No	[43], [46]
$H(4,4)$	No	[7]

4.2 Ovoids

Remark From Table 2, it follows, for example, that for $n \geq 4$ the polar space $Q^+(2n+1,q)$, with $q = p^h$ and $p \in \{2,3\}$, has no ovoid. For $n \geq 5$ the polar space $Q^+(2n+1,q)$, with $q = p^h$ and $p \in \{2,3,5,7\}$, has no ovoid.

For $n \geq 3$ the polar space $H(2n+1,q^2)$, with $q = p^h$ and $p \in \{2,3\}$, has no ovoid. For $n \geq 4$ the polar space $H(2n+1,q^2)$, with $q = 5^h$, has no ovoid.

Open problems Determine the existence or non-existence of ovoids in the following cases:

(a) $Q(6,q)$, for q odd with $q \neq 3^h, 5, 7$;

(b) $Q^+(7,q)$, for q odd with $q \equiv 1$ (mod 3) and not a prime;

$Q^+(2n+1,q)$, for $n > 3$, $q = p^h$, p prime, and

$$p^n \leq \binom{2n+p}{2n+1} - \binom{2n+p-2}{2n+1};$$

(c) $H(2n+1,q^2)$, for $n \geq 2$, $q = p^h$, p prime, and

$$p^{2n+1} \leq \binom{2n+p}{2n+1}^2 - \binom{2n+p-1}{2n+1}^2.$$

Table 2: Existence of ovoids

Polar space	Existence	References
$W_3(q)$, q even	Yes	[38]
$W_3(q)$, q odd	No	[38]
$W_n(q)$, $n = 2t + 1$ and $t > 1$	No	[43]
$Q(4, q)$	Yes	[22], [25], [32], [49]
$Q(6, 5)$ and $Q(6, 7)$	No	[30]
$Q(6, q)$, $q = 3^h$	Yes	[22], [42], [43]
$Q(2n, q)$, $n > 2$ and q even	No	[43]
$Q(2n, q)$, $n > 3$ and q odd	No	[13]
$Q^+(3, q)$	Yes	
$Q^+(5, q)$	Yes	[17]
$Q^+(7, q)$, q odd with q prime	Yes	See (*) in Table 1
or $q \equiv 0$ or $2 \pmod 3$		
$Q^+(2n + 1, q)$, with $q = p^h$, p prime, and	No	[6]
$p^n > \dbinom{2n + p}{2n + 1} - \dbinom{2n + p - 2}{2n + 1}$		
$Q^-(2n + 1, q)$, $n > 1$	No	[43]
$H(3, q^2)$	Yes	[31], [44], [49]
$H(2n, q^2)$, $n \geq 2$	No	[43]
$H(2n + 1, q^2)$, $n > 1$,	No	[29]
$q = p^h$, p prime, and		
$p^{2n+1} > \dbinom{2n + p}{2n + 1}^2 - \dbinom{2n + p - 1}{2n + 1}^2$		

5 *m*-Systems and partial *m*-systems of finite classical polar spaces

Definition 5.1 Let P be a finite classical polar space of rank r, with $r \geq 2$. A *partial m-system* of P, with $0 \leq m \leq r - 1$, is any set $\{\pi_1, \pi_2, \ldots, \pi_k\}$ of $k(\neq 0)$ m-dimensional subspaces of P such that no generator containing π_i has a point in common with $(\pi_1 \cup \pi_2 \cup \ldots \cup \pi_k) - \pi_i$, with $i = 1, 2, \ldots, k$. A partial 0-system of size k is also called a *partial ovoid*, or a *cap*, or a *k-cap*; a partial $(r - 1)$-system is also called a *partial spread*.

Theorem 5.2 (Shult and Thas [35]) *Let M be a partial m-system of the classical polar space P. Then*

$$\text{for } P = \mathsf{W}_{2n+1}(q), \qquad |M| \le q^{n+1} + 1,$$
$$\text{for } P = \mathsf{Q}(2n, q), \qquad |M| \le q^{n} + 1,$$
$$\text{for } P = \mathsf{Q}^{+}(2n + 1, q), \quad |M| \le q^{n} + 1,$$
$$\text{for } P = \mathsf{Q}^{-}(2n + 1, q), \quad |M| \le q^{n+1} + 1,$$
$$\text{for } P = \mathsf{H}(2n, q^{2}), \qquad |M| \le q^{2n+1} + 1,$$
$$\text{for } P = \mathsf{H}(2n + 1, q^{2}), \quad |M| \le q^{2n+1} + 1.$$

Definition 5.3 Let M be a partial m-system of the finite classical polar space P. If for $|M|$ the upper bound in the statement of Theorem 4.2 is reached, then M is called an m-*system* of P.

So for an m-system M we have in the respective cases:

$$\text{if } P = \mathsf{W}_{2n+1}(q), \qquad \text{then } |M| = q^{n+1} + 1,$$
$$\text{if } P = \mathsf{Q}(2n, q), \qquad \text{then } |M| = q^{n} + 1,$$
$$\text{if } P = \mathsf{Q}^{+}(2n + 1, q), \quad \text{then } |M| = q^{n} + 1,$$
$$\text{if } P = \mathsf{Q}^{-}(2n + 1, q), \quad \text{then } |M| = q^{n+1} + 1,$$
$$\text{if } P = \mathsf{H}(2n, q^{2}), \qquad \text{then } |M| = q^{2n+1} + 1,$$
$$\text{if } P = \mathsf{H}(2n + 1, q^{2}), \quad \text{then } |M| = q^{2n+1} + 1.$$

For $m = 0$, the m-system is an ovoid of P; for $m = r - 1$, with r the rank of P, the m-sytem M is a spread of P. The fact that $|M|$ is independent of m gives us the explanation why an ovoid and a spread of a finite classical polar space P have the same size.

Theorem 5.4 (Shult and Thas [35]) *Let M be an m-system of the finite classical polar space P of rank r, with $m < r - 1$. Then the number θ of $(m + 1)$-dimensional subspaces of P containing an element of M and a given point x of P not in an element of M is independent of the choice of x. In the respective cases the number θ is given as follows:*

$$\text{for } P = \mathsf{W}_{2n+1}(q), \qquad \theta = q^{n-m} + 1,$$
$$\text{for } P = \mathsf{Q}(2n, q), \qquad \theta = q^{n-m-1} + 1,$$
$$\text{for } P = \mathsf{Q}^{+}(2n + 1, q), \quad \theta = q^{n-m-1} + 1,$$
$$\text{for } P = \mathsf{Q}^{-}(2n + 1, q), \quad \theta = q^{n-m} + 1,$$
$$\text{for } P = \mathsf{H}(2n, q^{2}), \qquad \theta = q^{2n-2m-1} + 1,$$
$$\text{for } P = \mathsf{H}(2n + 1, q^{2}), \quad \theta = q^{2n-2m-1} + 1.$$

Remark If x is a point of P not in an element of the m-system M, then for $m < r-1$ and $P \ne \mathsf{W}_{2n+1}(q)$, Theorem 4.4 says that the tangent hyperplane of P at x contains exactly θ elements of M; for $m < r - 1$ and $P = \mathsf{W}_{2n+1}(q)$, the hyperplane $x^{\perp}$, that is, x^{η} with η the symplectic polarity defining P, contains exactly θ elements of M.

6 Intersections with hyperplanes

From now on, let P be a finite classical polar space of rank r, with $r \geq 2$, and let M be an m-system of P.

Theorem 6.1 (Shult and Thas [35]) *For $P \neq W_{2n+1}(q)$, let ζ be the number of elements of M contained in a hyperplane π which is not tangent to P; for $P = W_{2n+1}(q)$, let ζ be the number of elements of M contained in a hyperplane $x^{\perp}$, with x not in an element of M. Then*

$$
\begin{aligned}
&\text{for } P = W_{2n+1}(q), && \zeta = \theta = q^{n-m} + 1, \\
&\text{for } P = Q(2n,q) \text{ and} && \\
&\quad \pi \cap Q(2n,q) = Q^+(2n-1,q), && \zeta = \theta = q^{n-m-1} + 1, \\
&\text{for } P = Q^-(2n+1,q), && \zeta = \theta = q^{n-m} + 1, \\
&\text{for } P = H(2n,q^2), && \zeta = \theta = q^{2n-2m-1} + 1.
\end{aligned}
$$

Remarks (i) If $P = Q(2n,q)$ and $\pi \cap Q(2n,q) = Q^-(2n-1,q)$, then $\zeta = 0$ for $m = n - 1$, and ζ depends on the choice of π for $m < n - 1$.

(ii) If $P = Q^+(2n+1,q)$, then $\zeta = 0$ for $m = n$, and ζ depends on the choice of π for $m < n$.

(iii) If $P = H(2n+1,q^2)$, then $\zeta = 0$ for $n = m$, and ζ depends on the choice of π for $m < n$.

Theorem 6.2 (Shult and Thas [35]) *For $P \in \{W_{2n+1}(q), Q^-(2n+1,q), H(2n,q^2)\}$ we have $\zeta = \theta$; that is, any hyperplane contains either one or θ elements of M. Hence the union $\tilde{M}$ of the elements of M has two intersection numbers β_1, β_2 with respect to hyperplanes.*

The intersection numbers β_1, β_2. In the respective cases, the intersection numbers β_1, β_2 are given as follows:

(a) for $P = W_{2n+1}(q)$,

$$
\beta_1 = \frac{(q^{m+1} - 1)(q^n + 1)}{q - 1} - q^n, \quad \beta_2 = \frac{(q^{m+1} - 1)(q^n + 1)}{q - 1};
$$

(b) for $P = Q^-(2n+1,q)$,

$$
\beta_1 = \frac{(q^{m+1} - 1)(q^n + 1)}{q - 1} - q^n, \quad \beta_2 = \frac{(q^{m+1} - 1)(q^n + 1)}{q - 1};
$$

(c) for $P = H(2n,q^2)$,

$$
\beta_1 = \frac{(q^{2m+2} - 1)(q^{2n-1} + 1)}{q^2 - 1} - q^{2n-1}, \quad \beta_2 = \frac{(q^{2m+2} - 1)(q^{2n-1} + 1)}{q^2 - 1}.
$$

Corollary 6.3 *For $P \in \{W_{2n+1}(q), Q^-(2n+1,q), H(2n,q^2)\}$ any m-system defines a strongly regular graph and a linear projective two-weight code.*

Proof See Calderbank and Kantor [8]. $\qquad\qquad\qquad\qquad\qquad\qquad\square$

Remark For more details on these graphs and codes, see Section 9.

7   Intersections with generators

Theorem 7.1 (Shult and Thas [35]) *Let M be an m-system of the finite classical polar space P (over $\mathrm{GF}(q)$), and let $\tilde{M}$ be the union of the elements of M. Then, for any generator γ of P,*

$$|\gamma \cap \tilde{M}| = (q^{m+1} - 1)/(q - 1).$$

k-ovoids of polar spaces. Let P be a finite, not necessarily classical, polar space of rank r, with $r \geq 2$. Hence P is allowed to be a non-classical generalized quadrangle; see Payne and Thas [31]. A pointset K of P is called a *k-ovoid* of P if each generator of P contains exactly k points of K. A k-ovoid with $k = 1$ is an ovoid. For $r = 2$, k-ovoids were already introduced by Thas [45]. By Theorem 6.1, the union of all elements of an m-system of P (over $\mathrm{GF}(q)$) is a k-ovoid with $k = (q^{m+1} - 1)/(q - 1)$.

8   Bounds on partial m-systems and non-existence of m-systems

Theorem 8.1 (Blokhuis and Moorhouse [6], Moorhouse [29]) *If K is a k-cap of the finite classical polar space P, naturally embedded in $\mathrm{PG}(n, q)$ with $q = p^h$ and p prime, then*

$$k \leq \binom{p + n - 1}{n}^h + 1. \tag{8.1}$$

(i) *If P arises from a quadric in $\mathrm{PG}(n, q)$, then (8.1) can be improved to*

$$k \leq \left[\binom{p + n - 1}{n} - \binom{p + n - 3}{n} \right]^h + 1. \tag{8.2}$$

(ii) *If P arises from a quadric in $\mathrm{PG}(n, q)$ and if n and q are both even, then (8.2) can be improved to*

$$k \leq n^h + 1. \tag{8.3}$$

(iii) *If P arises from a Hermitian variety in $\mathrm{PG}(n, q^2)$, with $q = p^h$ and p prime, then (8.1) can be improved to*

$$k \leq \left[\binom{p + n - 1}{n}^2 - \binom{p + n - 2}{n}^2 \right]^h + 1. \tag{8.4}$$

Remark   Some results on ovoids in §3.2 were deduced from Theorem 7.1.

Theorem 8.2 (Shult and Thas [37]) *Consider in* $\mathsf{PG}(n,q)$, $n \geq 2$ *and* $q = p^h$ *with* p *prime, a set of* m*-dimensional subspaces* $\pi_1, \pi_2, \ldots, \pi_k$ *and a set of* $(n - m - 1)$*-dimensional subspaces* $\zeta_1, \zeta_2, \ldots, \zeta_k$, *with* $m \leq (n - 1)/2$, *where* $\pi_i \cap \zeta_i \neq \emptyset$ *and* $\pi_j \cap \zeta_i = \emptyset$ *for all* $i, j = 1, 2, \ldots, k$ *with* $i \neq j$. *Then*

$$k \leq \left(\frac{\binom{n+1}{m+1} + p - 2}{p - 1} \right)^h + 1. \tag{8.5}$$

Theorem 8.3 (Shult and Thas [37]) (i) *If* M *is a partial* m*-system of size* k *of the finite classical polar space* P, *naturally embedded in* $\mathsf{PG}(n,q)$ *with* $q = p^h$ *and* p *prime, then*

$$k \leq \left(\frac{\binom{n+1}{m+1} + p - 2}{p - 1} \right)^h + 1. \tag{8.6}$$

(a) *For* $P \in \{Q(n,q), Q^+(n,q), Q^-(n,q)\}$, *with* q *odd, and for* $P = W_n(q)$, *with* q *and* m *odd, the inequality (8.6) can be improved to*

$$k \leq \left[\left(\frac{\binom{n+1}{m+1} + p - 2}{p - 1} \right) - \left(\frac{\binom{n+1}{m+1} + p - 4}{p - 3} \right) \right]^h + 1. \tag{8.7}$$

(b) *For* P *arising from a Hermitian variety in* $\mathsf{PG}(n, q^2)$, *the inequality (8.6) can be improved to*

$$k \leq \left[\left(\frac{\binom{n+1}{m+1} + p - 2}{p - 1} \right)^2 - \left(\frac{\binom{n+1}{m+1} + p - 3}{p - 2} \right)^2 \right]^h + 1. \tag{8.8}$$

(ii) *If the finite classical polar space* P *admits an* m*-system, then*

(a) *for* $P = W_{2n+1}(q)$, $q = p^h$ *and with* m *even if* p *is odd,*

$$p^{n+1} \leq \left(\frac{\binom{2n+2}{m+1} + p - 2}{p - 1} \right); \tag{8.9}$$

(b) *for* $P = W_{2n+1}(q)$, $q = p^h$ *odd and* m *odd,*

$$p^{n+1} \leq \left(\frac{\binom{2n+2}{m+1} + p - 2}{p - 1} \right) - \left(\frac{\binom{2n+2}{m+1} + p - 4}{p - 3} \right); \tag{8.10}$$

(c) *for* $P = Q(2n, q)$, $q = 2^h$,

$$2^n \le \binom{2n+1}{m+1};$$

(8.11)

(d) *for* $P = Q^+(2n+1, q)$, $q = 2^h$,

$$2^n \le \binom{2n+2}{m+1};$$

(8.12)

(e) *for* $P = Q^-(2n+1, q)$, $q = 2^h$,

$$2^{n+1} \le \binom{2n+2}{m+1};$$

(8.13)

(f) *for* $P = Q(2n, q)$, $q = p^h$ *and* q *odd*,

$$p^n \le \binom{\binom{2n+1}{m+1} + p - 2}{p-1} - \binom{\binom{2n+1}{m+1} + p - 4}{p-3};$$

(8.14)

(f) *for* $P = Q^+(2n+1, q)$, $q = p^h$ *and* q *odd*,

$$p^n \le \binom{\binom{2n+2}{m+1} + p - 2}{p-1} - \binom{\binom{2n+2}{m+1} + p - 4}{p-3};$$

(8.15)

(g) *for* $P = Q^-(2n+1, q)$, $q = p^h$ *and* q *odd*,

$$p^{n+1} \le \binom{\binom{2n+2}{m+1} + p - 2}{p-1} - \binom{\binom{2n+2}{m+1} + p - 4}{p-3};$$

(8.16)

(h) *for* $P = H(2n, q^2)$, $q = p^h$,

$$p^{2n+1} \le \left(\binom{\binom{2n+1}{m+1} + p - 2}{p-1} \right)^2 - \left(\binom{\binom{2n+1}{m+1} + p - 3}{p-2} \right)^2;$$

(8.17)

(i) *for* $P = H(2n+1, q^2)$, $q = p^h$,

$$p^{2n+1} \le \left(\binom{\binom{2n+2}{m+1} + p - 2}{p-1} \right)^2 - \left(\binom{\binom{2n+2}{m+1} + p - 3}{p-2} \right)^2.$$

(8.18)

Theorem 8.4 (Shult and Thas [37]) (i) *Let M be a partial m-system of size k of* $W_{2n+1}(q)$, *with* $q = p^h$ *and* $m > 0$.
For p odd with m even and for $p = 2$,

$$k \le \left(\frac{\binom{2n+2}{m+1} - \binom{2n+2}{m-1} + p - 2}{p-1} \right)^h + 1.$$

For p odd with m odd,

$$k \le \left[\left(\frac{\binom{2n+2}{m+1} - \binom{2n+2}{m-1} + p - 2}{p-1} \right) - \left(\frac{\binom{2n+2}{m+1} - \binom{2n+2}{m-1} + p - 4}{p-3} \right) \right]^h + 1.$$

If M is a partial m-system of size k of $Q(2n, q)$, with $m > 0$ and $q = 2^h$,
then

$$k \le \left[\binom{2n}{m+1} - \binom{2n}{m-1} \right]^h + 1.$$

(ii) *If* $W_{2n+1}(q)$ *admits an m-system, $q = p^h$ and $m > 0$, then for p odd with m even and for $p = 2$,*

$$p^{n+1} \le \left(\frac{\binom{2n+2}{m+1} - \binom{2n+2}{m-1} + p - 2}{p-1} \right),$$

and for p odd with m odd,

$$p^{n+1} \le \left(\frac{\binom{2n+2}{m+1} - \binom{2n+2}{m-1} + p - 2}{p-1} \right) - \left(\frac{\binom{2n+2}{m+1} - \binom{2n+2}{m-1} + p - 4}{p-3} \right).$$

If $Q(2n, q)$ with q even admits an m-system, with $m > 0$, then

$$2^n \le \binom{2n}{m+1} - \binom{2n}{m-1}.$$

9 m'-systems arising from a given m-system

Here the constructions of m'-systems starting from a given m-system are surveyed. The cases "ovoids arising from a given ovoid" and "spreads arising from a given spread" are considered in many papers, in particular in the papers on ovoids and spreads mentioned in Section 3. Here mainly the other cases are dealt with.

9.1 The results in Shult and Thas [35]

Definition 9.1 The 0-*system* of the non-singular conic $Q(2,q)$ of $PG(2,q)$, respectively the non-singular elliptic quadric $Q^-(3,q)$ of $PG(3,q)$, is the set of all points of $Q(2,q)$, respectively $Q^-(3,q)$. The 0-*system* of the non-singular Hermitian curve $H(2,q^2)$ of $PG(2,q^2)$ is the set of all points of $H(2,q^2)$. The 0-*system* of $W_1(q)$ is the set of all points of $PG(1,q)$.

Theorem 9.2 *If $Q^-(2n+1,q)$ has an m-system, then also $Q(2n+2,q)$ and $Q^+(2n+3,q)$ have an m-system; if $Q(2n+2,q)$ has an m-system, then also $Q^+(2n+3,q)$ has an m-system.*

Corollary 9.3 *Putting $(n,m) = (1,0)$, respectively $(n,m) = (0,0)$, Theorem 8.2 yields that $Q(4,q)$, $Q^+(5,q)$, $Q^+(3,q)$ admit an ovoid.*

Theorem 9.4 *The polar space $Q(2n,q)$, q even, has an m-system if and only if the polar space $W_{2n-1}(q)$ has an m-system.*

Theorem 9.5 *Let S_1 and S_2 be spreads of $Q^+(7,q)$, where the generators of S_1 and the generators of S_2 belong to different families of generators of $Q^+(7,q)$. Then for each $\xi_i \in S_1$ there is exactly one $\eta_j \in S_2$ with $\xi_i \cap \eta_j = \pi_{ij}$ a plane. Also, the q^3+1 planes π_{ij} form a 2-system of $Q^+(7,q)$.*

Remark If $Q^+(4n+3,q)$ admits a $2n$-system, then it admits a spread. This spread is obtained by considering all generators of a given family of generators of $Q^+(4n+3,q)$, containing an element of the $2n$-system.

Theorem 9.6 (i) *If $H(2n,q^2)$ admits an m-system M, then $Q^-(4n+1,q)$ admits a $(2m+1)$-system M'.*
    (ii) *If $H(2n+1,q^2)$ admits an m-system M, then $Q^+(4n+3,q)$ admits a $(2m+1)$-system M'.*

Corollary 9.7 *Putting $(n,m) = (1,0)$, Theorem 8.6 yields that $Q^-(5,q)$ and $Q^+(7,q)$ admit a 1-system.*

Theorem 9.8 (i) *If $Q(2n,q^2)$, with $n \geq 1$ and q odd, admits an m-system M, then $Q^+(4n+1,q)$ admits a $(2m+1)$-system M'. If $Q(2n,q^2)$, with $n \geq 1$ and q even, admits an m-system M, then $Q(4n,q)$, and hence also $Q^+(4n+1,q)$, admits a $(2m+1)$-system M'.*
    (ii) *If $Q^-(2n+1,q^2)$, with $n \geq 1$, admits an m-system M, then $Q^-(4n+3,q)$ admits a $(2m+1)$-system M'.*

Corollary 9.9 *Putting $(n,m) = (1,0)$, Theorem 8.8 yields that $Q^+(5,q)$ and $Q^-(7,q)$ admit a 1-system, and that $Q(4,q)$ with q even admits a spread.*

Theorem 9.10 *If $H(2n,q^2)$, $n \geq 1$, admits an m-system M, then $W_{4n+1}(q)$ admits a $(2m+1)$-system M'.*

Corollary 9.11 *Putting $(n,m) = (1,0)$, Theorem 8.10 implies that $W_5(q)$ admits a 1-system.*

Theorem 9.12 *If $H(l-1,q^{2n})$, n odd, admits an m-system M, then $H(nl - 1, q^2)$ admits a partial $(mn + n - 1)$-system of size $q^{ln} + 1$ when l is odd and $q^{(l-1)n} + 1$ when l is even; for l odd this partial $(mn + n - 1)$-system is an $(mn + n - 1)$-system of $H(nl - 1, q^2)$.*

Corollary 9.13 *Putting $(l,m) = (3,0)$, Theorem 8.12 implies that $H(3n - 1, q^2)$, n odd, admits an $(n-1)$-system.*

9.2 The results in Shult and Thas [36]

Theorem 9.14 *If $H(n,q^2)$, $n \geq 2$, admits an m-system M, then $W_{2n+1}(q)$ admits a partial $(2m+1)$-system M' of size $q^n + 1$ if n is odd and $q^{n+1} + 1$ if n is even; so for n even there arises a $(2m+1)$-system of $W_{2n+1}(q)$.*

Corollary 9.15 *Putting $(n,m) = (2,0)$, Theorem 8.14 implies that $W_5(q)$ admits a 1-system.*

Theorem 9.16 *Assume that we do not have both q even and n odd.*

(i) If $Q^\epsilon(n-1,q^e)$, n even and $\epsilon \in \{+,-\}$, admits an m-system M, then $Q^\epsilon(ne - 1, q)$ admits a partial $(me + e - 1)$-system M' of size $q^{e(n-2)/2} + 1$ for $\epsilon = +$ and of size $q^{en/2} + 1$ for $\epsilon = -$; hence, for $\epsilon = -$ there arises a $(me + e - 1)$-system of $Q^-(ne - 1, q)$.

(ii) If $Q(n-1,q^e)$, n odd, admits an m-system M, then for e odd $Q(ne-1,q)$ admits a partial $(me + e - 1)$-system M' of size $q^{e(n-1)/2} + 1$, and for e even $Q^\epsilon(ne - 1, q)$ admits a partial $(me + e - 1)$-system M' of size $q^{e(n-1)/2} + 1$ for any $\epsilon \in \{+,-\}$; hence for $e = 2$ and $\epsilon = +$ there arises a $(2m + 1)$-system of $Q^+(2n - 1, q)$ (n odd).

Corollary 9.17 *If q is not even and n not odd together, then, putting $m = 0$, Theorem 8.16 implies that $Q^-(ne - 1, q)$ admits an $(e - 1)$-system and that $Q^+(2n - 1, q)$, n odd, admits a 1-system.*

Remark If n is odd with q even, then the trace map applied to $Q(n-1,q^e)$ yields a singular quadric Q' with a $(e-2)$-dimensional singular space. Without any doubt any m-system of $Q(n-1,q^e)$ will define a partial m'-system of the non-singular base of Q'.

9.3 The results in Hamilton and Quinn [16]

Theorem 9.18 *If* $W_{2n-1}(q^e)$ *admits an* m-*system* M, *then* $W_{2ne-1}(q)$ *admits an* $(me + e - 1)$-*system.*

Corollary 9.19 *Putting* $(n, m) = (1, 0)$, *Theorem 8.18 implies that* $W_{2e-1}(q)$ *admits an* $(e - 1)$-*system.*

9.4 Final remark

Many infinite classes of examples can be constructed using and combining the results in Sections 8.1, 8.2 and 8.3.

10 m-systems, strongly regular graphs and linear projective two-weight codes

10.1 Connections between the structures

Let V be a set of points in $PG(n, q)$ such that for any hyperplane π of $PG(n, q)$ the condition $|\pi \cap V| \in \{\beta_1, \beta_2\}$, with $\beta_1 \neq \beta_2$, is satisfied. Let $PG(n, q)$ be embedded in $PG(n+1, q)$ and call two distinct points x, y of $PG(n+1, q) - PG(n, q)$ adjacent if and only if the line xy contains a point of V. With such an adjacency $PG(n + 1, q) - PG(n, q)$ becomes a strongly regular graph. This is due to Calderbank and Kantor [8].

Let $V = \{p_1, p_2, \ldots, p_s\}$, with $p_i = (x_{i0}, x_{i1}, \ldots, x_{in})$ and $i = 1, 2, \ldots, s$. Then, by Calderbank and Kantor [8], the matrix

$$
G = \begin{bmatrix}
x_{10} & x_{20} & \cdots & x_{s0} \\
x_{11} & x_{21} & \cdots & x_{s1} \\
\vdots & \vdots & & \vdots \\
x_{1n} & x_{2n} & \cdots & x_{sn}
\end{bmatrix}
$$

generates a linear projective $[s, n + 1]$-code C over $GF(q)$ whose codewords can only have weights $s - \beta_1$ and $s - \beta_2$. Recall that a linear code C is projective if and only if any two columns of a generator matrix are linearly independent, that is, if and only if the minimum weight of the dual code $C^\perp$ is at least 3. Conversely, any linear projective two-weight $[s, n + 1]$-code C over $GF(q)$ defines a set V of s points in $PG(n, q)$, which has two intersection numbers with respect to hyperplanes.

In Section 5 it was mentioned that for any m-system M of the polar space $P \in \{W_{2n-1}(q), Q^-(2n + 1, q), H(2n, q^2)\}$ the union $\tilde{M}$ of the elements of M is a set with two intersection numbers β_1, β_2 with respect to hyperplanes.

10.2 The theorems of Hamilton and Mathon [14]

Expressing that the strongly regular graph arising from an m-system of the polar space $P \in \{W_{2n+1}(q), Q^-(2n+1,q), H(2n,q^2)\}$ has $\lambda \geq 0$ one obtains.

Theorem 10.1 *For any m-system of* $W_{2n+1}(q), Q^-(2n+1,q)$ *or* $H(2n,q^2)$, *the inequality* $n \leq 2m+1$ *is satisfied.*

Corollary 10.2 *For any m-system of* $Q(2n+2,q), q$ *even, the bound* $n \leq 2m+1$ *holds.*

10.3 The theorems of Hamilton and Quinn [16]

Theorem 10.3 *Let* M_i *be an* m_i-*system of* $W_{2n+1}(q), i = 1, 2, \ldots, k$, *for some integer* $k > 1$. *For* $i = 1, 2, \ldots, k$ *define*

$$a_i = \frac{(q^{m_i+1}-1)(q^n+1)}{q-1}.$$

(i) *If, for all* $i \neq j$, M_i *and* M_j *are disjoint, that is,* $\tilde{M}_i \cap \tilde{M}_j = \emptyset$, *then the set* $\tilde{M}_1 \cup \tilde{M}_2 \cup \ldots \cup \tilde{M}_k$ *has two intersection numbers* $a_1 + a_2 + \cdots + a_k$ *and* $a_1 + a_2 + \cdots + a_k - q^n$ *with respect to hyperplanes in* $PG(2n+1,q)$.

(ii) *Suppose* M_i *is contained by* M_{i+1}, *that is, every element of* M_i *is a subspace of a unique element of* $M_{i+1}, i = 1, 2, \ldots, k-1$. *Then*

(a) *if* k *is even, the set*

$$\mathcal{K} = (\tilde{M}_k - \tilde{M}_{k-1}) \cup (\tilde{M}_{k-2} - \tilde{M}_{k-3}) \cup \ldots \cup (\tilde{M}_2 - \tilde{M}_1)$$

has two intersection numbers $a_k - a_{k-1} + a_{k-2} - a_{k-3} + \cdots + a_2 - a_1$ *and* $a_k - a_{k-1} + a_{k-2} - a_{k-3} + \cdots + a_2 - a_1 - q^n$ *with respect to hyperplanes in* $PG(2n+1,q)$;

(b) *if* k *is odd, the set*

$$\mathcal{K} = (\tilde{M}_k - \tilde{M}_{k-1}) \cup (\tilde{M}_{k-2} - \tilde{M}_{k-3}) \cup \ldots \cup (\tilde{M}_3 - \tilde{M}_2) \cup \tilde{M}_1$$

has two intersection numbers $a_k - a_{k-1} + a_{k-2} - a_{k-3} + \cdots + a_3 - a_2 + a_1$ *and* $a_k - a_{k-1} + a_{k-2} - a_{k-3} + \cdots + a_3 - a_2 + a_1 - q^n$ *with respect to hyperplanes in* $PG(2n+1,q)$.

Theorem 10.4 *Let* M_i *be an* m_i-*system of* $Q^-(2n+1,q), i = 1, 2, \ldots, k$, *for some integer* $k > 1$. *For* $i = 1, 2, \ldots, k$ *define* a_i *as in Theorem 10.3. Then the same conclusions hold.*

Theorem 10.5 *Let* M_i *be an* m_i-*system of* $H(2n,q^2), i = 1, 2, \ldots, k$, *for some integer* $k > 1$. *For* $i = 1, 2, \ldots, k$ *define*

$$a_i = \frac{(q^{2m_i+2}-1)(q^{2n-1}+1)}{q^2-1}.$$

(i) *If for all $i \neq j$ M_i and M_j are disjoint, then the set $\tilde{M}_1 \cup \tilde{M}_2 \cup \ldots \cup \tilde{M}_k$ has two intersection numbers $a_1 + a_2 + \cdots + a_k$ and $a_1 + a_2 + \cdots + a_k - q^{2n-1}$ with respect to hyperplanes in $\mathrm{PG}(2n, q^2)$.*

(ii) *Suppose M_i is contained by M_{i+1}, $i = 1, 2, \ldots, k-1$. Then*

(a) *if k is even, the set*

$$\mathcal{K} = (\tilde{M}_k - \tilde{M}_{k-1}) \cup (\tilde{M}_{k-2} - \tilde{M}_{k-3}) \cup \ldots \cup (\tilde{M}_2 - \tilde{M}_1)$$

has two intersection numbers $a_k - a_{k-1} + a_{k-2} - a_{k-3} + \cdots + a_2 - a_1$ and $a_k - a_{k-1} + a_{k-2} - a_{k-3} + \cdots + a_2 - a_1 - q^{2n-1}$ with respect to hyperplanes in $\mathrm{PG}(2n, q^2)$;

(b) *if k is odd, the set*

$$\mathcal{K} = (\tilde{M}_k - \tilde{M}_{k-1}) \cup (\tilde{M}_{k-2} - \tilde{M}_{k-3}) \cup \ldots \cup (\tilde{M}_3 - \tilde{M}_2) \cup \tilde{M}_1$$

has two intersection numbers $a_k - a_{k-1} + a_{k-2} - a_{k-3} + \cdots + a_3 - a_2 + a_1$ and $a_k - a_{k-1} + a_{k-2} - a_{k-3} + \cdots + a_3 - a_2 + a_1 - q^{2n-1}$ with respect to hyperplanes in $\mathrm{PG}(2n, q^2)$.

Examples (a) There are no examples of an m_1-system and an m_2-system of the finite classical polar spaces $P \in \{W_{2n+1}(q), Q^-(2n+1, q), H(2n, q^2)\}$ which are disjoint.

(b) There exists a chain of $(st - s_i - 1)$-systems of $Q^-(2st - 1, q)$, q even and $i = 1, 2, \ldots, k$, with each element contained by the previous one in the chain.

11 m-systems and maximal arcs

11.1   Maximal arcs

In a finite projective plane of order q, a $\{k; d\}$-*arc* K is a non-empty proper subset of k points of the plane such that some line of the plane meets K in d points, but no line meets K in more than d points. For given q and d, the size of K cannot exceed $qd - q + d$. If equality occurs the set is called a *maximal arc*. Equivalently, a maximal arc can be defined as a non-empty, proper subset of points of the plane such that every line meets the set in either 0 or d points for some d. The integer d is known as the *degree* of the maximal arc. For example, any point of a projective plane of order q is a maximal $\{1; 1\}$-arc in that plane, and the complement of any line is a maximal $\{q^2; q\}$-arc. These are known as the *trivial* maximal arcs.

If K is a maximal $\{qd - q + d; d\}$-arc, then the set of lines external to K in the plane is a maximal $\{q(q - d + 1)/d; q/d\}$-arc in the dual plane, called the *dual* of K. It follows that a necessary condition for the existence of a maximal $\{qd - q + d; d\}$-arc in a projective plane of order q is that d divides q. But it is not sufficient. Ball, Blokhuis and Mazzocca [4] proved that no non-trivial maximal arcs exist in $PG(2, q)$ for q odd; see also Ball and Blokhuis [3] for an easier proof. In $PG(2, q)$, q even, Denniston [11] has given a construction of maximal $\{k; d\}$-arcs for all d dividing q. In Thas [39, 41] other constructions of maximal arcs in Desarguesian planes are given, and further maximal arcs in non-Desarguesian planes are constructed. No other maximal arcs in Desarguesian planes are known.

From maximal arcs partial geometries and strongly regular graphs can be constructed, see e.g. Thas [39].

11.2 The constructions of Thas

Theorem 11.1 *Let* $Q^- = Q^-(2n + 1, q)$ *be a non-singular elliptic quadric in* $PG(2n + 1, q)$, $n > 0$, *and let* $\overline{S}$ *be a spread of* Q^-. *Suppose there exists an n-spread* $S = \{\pi_1, \pi_2, \dots, \pi_{q^{n+1}+1}\}$ *of* $PG(2n+1, q)$ *such that* $\overline{S} = \{Q^- \cap \pi_1, Q^- \cap \pi_2, \dots, Q^- \cap \pi_{q^{n+1}+1}\}$. *Embed* $PG(2n+1, q)$ *as a hyperplane* Σ *in* $PG(2n+2, q)$ *and choose some point* $x \in PG(2n+2, q) - \Sigma$. *Let* K *be the set of all points not in* Σ *of the cone with vertex* x *and base* Q^-. *Then* K *is a degree* q^n *maximal arc in the projective translation plane* $\pi(S)$ *of order* q^{n+1} *determined by the spread* S.

Remarks (i) For the definition of an n-spread S of $PG(2n + 1, q)$ and the translation plane defined by S, see Section 8 of Thas [48].

(ii) Theorem 11.1 also holds if for $n = 1$ the quadric Q^- is replaced by any ovoid of $PG(3, q)$; see Thas [39].

(iii) Blokhuis, Hamilton and Wilbrink [5] prove that for q odd such a pair $(S, \overline{S})$ does not exist.

(iv) Hamilton and Penttila [15] prove that for S regular (see Section 8 of Thas [48]) the maximal arc K is a Denniston maximal arc of the Desarguesian plane $\pi(S)$.

(v) The spread S is always symplectic, that is, S is a spread of the polar space $W_{2n+1}(q)$ defined by $Q^-(2n + 1, q)$. Also, $\overline{S}$ is an $(n - 1)$-system of $W_{2n+1}(q)$.

(vi) Let q be even, let $Q(2n+2, q)$ be a non-singular quadric of $PG(2n+2, q)$ and let $Q^-(2n + 1, q)$ be contained in $Q(2n + 2, q)$. If S^* is any spread of $Q(2n + 2, q)$, then S^* induces a spread $\overline{S}$ of $Q^-(2n + 1, q)$. By projection from the nucleus of $Q(2n + 2, q)$ onto the hyperplane $PG(2n + 1, q)$ containing $Q^-(2n+1, q)$, the spread S^* yields a spread S of $PG(2n+1, q)$ with the desired property. In such a way all suitable pairs $(S, \overline{S})$ are obtained.

11.3   The construction of Hamilton and Quinn [16]

Theorem 11.2 *Let M be an m-system of the polar space $W_{2n+1}(q)$ in $PG(2n+1, q)$, $n > 0$. Suppose there exists a spread S of $W_{2n+1}(q)$ such that M is contained by S. Embed $PG(2n+1, q)$ in $PG(2n+2, q)$ and choose some point $x \in PG(2n+2, q) - PG(2n+1, q)$. Let K be the set of all points not in $PG(2n+1, q)$ of the cone with vertex x and base $\tilde{M}$, with $\tilde{M}$ the set of all points contained in elements of M. Then K is a degree q^{m+1} maximal arc in the translation plane $\pi(S)$ of order q^{n+1} defined by S.*

Remark   Consider an $(n-1)$-system of $Q^-(2n+1, q^s)$, q even, contained by a spread of the associated $W_{2n+1}(q^s)$, as in Theorem 11.1. By Remark (v) of Section 11.2 this $(n-1)$-system is also an $(n-1)$-system of $W_{2n+1}(q)$. Hence by Theorem 8.18 there exists an $(ns-1)$-system of $W_{2ns+2s-1}(q)$ contained by a spread of $W_{2ns+2s-1}(q)$. Unfortunately, but expected, the translation plane and maximal arc thus obtained are isomorphic to the original translation plane and maximal arc. So here nothing new is constructed.

Consider a non-singular quadric $Q^+(4n-1, q)$, $n \geq 2$ and q even, and let $Q(4n-2, q)$ be a non-singular parabolic quadric on $Q^+(4n-1, q)$. Let y be the nucleus of $Q(4n-2, q)$. Now we project $Q(4n-2, q)$ from y onto a hyperplane $PG(4n-3, q)$ of $PG(4n-2, q)$, with $y \notin PG(4n-3, q)$. Then the subspaces of the polar space $Q(4n-2, q)$ are projected onto the subspaces of a symplectic polar space $W_{4n-3}(q)$. Let S be a spread of $W_{4n-3}(q)$. With S there corresponds a spread S' of $Q(4n-2, q)$. The generators of a chosen family of generators of $Q^+(4n-1, q)$, which contain the elements of S' constitute a spread $\hat{S}$ of $Q^+(4n-1, q)$. Now considering a non-singular parabolic quadric $Q^*(4n-2, q)$ on $Q^+(4n-1, q)$ and intersecting it with the elements of $\hat{S}$, there arises a spread $S^{*'}$ of $Q^*(4n-2, q)$. Projecting again from the nucleus y^* of $Q^*(4n-2, q)$, there arises a spread S^* of some $W^*_{4n-3}(q)$. Such a spread S^* is called a *cousin* of S. Also, the projective translation plane $\pi(S^*)$ of order q^{2n-1} arising from S^* is called a *cousin* of the projective translation plane $\pi(S)$. This construction and these notations are due to Kantor [20, 21].

Let S be a spread of $W_{4n-3}(q)$, $n \geq 2$ and q even. Then four cases are distinguished:

1. $y^* = y$;

2. $y^* \neq y$, $y^* \in y^\eta$, with η the symplectic polarity defining $W_{4n-3}(q)$;

3. $y^* \neq y$ and the line yy^* intersects $Q^+(4n-1, q)$ in two distinct points;

4. $y^* \neq y$ and $yy^* \cap Q^+(4n-1, q) = \emptyset$.

Kantor [20, 21] proves that, for S regular, spreads S^* corresponding to different classes yield non-isomorphic translation planes $\pi(S^*)$ of order q^{2n-1}; notice that spreads corresponding to the same class do not necessarily yield isomorphic translation planes. It follows that for S regular and $y^* \neq y$ the plane $\pi(S^*)$ is always non-Desarguesian.

Theorem 11.3 *Let M be an m-system of a polar space $Q^-(4n-3,q)$ in $PG(4n-3,q)$, $n \geq 2$ and q even. Suppose that the associated symplectic polar space $W_{4n-3}(q)$ admits a spread S such that M is contained by S. Then the m-system gives rise to degree q^{m+1} maximal arcs in (at least) q of the projective planes arising from the fourth cousins of S.*

Corollary 11.4 *Let s, t be positive integers, with $t > 1$, such that st is odd. Then there exist degree $q^{s(t-1)}$ maximal arcs in (at least) q of the fourth cousins of $PG(2, q^{st})$, q even.*

Remark In the case $s = 1$ these are Thas maximal arcs. However, for $s > 1$, the maximal arcs are new. This follows since the fourth cousins of the regular $(st - 1)$-spread of $PG(2st - 1, q)$, q even, are symplectic with kernel $GF(q)$; for the definition of kernel we refer to Kallaher [19]. But the only maximal arcs known in such planes are Thas maximal arcs which have different degree to those of the corollary. Notice that the procedure can also be applied to non-Desarguesian planes of order q^{st}. But the isomorphism problem for fourth cousins has been solved by Kantor [20, 21] only in the Desarguesian case, so that only in this case the maximal arcs can be identified as new.

12 Partial m-systems, m-systems, and the BLT-property

12.1 (Partial) m-systems and the BLT-property

A *BLT set* is a non-empty set $\mathcal{B}$ of disjoint lines of $W_3(q)$, q odd, with the property that every line of $W_3(q)$ which is not a member of $\mathcal{B}$ meets non-trivially exactly either two or none of the lines of $\mathcal{B}$. The dual concept in the generalized quadrangle $Q(4,q)$, which is the dual of the generalized quadrangle $W_3(q)$, is called a *dual BLT set*. BLT sets play a key role in the theory of translation planes and in the theory of generalized quadrangles. In Bader, Lunardon and Thas [2] it is proved that from any flock of a quadratic cone K with vertex x in $PG(3,q)$, that is, a partition of $K \setminus \{x\}$ into q disjoint conics, there arise q so-called "derived" flocks. Crucial in the construction is a certain pointset of size $q+1$ on $Q(4,q)$, the dual in $W_3(q)$ of which was subsequently called BLT set in a paper by Kantor [23]. On the other hand De Soete and Thas [10], in a paper on characterizations of generalized quadrangles of order $(s, s+2)$, introduced already the dual BLT sets, which they called $\{0,2\}-sets$.

More generally, one may define a *BLT set* of $PG(m,q)$'s as a non-empty collection $\mathcal{B}$ of disjoint totally singular m-dimensional spaces of a polar space P, having the property that each line of P not contained in a member of $\mathcal{B}$ meets non-trivially exactly either zero or two members of $\mathcal{B}$; see Shult and Thas [36].

Note that a BLT set of points of P is a subset of the points of P with each line of P containing either zero or two points of this subset. Examples are the hemisystem of $Q^-(5,3)$, see Section 9.5 of Thas [48], and the union of two disjoint ovoids of $H(3,q^2)$, see, for example, Hamilton and Quinn [16].

Theorem 12.1 (Thas [46], see Knarr [24]) *Let P be a finite generalized quadrangle. Then a BLT set of lines exists only in the cases that $P = W_3(q)$, q odd, or $P = Q^-(5,q)$, q odd, and has $q+1$ or q^2+1 lines in the respective cases.*

Remarks (i) For $W_3(q)$, q odd, many non-isomorphic BLT sets are known. They lead to new generalized quadrangles, new projective planes, new ovoids of $Q^+(5,q)$ and new ovoids of $Q(4,q)$; see, for example, Bader, Lunardon and Thas [2], Thas [47, 48].

(ii) For $Q^-(5,q)$, q odd, a unique example is known. This example can be constructed as follows. Let π and π' be two disjoint planes of $PG(5,q^2)$ which are conjugate with respect to the quadratic extension $GF(q^2)$ of $GF(q)$. Let C be a (non-singular) conic of π and let C' be the conic of π' consisting of the points conjugate to those of C. Joining the points of C to their conjugates, there arise q^2+1 lines of $PG(5,q)$. For q odd these lines are contained in a $Q^-(5,q)$ and form a BLT set. For q odd these lines are also contained in a $Q^+(5,q)$. Under the Klein correspondence the points on these lines are the images of all tangent lines of some $Q^-(3,q)$.

Theorem 12.2 (Shult and Thas [36]) *A BLT set of $PG(m,q)$'s, with $m > 0$, of a polar space P can exist only if $m = 1$ and either $P = W_3(q)$, q odd, or $Q^-(5,q)$, q odd.*

We say that a non-empty collection $\mathcal{B}$ of disjoint totally singular $PG(m,q)$'s of a polar space P possesses the *BLT property* if and only if there is no line of P meeting three distinct members of $\mathcal{B}$ non-trivially. In Shult and Thas [36] it is shown that from certain partial m-systems possessing the BLT property generalized quadrangles can be constructed. The following existence and non-existence results are taken from Shult and Thas [36].

Existence results

(a) Under the field reduction $Q(2,q^e) \longrightarrow \overline{Q}(3e-1,q)$ with q odd, where $\overline{Q}$ is of type Q^+ or Q^- when e is even and of type Q when e is odd, the points of $Q(2,q^e)$ give rise to a partial $(e-1)$-system of size q^e+1 of $\overline{Q}(3e-1,q)$ having the BLT property. In this case there arises the generalized quadrangle $Q(4,q^e)$.

(b) Similarly, under the field reduction $Q^-(3,q^e) \longrightarrow Q^-(4e-1,q)$, the points of $Q^-(3,q^e)$ give rise to an $(e-1)$-system of $Q^-(4e-1,q)$ with the BLT property. In this case there arises the generalized quadrangle $Q^-(5,q^e)$.

(c) The field reduction $H(2, q^2) \longrightarrow W_5(q)$ with q odd, provides a 1-system with the BLT property of $W_5(q)$. The resulting generalized quadrangle is $H(4, q^2)$.

Non-existence result

Let M be a partial $(n-1)$-system of size $q^m + 1$ of a polar space $\overline{Q}(2n + m - 1, q)$, $\overline{Q}$ any non-singular quadric, having the BLT property. Then from the theory of translation generalized quadrangles, see Payne and Thas [31], it follows that either $n = m$ or $ma = n(a+1)$ with a odd. If q is even, then necessarily either $n = m$ or $m = 2n$.

Remark If the partial m-system defines a generalized quadrangle, then further conditions can be derived; see Shult and Thas [36].

12.2 m-systems having the BLT property

Only the following m-systems can have the BLT property; see Shult and Thas [36].

(a) M is a 1-system of $q^3 + 1$ lines in $W_5(q)$, q odd. The corresponding generalized quadrangle has order (q^2, q^3). One such system is known; the resulting generalized quadrangle is $H(4, q^2)$.

(b) M is a 1-system of $q^2 + 1$ lines in $Q^+(5, q)$, q odd. The corresponding generalized quadrangle has order (q^2, q^2). Shult and Thas [35] prove that exactly one such 1-system exists; the resulting generalized quadrangle is $Q(4, q^2)$.

(c) M is a 2-system of $q^4 + 1$ planes in $Q^+(9, q)$, q odd. The corresponding generalized quadrangle has order (q^3, q^4). No example is known.

(d) M is an $(r-2)$-system of $q^{2r-2}+1$ $(r-2)$-dimensional spaces in $Q^-(4r - 5, q)$, $r \geq 3$. For each r one such system is known; the resulting generalized quadrangle is $Q^-(5, q^{r-1})$.

(e) M is a 2-system of $q^9 + 1$ planes in $H(9, q^2)$. The corresponding generalized quadrangle has order (q^6, q^9). No example is known.

13   (Partial) m-systems and SPG reguli

An *SPG regulus* of $PG(n, q)$ is a set R of m-dimensional subspaces $\pi_1, \pi_2, \cdots,$ π_r, $r > 1$, of $PG(n, q)$, satisfying:

(SPG1) $\pi_i \cap \pi_j = \emptyset$ for all $i \neq j$.

(SPG2)   If $PG(m+1, q)$ contains $\pi_i \in R$, then it has a point in common
with either 0 or α ($\alpha > 0$) spaces in $R - \{\pi_i\}$. If $PG(m+1, q)$ has no
point in common with $\pi_j \in R$ for all $j \neq i$, then it is called a *tangent*
$(m+1)$-*space* of R at π_i.

(SPG3)   If the point x of $PG(n, q)$ is not contained in an element of R, then
it is contained in a constant number θ ($\theta \geq 0$) of tangent $(m+1)$-spaces
of R.

SPG reguli were introduced by Thas [44]. He shows that each SPG regulus
gives rise to a semi-partial geometry, hence also to a strongly regular graph.

Theorem 13.1 (Luyckx [26]) *If* $P \in \{W_{2n+1}(q), Q^-(2n+1, q), H(2n, q^2)\}$,
then all m-systems of P are SPG *reguli of the ambient space of P.*

14   Generalized hexagons and 1-systems

Consider the classical generalized hexagon $H(q)$ of order q embedded in
$Q(6, q)$; for the definition of generalized hexagon see Van Maldeghem [50]. A
spread of $H(q)$ is a set S of lines of $H(q)$, of size $q^3 + 1$, any two of which are
opposite in $H(q)$. The generalized hexagon $H(q)$ always admits a spread, see
Thas [42].

Theorem 14.1 (Shult and Thas [35]) *Any spread S of* $H(q)$ *is a 1-system
of* $Q(6, q)$.

15   Characterizations, classifications and small parameters

The following uniqueness theorems on m-systems were proved.

Theorem 15.1 (Shult and Thas [35]) *Up to isomorphism the polar space*
$Q^+(5, q)$, *q odd, admits exactly one 1-system.*

For small parameters several results were obtained, mostly with the aid of
a computer. Here we just mention one recent theorem on symplectic polar
spaces over $GF(2)$.

Theorem 15.2 (Hamilton and Mathon [14]) (i) *Up to isomorphism, the
polar space* $W_5(2)$ *admits a unique 1-system and a unique spread. Hence all
symplectic 2-spreads of* $PG(5, 2)$ *are regular.*

(ii) *Up to isomorphism* $W_7(2)$ *admits a unique 1-system, a unique 2-system
and a unique spread. Hence all symplectic 3-spreads of* $PG(7, 2)$ *are regular.*

(iii) *The polar space* $W_9(2)$ *admits no 1-systems and no 2-systems. Up to
isomorphism* $W_9(2)$ *admits exactly two spreads and ten 3-systems.*

A 1-system M of the polar space $Q(6, q)$ is said to be *locally Hermitian* at a line L of M if for any line $N \in M - \{L\}$, the regulus on $Q(6, q)$ defined by L and N, that is, the regulus containing L of the hyperbolic quadric $\langle L, N \rangle \cap Q(6, q)$, is entirely contained in M. Such 1-systems consist of q^2 reguli through the line L. The 1-system M is said to be *Hermitian* or *classical* if it is locally Hermitian at every line of M. A *flock* of a quadratic cone LC, with line vertex L and base the non-singular conic C, of $PG(4, q)$ is a partition of $LC - L$ in q^2 mutually disjoint conics, with the extra condition that any two disjoint conics of the partition generate $PG(4, q)$. If q is odd and the planes $\pi_1, \pi_2, \cdots, \pi_{q^2}$ of the elements of the flock mutually intersect in internal or external points of the cone, then we speak of an *i-flock* or *e-flock*.

Assume that q is odd, so a polarity $\perp$ is defined by $Q(6, q)$. Let M be a 1-system of $Q(6, q)$ which is locally Hermitian at some line L. The q^2 hyperbolic quadrics, the reguli through L of which constitute M, will be denoted by $Q_i^+(3, q), i = 1, 2, \cdots, q^2$. Let $C_i =< Q_i^+(3, q) >^\perp \cap Q(6, q)$ and let $LC = L^\perp \cap Q(6, q), i = 1, 2, \cdots, q^2$. Then $\{C_1, C_2, \cdots, C_{q^2}\}$ is an *i*-flock of the cone LC. Conversely, any *i*-flock of LC defines a locally Hermitian 1-system of $Q(6, q)$ at L. Hence we have the following theorem.

Theorem 15.3 (Luyckx and Thas [27]) *For q odd, locally Hermitian 1-systems at L of* $Q(6, q)$ *and i-flocks of LC are equivalent objects.*

Theorem 15.4 (Luyckx and Thas [27]) *Let L be a line of* $Q(6, q)$ *and let* $LC = L^\perp \cap Q(6, q)$, *with q odd. Further, let π be the plane containing C and assume that $C' \subseteq \pi$ is a non-singular conic with the property that all of its points are internal points of C. Then any rational normal cubic scroll* $\mathcal{R}^3$ *with directrix line L and $C' \subseteq \mathcal{R}^3$, uniquely defines an i-flock of LC, and consequently also a locally Hermitian 1-system of* $Q(6, q)$.

Remark For the definition of a rational normal cubic scroll $\mathcal{R}^3$, see Semple and Roth [33]. The elements of the *i*-flock in Theorem 14.4 are the intersections of LC with the planes of the conics on $\mathcal{R}^3$.

Let M be a 1-system of $Q(6, q)$, q odd, which is locally Hermitian at $L \in M$. For $x \in L$ arbitrary we consider a 4-dimensional space $\gamma \subseteq x^\perp - \{x\}$; so $\gamma \cap Q(6, q) = Q(4, q)$. Denote the q^2 reguli through L which constitute M by $R_1, R_2, \cdots, R_{q^2}$; the opposite regulus of R_i will be denoted by $\overline{R}_i$. Each $R_i, i = 1, 2, \cdots, q^2$, has a unique transversal $N_i \in \overline{R}_i$ through x, which intersects γ in a point n_i. If l denotes the point $L \cap \gamma$, then $\{l, n_1, n_2, \cdots, n_{q^2}\}$ is an ovoid of $Q(4, q)$. This ovoid will be called *the projection of M from x along reguli*. The 1-system M is called *semi-classical* if for each $x \in L$ the projection of M from x along reguli is the classical ovoid of $Q(4, q)$, that is, an elliptic quadric.

Theorem 15.5 (Luyckx and Thas [27]) *For q odd, any locally Hermitian non-Hermitian 1-system of* $Q(6, q)$ *is semi-classical if and only if its associated i-flock arises from a rational normal cubic scroll* $\mathcal{R}^3$.

Corollary 15.6 (Luyckx and Thas [27]) *For q odd, all locally Hermitian non-Hermitian 1-systems of* $Q(6, q)$ *are known.*

References

[1] L. Bader, W.M. Kantor, and G. Lunardon, Symplectic spreads from twisted fields, *Boll. Un. Mat. Ital. A* **8** (1994), 383–389.

[2] L. Bader, G. Lunardon, and J.A. Thas, Derivation of flocks of quadratic cones, *Forum Math.* **2** (1990), 163–174.

[3] S. Ball and A. Blokhuis, An easier proof of the maximal arcs conjecture, *Proc. Amer. Math. Soc.* **126** (1998), 3377–3380.

[4] S. Ball, A. Blokhuis, and F. Mazzocca, Maximal arcs in Desarguesian planes of odd order do not exist, *Combinatorica* **17** (1997), 31–47.

[5] A. Blokhuis, N. Hamilton, and H. Wilbrink, On the non-existence of Thas maximal arcs in odd order projective planes, *European J. Combin.* **19** (1998), 413–417.

[6] A. Blokhuis and G.E. Moorhouse, Some p-ranks related to orthogonal spaces, *J. Algebraic Combin.* **4** (1995), 295–316.

[7] A.E. Brouwer, Private communication, 1981.

[8] A.R. Calderbank and W.M. Kantor, The geometry of two-weight codes, *Bull. London Math. Soc.* **18** (1986), 97–122.

[9] J.H. Conway, P.B. Kleidman, and R.A. Wilson, New families of ovoids in O_8^+, *Geom. Dedicata* **26** (1988), 157–170.

[10] M. De Soete and J.A. Thas, A characterization theorem for the generalized quadrangle $T_2^*(O)$ of order $(s, s + 2)$, *Ars Combin.* **17** (1984), 225–242.

[11] R.H.F. Denniston, Some maximal arcs in finite projective planes, *J. Combin. Theory* **6** (1969), 317–319.

[12] R.H. Dye, Partitions and their stabilizers for line complexes and quadrics, *Ann. Mat. Pura Appl.* **114** (1977), 173–194.

[13] A. Gunawardena and G.E. Moorhouse, The non-existence of ovoids in $O_9(q)$, *European J. Combin.* **18** (1997), 171–173.

[14] N. Hamilton and R. Mathon, Existence and non-existence of m-systems of polar spaces, *European J. Combin.*, to appear.

[15] N. Hamilton and T. Penttila, Groups of maximal arcs, *J. Combin. Theory Ser. A*, to appear.

[16] N. Hamilton and C.T. Quinn, *m*-Systems of polar spaces and maximal arcs in projective planes, *Bull. Belg. Math. Soc. Simon Stevin* **7** (2000), 237-248.

[17] J.W.P. Hirschfeld, *Finite Projective Spaces of Three Dimensions*, Oxford University Press, Oxford, 1985.

[18] J.W.P. Hirschfeld and J.A. Thas, *General Galois Geometries*, Oxford University Press, Oxford, 1991.

[19] M.J. Kallaher, Translation planes, *Handbook of Incidence Geometry*, North-Holland, Amsterdam, 1995, 137-192.

[20] W.M. Kantor, Spreads, translation planes and Kerdock sets. I, *SIAM J. Algebraic Discrete Methods* **3** (1982), 151-165.

[21] W.M. Kantor, Spreads, translation planes and Kerdock sets. II, *SIAM J. Algebraic Discrete Methods* **3** (1982), 308-318.

[22] W.M. Kantor, Ovoids and translation planes, *Canad. J. Math.* **34(5)** (1982), 1195-1203.

[23] W.M. Kantor, Generalized quadrangles, flocks, and BLT sets, *J. Combin. Theory Ser. A* **58** (1991), 153-157.

[24] N. Knarr, A geometric construction of generalized quadrangles from polar spaces of rank three, *Results Math.* **21(3-4)** (1992), 332-344.

[25] H. Lüneburg, *Die Suzukigruppen und ihre Geometrien, Lecture Notes in Math.* **10** , Springer, Berlin, 1965.

[26] D. Luyckx, *m*-Systems and SPG reguli, preprint.

[27] D. Luyckx and J.A. Thas, Flocks and locally hermitian 1-systems of $Q(6,q)$, in *Finite Geometries*, Kluwer Academic Publishers, Norwell, 2001, 257-275.

[28] G.E. Moorhouse, Root lattice constructions of ovoids, in *Finite Geometry and Combinatorics (Deinze, 1992)*, Cambridge University Press, Cambridge, 1993, 269-275.

[29] G.E. Moorhouse, Some *p*-ranks related to Hermitian varieties, *J. Statist. Plann. Inference* **56** (1996), 229-241.

[30] C.M. O'Keefe and J.A. Thas, Ovoids of the quadric $Q(2n,q)$, *European J. Combin.* **16** (1995), 87-92.

[31] S.E. Payne and J.A. Thas, *Finite Generalized Quadrangles*, Pitman, London, 1984.

[32] T. Penttila and B. Williams, Ovoids in parabolic spaces. *Geom. Dedicata* **82** (2000), 1–19.

[33] J.G. Semple and L. Roth, *Introduction to Algebraic Geometry*, Oxford University Press, New York, 1985.

[34] E.E. Shult, A sporadic ovoid in $\Omega^+(8,7)$, *Algebras, Groups and Geometries* **2** (1985), 495–513.

[35] E.E. Shult and J.A. Thas, m-systems of polar spaces, *J. Combin. Theory Ser. A* **68** (1994), 184–204.

[36] E.E. Shult and J.A. Thas, Constructions of polygons from buildings, *Proc. London Math. Soc.* **71** (1995), 397–440.

[37] E.E. Shult and J.A. Thas, m-systems and partial m-systems of polar spaces, *Des. Codes Cryptogr.* **8** (1996), 229–238.

[38] J.A. Thas, Ovoidal translation planes, *Arch. Math. (Basel)* **23** (1972), 110–112.

[39] J.A. Thas, Construction of maximal arcs and partial geometries, *Geom. Dedicata* **3** (1974), 61–64.

[40] J.A. Thas, Two infinite classes of perfect codes in metrically regular graphs, *J. Combin. Theory Ser. B* **23** (1977), 236–238.

[41] J.A. Thas, Construction of maximal arcs and dual ovals in translation planes, *European J. Combin.* **1** (1980), 189–192.

[42] J.A. Thas, Polar spaces, generalized hexagons and perfect codes, *J. Combin. Theory Ser. A* **29** (1980), 87–93.

[43] J.A. Thas, Ovoids and spreads of finite classical polar spaces, *Geom. Dedicata* **10** (1981), 135–144.

[44] J.A. Thas, Semi-partial geometries and spreads of classical polar spaces, *J. Combin. Theory Ser. A* **35** (1983), 58–66.

[45] J.A. Thas, Interesting pointsets in generalized quadrangles and partial geometries, *Linear Algebra Appl.* **114/115** (1989), 103–131.

[46] J.A. Thas, Old and new results on spreads and ovoids of finite classical polar spaces, *Combinatorics '90 (Gaeta, 1990)*, North-Holland, Amsterdam, 1992, 529–544.

[47] J.A. Thas, Generalized polygons, *Handbook of Incidence Geometry*, North-Holland, Amsterdam, 1995, 383–431.

[48] J.A. Thas, Projective geometry over a finite field, *Handbook of Incidence Geometry*, North-Holland, Amsterdam, 1995, 295–347.

[49] J.A. Thas and S.E. Payne, Spreads and ovoids in finite generalized quadrangles, *Geom. Dedicata* **52** (1994), 227–253.

[50] H. Van Maldeghem, *Generalized Polygons*, Birkhäuser, Basel, 1998.

Department of Pure Mathematics
and Computer Algebra
Ghent University
Krijgslaan 281
B-9000 Gent
Belgium
jat@cage.rug.ac.be

List colourings of graphs

Douglas R. Woodall

Abstract

A list colouring of a graph is a colouring in which each vertex v receives a colour from a prescribed list $L(v)$ of colours. This paper about list colourings can be thought of as being divided into two parts. The first part, comprising Sections 1, 2 and 6, is about proper colourings, in which adjacent vertices must receive different colours. It is a survey of known conjectures and results with few proofs, although Section 6 discusses several different *methods* of proof. Section 1 is intended as a first introduction to the concept of list colouring, and Section 2 discusses conjectures and results, mainly about graphs for which "ch = χ". The other part of the paper, comprising Sections 3, 4 and 5, is about improper or defective colourings, in which a vertex is allowed to have some neighbours with the same colour as itself, but not too many. Although still written mainly as a survey, this part of the paper contains a number of new proofs and new conjectures. Section 3 is about subcontractions, and includes conjectures broadly similar to Hadwiger's conjecture. Section 4 is about planar and related graphs. Section 5 is also about planar and related graphs, but this time with additional constraints imposed on the lists.

1 Introduction

List colourings of graphs were introduced independently by Vizing [61], by Erdős, Rubin and Taylor [15], and, from a slightly different perspective, by Levow [43]. Two excellent articles on them are those by Alon [1] and Tuza [59], and the second of these has been updated by Kratochvíl, Tuza and Voigt [42]. Readers who are familiar with the content of those articles may prefer to skip this section, which is intended as an introduction to list colouring for readers who are unfamiliar with the concept.

1.1 The basic concept

A *vertex-colouring*, or just *colouring*, of a finite simple graph G is an assignment of a colour to each vertex of G. A colouring is *proper* if adjacent vertices always get different colours. A graph is k-*colourable* if it has a proper colouring using at most k different colours.

The premise in this classical situation is that we have a palette of k colours available, and any vertex can receive any one of the k colours, subject only to the restriction that adjacent vertices must get different colours. In contrast, the premise in list-colouring is that every vertex has its own palette of colours, and a vertex may be coloured only with a colour from its own palette. These palettes are called *lists*—somewhat incongruously, since the term *list* is usually taken to imply the existence of an ordering, whereas the palettes are simply unordered sets of colours: in other words, they are *unordered lists*.

A *list-assignment* L to (the vertices of) G is the assignment of a "list" (set) $L(v)$ of colours to every vertex v of G; and a *k-list-assignment* is a list-assignment such that $|L(v)| \geq k$ for every vertex v. If L is a list-assignment to G, then an *L-colouring* of G is a colouring (not necessarily proper) in which each vertex receives a colour from its own list; we talk loosely of *colouring G from its lists*. The graph G is *k-list-colourable*, or *k-choosable*, if it is properly L-colourable for every k-list-assignment L to G.

The *chromatic number* $\chi(G)$ of G is the smallest number k such that G is k-colourable. The *list chromatic number* or *choosability* ch(G) of G, sometimes written $\chi_{\text{list}}(G)$ or $\chi_l(G)$, is the smallest number k such that G is k-choosable. It is evident that ch$(G) \geq \chi(G)$, since if $k < \chi(G)$ then G is not L-colourable when every vertex v of G is given the *same* list $L(v)$ of k colours.

On meeting these ideas for the first time, one might think that ch$(G) = \chi(G)$ always, since if one wants to colour adjacent vertices with different colours, then it might appear that having different lists on different vertices could only make the task easier. Indeed, there are many situations in which it does make the task easier. For example, if G is a circuit of odd length, or a complete graph, with maximum degree Δ, then it is easy to see that $\chi(G) = \Delta + 1$. It follows that ch$(G) \geq \Delta + 1$, since, in particular, if one gives every vertex of G the *same* list of Δ colours, then one cannot colour G properly from these lists. However, this is the *only* case in which the colouring is not possible: if one gives every vertex v a list $L(v)$ of Δ colours, and the lists are not all identical, then G has a proper L-colouring. To see this, choose an edge uv such that $L(u) \neq L(v)$, colour u with a colour from $L(u) \setminus L(v)$, and then colour the remaining vertices in order round the circuit, or in an arbitrary order if G is complete, but leaving v until last; it is not difficult to see that this can be done in such a way as to form a proper L-colouring of G, as required.

However, there are situations in which having different lists can make the colouring impossible. The simplest examples are the complete-bipartite graphs.

1.2 Complete-bipartite graphs

If one assigns lists $\{a, b\}$ and $\{c, d\}$ to the vertices in one partite class of $K_{2,4}$, and lists $\{a, c\}$, $\{a, d\}$, $\{b, c\}$ and $\{b, d\}$ to the vertices in the other class, then the vertices cannot be coloured from these lists. In a similar way, if one assigns lists $\{a, b\}$, $\{a, c\}$ and $\{b, c\}$ to the three vertices in each class of $K_{3,3}$, then there is no colouring from these lists. It follows that these graphs are not 2-choosable, despite being bipartite and hence 2-colourable; that is, ch$(G) > \chi(G) = 2$ for these two graphs. It is not difficult to see that ch$(G) = 3$

in each case.

These constructions can easily be generalized to give examples of graphs G such that $\mathrm{ch}(G) > k$ and $\chi(G) = 2$ for arbitrary integers $k \geq 2$. One such example is the graph K_{k,k^k}, where the k vertices of the first class are given k disjoint lists of k colours each, and the lists on the k^k vertices of the second class are all the possible systems of distinct representatives of the first k lists.

A smaller example [15] is $K_{r,r}$, with $r = \binom{2k-1}{k}$, where the lists given to the r vertices of each partite class consist of the r different k-subsets of a set of $2k-1$ colours. In this case, for any $k-1$ of the colours, there is a vertex in each class that has none of those colours in its list. It follows that in any colouring from these lists, whether proper or not, at least k colours must be used on each class; and since there are fewer than $2k$ colours in total, the colouring cannot be proper.

The question of how large the choosability of a complete-bipartite graph can be, in terms of its number of vertices, was first posed by Erdős, Rubin and Taylor [15], who proved that $\mathrm{ch}(K_{r,r}) = \log_2 r + o(\log r)$ as $r \to \infty$. They also showed that $\mathrm{ch}(K_{7,7}) \geq 4$, by exhibiting a list-assignment based on the Fano configuration (reproduced in [29] and in [59, Example 0.3]); in contrast, Hanson, MacGillivray and Toft [29] have shown that $\mathrm{ch}(G) \leq 3$ for every bipartite graph G with up to 13 vertices. For further results about $\mathrm{ch}(K_{p,q})$ see [59, Section 1.2].

1.3 The 6-, 5- and 4-colour theorems for planar graphs

It follows from the above examples that the choosability $\mathrm{ch}(G)$ of a graph G is not always equal to its chromatic number $\chi(G)$, and indeed that there is no general upper bound for $\mathrm{ch}(G)$ in terms of $\chi(G)$. Thus every question that has ever been asked about graph colourings can be asked again about choosability.

If one takes a theorem about colourings, and one changes "colouring" to "choosability", there are basically three different things that can happen:

(i) the theorem remains true and the proof still works;

(ii) the theorem remains true but the proof does not work;

(iii) the theorem becomes false.

These three possibilities are illustrated rather nicely by the 6-colour theorem, the 5-colour theorem and the 4-colour theorem for planar graphs.

Suppose one has proved, presumably by using Euler's theorem, that every planar graph contains a vertex with degree at most 5. Then the 6-colour theorem follows easily by the following argument. Given a hypothetical minimal non-6-colourable planar graph G, choose a vertex v with degree at most 5; colour (properly) the vertices of $G - v$ with six colours, which is possible by the minimality of G; and then at least one of the six colours is not present on the five or fewer neighbours of v and so can be used on v. This gives a 6-colouring of G, which is a contradiction. The same argument works equally well to prove that every planar graph is 6-choosable: when the time comes to colour v, it makes no difference whether the six colours that are potentially available for v are the same as the ones that were available for the other vertices; all that matters is that at least one colour that is potentially available for v has not been used on any neighbour of v.

In contrast, the usual proof of the 5-colour theorem involves refinements to the above argument that do not extend to choosability. If the vertex v in this argument has five neighbours with five different colours, then there is no colour that one can give to v. So one must ensure that at least two neighbours of v have the same colour, either by using a Kempe interchange of colours to make two colours the same, or by identifying two nonadjacent neighbours of v before colouring $G-v$ in order to ensure that they get the same colour from the outset. Neither of these tricks will work for choosability. Kempe interchanges will not work because they might require a vertex to be recoloured with a colour that is not present in its list; and if all the neighbours of v have different lists then it is not possible to identify two of them in any helpful way. However, the result remains true. Thomassen [58] gave a remarkably simple and elegant proof of the fact that every planar graph is 5-choosable. Thomassen's proof does not use Euler's formula or Kempe interchanges, and it is arguably shorter than the usual proof of the 5-colour theorem, despite proving a much stronger result. The proof of Theorem 5.2 of the present paper uses substantially the same method, which is described in Section 6.1.3 as the *boundary method*.

Finally, there is the 4-colour theorem itself. Here the choosability analogue is false. The first example of a non-4-choosable planar graph was given by Voigt [62], and further examples were given by Gutner [22] and Mirzakhani [45]; these last two examples are even 3-colourable.

Further results about planar graphs are given in Sections 4, 5 and 6.1.1; Section 6.1.1 also discusses graphs in other surfaces.

2   Graphs for which ch $= \chi$

At the present time it is not at all clear for which graphs G it is true that $\mathrm{ch}(G) = \chi(G)$. Rubin (see [15]) characterized 2-choosable bipartite graphs, that is, graphs G such that $\mathrm{ch}(G) = \chi(G) = 2$. But, as Tuza [59] remarks, it seems hopeless to find a characterization of all graphs G for which $\mathrm{ch}(G) = \chi(G)$. However, there are certain classes of graphs for which this equation is conjectured to hold.

2.1   List-colouring conjectures

To state these conjectures, we need the concepts of edge-choosability and total choosability. These are natural analogues of (vertex-)choosability, and we now define them. A (proper) *total colouring* of a multigraph H is an assignment of a colour to every vertex and every edge of H in such a way that no two adjacent vertices or adjacent edges have the same colour, and no vertex has the same colour as an edge incident with it. The *total chromatic number* $\chi''(H)$ of H is the smallest integer k such that H has a total colouring using k colours. The *total choosability* or *list total chromatic number* $\text{ch}''(H)$ of H is the smallest integer k such that whenever every vertex and every edge of H is given a list of at least k colours, there exists a total colouring of H in which every vertex and every edge receives a colour from its own list. The *edge chromatic number* (or *chromatic index*) $\chi'(H)$ and the *edge choosability* $\text{ch}'(H)$ are defined similarly in terms of colouring edges alone.

Let $T(H)$ denote the *total graph* of H, which has a vertex corresponding to every vertex and every edge of H, with an edge joining two vertices of $T(H)$ whenever the corresponding elements of H are required to be coloured differently in a total colouring of H. The *line graph* $L(H)$ of H is defined analogously with respect to edge-colourings; it is the subgraph of $T(H)$ induced by the vertices representing edges of H. In view of these definitions it is possible, and it is sometimes useful, to think of edge-colourings and total colourings not so much as new types of colourings, but rather as vertex-colourings of restricted classes of graphs, namely line graphs and total graphs. In particular, $\chi'(H) = \chi(L(H))$, $\chi''(H) = \chi(T(H))$, $\text{ch}'(H) = \text{ch}(L(H))$, and $\text{ch}''(H) = \text{ch}(T(H))$.

The following conjecture seems to have been made independently by Vizing, by Gupta, and by Albertson and Collins; see [9, 25]. It used to be known as *the list-colouring conjecture*, abbreviated *LCC*, but a more specific name now seems appropriate.

The List-Edge-Colouring Conjecture (LECC) *For every multigraph H,* $\text{ch}'(H) = \chi'(H)$.

An equivalent formulation of the LECC is that $\text{ch}(G) = \chi(G)$ for every graph G that is the line graph of a multigraph H. Since every line graph is *claw-free* (that is, it does not have $K_{1,3}$ as an induced subgraph), the following conjecture, due to Gravier and Maffray [20], would imply the LECC.

The List-Colouring Conjecture for Claw-Free Graphs *For every claw-free graph G,* $\text{ch}(G) = \chi(G)$.

The analogous conjecture to the LECC for total colourings was made independently and almost simultaneously by Borodin, Kostochka and Woodall [6], by Juvan, Mohar and Škrekovski [35] and by Hilton and Johnson [31].

The List-Total-Colouring Conjecture (LTCC) *For every multigraph H,* $\text{ch}''(H) = \chi''(H)$.

An equivalent formulation of the LTCC is that $\mathrm{ch}(G) = \chi(G)$ for every graph G that is the total graph of a multigraph H.

The *square* G^2 of a graph G is the graph with the same vertex-set as G in which two vertices are adjacent if their distance apart in G is at most 2. Note that if G is obtained by placing a vertex in the middle of every edge of a multigraph H, then $G^2 = T(H)$. Thus the following conjecture (the LSCC), made by Kostochka and Woodall [36], implies the LTCC; indeed, the LTCC is equivalent to the special case of the LSCC for bipartite graphs in which every vertex in one partite set has degree 2.

The List-Square-Colouring Conjecture (LSCC)   *For every graph G,* $\mathrm{ch}(G^2) = \chi(G^2)$.

Finally, Ohba [46] proved that for every graph G there exists an integer n_0 such that

$$\mathrm{ch}(G + K_n) = \chi(G + K_n) \quad \text{for every } n \geq n_0, \tag{2.1}$$

where $+$ denotes "join"; and he made the following conjecture, which would imply that this is true with $n_0 = \max\{0, |V(G)| - 2\chi(G) - 1\}$.

Ohba's Conjecture [46]   *If* $|V(G)| \leq 2\chi(G) + 1$, *then* $\mathrm{ch}(G) = \chi(G)$.

Before discussing what is known about these conjectures, we shall consider three more conjectures.

2.2   The $(a:b)$-choosability conjectures

Let F and G be (simple) graphs such that $V(G) = \{v_1, \ldots, v_n\}$. We say that F is an *inflation* of G if $V(F)$ can be written as a disjoint union $V(F) = V_1 \cup \ldots \cup V_n$ in such a way that if $x \in V_i$ and $y \in V_j$ then $xy \in E(F)$ if and only if $i = j$ or $v_i v_j \in E(G)$. (So to *inflate* a graph is to replace each vertex by a complete graph.) If $|V_i| = t$ for all i then we write $F = G_{(t)}$ and call it a *uniform inflation* of G. (Another way of looking at $G_{(t)}$ is as the *lexicographic product* or *composition* $G[K_t]$ of G and K_t, also called the *wreath product* $G * K_t$.)

Following Erdős, Rubin and Taylor [15], we say that a graph is $(a:b)$-*choosable* if, whenever each vertex is assigned a list of a colours, we can give each vertex a set of b colours from its list in such a way that adjacent vertices get disjoint sets of colours; so $(a:1)$-*choosable* means the same as a-*choosable*. It is easy to see that G is $(a:b)$-choosable if $G_{(b)}$ is a-choosable, and Kostochka and Woodall [36] conjectured that the converse holds; this is the $(a:b)$-*choosability equivalence conjecture*, below. Erdős, Rubin and Taylor [15] asked whether, for $a, b, t \in \mathbb{N}$, every graph that is $(a:b)$-choosable is necessarily $(ta:tb)$-choosable. The only pair (a, b) for which this is known to be true [60] is $(a, b) = (2, 1)$. Nevertheless it is widely believed to be true for all (a, b). This is the first of the following conjectures, which appeared in this form in [36].

The Weak $(a:b)$-Choosability Conjecture (Weak $(a:b)$-CC). *For all $a, b, t \in \mathbb{N}$, if a graph G is $(a:b)$-choosable, then G is $(ta:tb)$-choosable.*

The Strong $(a:b)$-Choosability Conjecture (Strong $(a:b)$-CC). *For all $a, b, t \in \mathbb{N}$, if a graph G is $(a:b)$-choosable, then $G_{(t)}$ is $(ta:b)$-choosable.*

The $(a:b)$-Choosability Equivalence Conjecture $((a:b)$-CEC). *For all $a, b \in \mathbb{N}$, a graph G is $(a:b)$-choosable if and only if $G_{(b)}$ is a-choosable.*

It is easy to see that the strong $(a:b)$-CC implies the weak $(a:b)$-CC, and if the $(a:b)$-CEC is true then the other two conjectures are equivalent. For certain families of graphs satisfying ch $= \chi$, all three conjectures are true.

Theorem 2.1 ([36]) *Let G be a graph such that $\mathrm{ch}(G_{(t)}) = \chi(G_{(t)})$ for all $t \in \mathbb{N}$. Then all three $(a:b)$-choosability conjectures hold for G.*

The classes of line graphs, claw-free graphs and squares are all closed under uniform inflations. Thus, by Theorem 2.1, the truth of the $(a:b)$-choosability conjectures for these classes of graphs would follow from the truth of the LECC, the list-colouring conjecture for claw-free graphs, and the LSCC, respectively. In contrast, the class of total graphs is not closed under uniform inflations, and so the truth of the LTCC would not apparently imply the truth of the $(a:b)$-choosability conjectures for total graphs. This is why it seemed useful to us to formulate the LSCC, as a stronger version of the LTCC which does have implications for the $(a:b)$-choosability conjectures.

2.3 Theorems proving that ch $= \chi$

Within the last few years, great impetus has been given to the study of choosability by the papers of Alon and Tarsi [2] and Galvin [18], which introduced two new methods of proof, the *Alon–Tarsi method* and the *kernel method*; see Sections 6.2.1 and 6.2.2. However, we start by describing results that can be proved without the use of these methods.

One easy result is that $\mathrm{ch}(G) = \chi(G) = \omega(G)$ for every interval graph G, where $\omega(G)$ is the order of a largest clique in G. I do not recall seeing this result in the literature, but a proof is sketched near the start of Section 6.2.

Both Vizing [61] and Erdős, Rubin and Taylor [15] proved that $\mathrm{ch}(G) = \chi(G) = 2$ if G is a circuit of even length, and (as already remarked) Rubin went further and characterized graphs for which $\mathrm{ch}(G) = \chi(G) = 2$. It is easy to see that $\mathrm{ch}(G) = \chi(G) = 3$ if G is a circuit of odd length, so that $\mathrm{ch}(G) = \chi(G)$ for all circuits.

By a simple application of Hall's theorem (Section 6.1.2), Erdős, Rubin and Taylor [15] proved that the complete k-partite graph $K_{2,2,\dots,2}$ is k-choosable, thereby showing that $\mathrm{ch}(G) = \chi(G)$ if G has a $\chi(G)$-colouring in which every colour class has at most two vertices. This holds, in particular, if G does not contain a set of three independent vertices. Gravier and Maffray [21] extended this by showing that $\text{ch}(G) = \chi(G)$ if $\chi(G) \geq 3$ and G has a $\chi(G)$-colouring in which no colour class has more than three vertices and at most two colour classes have three vertices. They used this to prove that $\text{ch}(G) = \chi(G)$ if G is claw-free and $|V(G)| \leq 2\chi(G) + 2$; this shows that Ohba's conjecture is true (and not sharp) for claw-free graphs. Enomoto, Ohba, Ota and Sakamoto [14] pointed out that the complete k-partite graph $K_{4,2,\ldots,2}$ is not k-choosable if k is even, thereby showing that the upper bound on $|V(G)|$ in Ohba's conjecture is sharp in general. They also proved Ohba's conjecture in the case when at most one colour class has more than two vertices.

Gravier and Maffray [20] proved the list-colouring conjecture for elementary claw-free graphs containing no K_4, where a graph is *elementary* if its edges can be coloured with two colours in such a way that every chordless path of length two has its two edges coloured differently.

Ohba [46] proved that if G is a graph that has a $\chi(G)$-colouring in which the second-largest colour class has α_2 vertices, and if $(\alpha_2 - 1)|V(G)| \leq \alpha_2\chi(G)$, then $\text{ch}(G) = \chi(G)$. From this it is easy to see that (2.1) holds (for every graph G) with $n_0 = \max\{0, (\alpha_2 - 1)|V(G)| - \alpha_2\chi(G)\}$.

Kostochka and Woodall [36] proved that $\text{ch}(G) = \chi(G)$ if G is an inflation of a graph with at most five vertices or is the inflation of the square of a graph with at most seven vertices; this proves all the $(a : b)$-choosability conjectures for such graphs, and it proves the LSCC for every inflation of a graph with at most seven vertices. All of the proofs mentioned in the last four paragraphs use Hall's theorem (Section 6.1.2) to deal with some special cases, but some of them require a great deal of ingenuity in addition.

An easy inductive argument proves the LECC, LTCC and LSCC for any multigraph whose underlying simple graph G_0 is a forest; indeed, if G is such a multigraph then $\text{ch}'(G) = \chi'(G) = \Delta(G)$, $\text{ch}''(G) = \chi''(G) = \Delta(G) + 1$ and $\text{ch}(G^2) = \chi(G^2) = \Delta(G_0) + 1$, where Δ denotes maximum degree. Because $T(K_3) = L(K_4)$ (the octahedron), if G is a multigraph with underlying simple graph K_3 then there is a multigraph H with at most four vertices such that $T(G) = L(H)$, and so the truth of the LECC for H (proved in [47]) implies the truth of the LTCC for G; it follows that the LTCC holds for multigraphs with at most three vertices.

Juvan, Mohar and Thomas [34] proved the LECC for series-parallel graphs, but not for series-parallel multigraphs; specifically, they proved that $\text{ch}'(G) = \chi'(G) = \Delta(G)$ for a series-parallel simple graph G. This proves the LECC also for (simple) outerplanar graphs, since outerplanar graphs are series-parallel. The LTCC is also known to be true for outerplanar graphs [33].

The first of the two new methods of proof mentioned above was the method of Alon and Tarsi [2]; see Section 6.2.1. Almost as soon as this method became available, Fleischner and Stiebitz [16] used it to prove that if G is a 4-regular graph on $3n$ vertices whose edges can be decomposed into a hamiltonian circuit and n pairwise vertex-disjoint triangles, then $\text{ch}(G) = \chi(G) = 3$. They stated only that $\chi(G) = 3$, but their proof shows also that $\mathrm{ch}(G) = 3$.

Ellingham and Goddyn [13] explored the Alon–Tarsi method in greater depth and used it to prove that if G is a 2-connected 3-regular planar graph then $\mathrm{ch}'(G) = \chi'(G) = 3$. More generally, they proved that if G is a d-regular planar multigraph then $\mathrm{ch}'(G) = d$ if and only if $\chi'(G) = d$. This proves all the $(a:b)$-choosability conjectures for line graphs of edge-d-colourable d-regular planar multigraphs.

Häggkvist and Janssen [26] used the Alon–Tarsi method to prove, among other results, that $\mathrm{ch}'(K_n) \le n$ for every n, which implies that $\mathrm{ch}'(K_n) = \chi'(K_n)$ when n is odd.

A very recent result of Prowse and Woodall [49] is that $\mathrm{ch}(G) = \chi(G)$, and all the $(a:b)$-choosability conjectures hold, if G is a power of a circuit, $G = C_n^p$; that is, the vertices of G are $v_1, \ldots, v_n$, and each vertex v_i is adjacent to $v_{i-p}, \ldots, v_{i-1}, v_{i+1}, \ldots, v_{i+p}$, where subscripts are taken module n.

The other new method of proof mentioned above is the kernel method; see Section 6.2.2. This was developed by Galvin [18], who used it to prove that $\mathrm{ch}'(G) = \chi'(G)$ for every bipartite multigraph G. This proves the LECC for bipartite multigraphs, and it also proves all the $(a:b)$-choosability conjectures for their line graphs. It also implies that $\mathrm{ch}''(G) \le \Delta(G) + 2$, which proves the LTCC for any bipartite multigraph G for which $\chi''(G) > \Delta(G) + 1$.

Quite apart from the new method that it contained, another interesting feature of Galvin's proof is that he proved that $\mathrm{ch}'(G) = \chi'(G)$ directly, without using (or even mentioning) the well-known fact that $\chi'(G) = \Delta(G)$ when G is bipartite. Most other theorems of this type have been proved by finding some sort of formula for $\chi'(G)$, and then proving that G is list-colourable from lists of this size. However, Plantholt and Tipnis [48] used a similar approach to Galvin's to prove the LECC for every multigraph whose underlying simple graph is of the form "bipartite plus one edge", and also for every multigraph containing a vertex v with degree at most 6 such that $G - v$ is bipartite.

Alternative presentations of Galvin's proof have been given by Slivnik [57] and by Borodin, Kostochka and Woodall [6]. These last authors extended the method to prove that if each edge uw of G is given a list of at least $\max\{d(u), d(w)\}$ colours, then the edges of G can be coloured from these lists. They further used this to prove the sharp result that $\mathrm{ch}'(G) \le \lfloor \frac{3}{2}\Delta(G) \rfloor$ for every multigraph G, which is the choosability analogue of the classic theorem of Shannon [53] that $\chi'(G) \le \lfloor \frac{3}{2}\Delta(G) \rfloor$ for every multigraph G. (The conjecture that $\mathrm{ch}''(G) \le \lfloor \frac{3}{2}\Delta(G) \rfloor$ if $\Delta(G) \ge 4$ is stated in [6] but remains unproved.) More relevantly, in the context of the present conjectures, various results are proved in [6] about the choosability of a graph G embedded in a surface of nonnegative characteristic, of which the simplest to state is that if $\Delta(G) \ge 12$ then $\mathrm{ch}'(G) = \chi'(G) = \Delta(G)$ and $\mathrm{ch}''(G) = \chi''(G) = \Delta(G) + 1$.

A multigraph is called *line-perfect* if its line graph is perfect. By a result of Maffray [44], a multigraph is line-perfect if and only if every block of it is bipartite or has underlying simple graph of the form K_4 or $K_{1,1,p}$. Peterson and Woodall [47] used the results of [18] and [6] to prove the LECC for line-perfect multigraphs. Woodall [66] extended this to multigraphs in which every block is line-perfect or a multicircuit, where a *multicircuit* is a multigraph whose underlying simple graph is a circuit (a connected 2-regular graph). It follows that the $(a : b)$-choosability conjectures all hold for the line graphs of such multigraphs.

Kostochka and Woodall [37, 38] proved the LTCC for multicircuits. It may give some indication of the different levels of difficulty of the two conjectures to note that the proof of the LECC for multicircuits takes about a page, while the proof of the LTCC for multicircuits takes two 20-page papers (one based on the kernel method and one based on the Alon–Tarsi method). Another difference is that while the truth of the LECC for multicircuits implies the truth of all the $(a : b)$-choosability conjectures for their line graphs, we have signally failed to prove the truth of the $(a : b)$-choosability conjectures for total graphs of multicircuits in general, although we did prove it for a fairly wide class of multicircuits of even order in [36].

3   Subcontractions and defective choosability; analogues of Hadwiger's conjecture

A graph H is a *subcontraction* or *minor* of a graph G if one can form an isomorphic copy of H from G by contracting edges and deleting edges and vertices. A graph G is called H-*minor-free* if it does not have H as a minor. Hadwiger's conjecture [23, 24] is that every K_{r+1}-minor-free graph is r-colourable; this is now known to be true for $r \leq 5$ [52].

The analogous statement cannot hold for choosability, since, as we have already remarked, there are planar (hence, K_5-minor-free) graphs that are not 4-choosable. However, there is some evidence that complete-bipartite graphs play a similar role in choosability to the role played by complete graphs in the theory of ordinary colourings.

For example, a famous theorem of Hajós [27] says that every graph that is not r-colourable can be obtained from K_{r+1} by a sequence of three types of operations. The analogous statement for choosability is false, the simplest counterexamples being the complete-bipartite graphs. However, Gravier [19] has proved that every non-r-choosable graph can be obtained from non-r-choosable complete-bipartite graphs by a sequence of essentially the same three types of operations (with one minor and natural change, namely, that nonadjacent vertices may only be identified if they have the same list in some r-list-assignment L for which the graph is not L-choosable).

Tables 1 and 2 in Section 4.1 also suggest (rather superficially) that $K_{r,s}$-minor-free graphs may behave better than K_r-minor-free graphs with respect to defective choosability, a concept that we now define.

In a vertex-coloured graph, the *defect* $\mathrm{def}(v)$ of a vertex v is the number of vertices adjacent to v that have the same colour as v; so a colouring is proper if every vertex has defect 0. A graph G is $(k, d)^*$-*colourable* if its vertices can be coloured with k colours in such a way that no vertex has defect greater than d. If L is a list-assignment to G, then an $(L, d)^*$-*colouring* is an L-colouring in which no vertex has defect greater than d, and G is $(L, d)^*$-*choosable* if it has an $(L, d)^*$-colouring. Finally, G is $(k, d)^*$-*choosable* if it is $(L, d)^*$-choosable whenever L is a k-list-assignment. Note that $(k, 0)^*$-*colourable* means the same as (properly) k-*colourable*, and $(k, 0)^*$-*choosable* means the same as k-*choosable*.

Section 3.1 contains some conjectures about the choosability of complete-bipartite graphs that are in some sense analogous to Hadwiger's conjecture, and Section 3.2 contains some more general conjectures based on an earlier conjecture of Woodall [65]. In these sections the term "conjecture" is used in the weak sense, to denote a result that may well be true rather than one that I strongly believe to be true.

3.1 $K_{r,s}$-minor-free graphs

In the following conjectures, the D conjectures are the "defect" versions of the C conjectures; r and s are arbitrary positive integers.

Conjecture $C(r, s, \chi)$ *Every $K_{r,s}$-minor-free graph is $(r + s - 1)$-colourable.*

Conjecture $C(r, s, \mathrm{ch})$ *Every $K_{r,s}$-minor-free graph is $(r + s - 1)$-choosable.*

Conjecture $D(r, s, \chi)$ *Every $K_{r,s}$-minor-free graph is $(r + s - 1 - d, d)^*$-colourable whenever $0 \le d \le s - 1$.*

Conjecture $D(r, s, \mathrm{ch})$ *Every $K_{r,s}$-minor-free graph is $(r + s - 1 - d, d)^*$-choosable whenever $0 \le d \le s - 1$.*

Since every $K_{r,s}$-minor-free graph is K_{r+s}-minor-free, Conjecture $C(r, s, \chi)$ would follow from Hadwiger's conjecture. It can be thought of as a variant of Hadwiger's conjecture that is sufficiently weak that the choosability analogue, Conjecture $C(r, s, \mathrm{ch})$, might be true. The complete graph K_{r+s-1} shows that both these conjectures are best possible.

In the rest of this Section we shall prove these conjectures for $r \le 2$. (The only other case where much is known is for $K_{3,3}$-minor-free graphs, which are discussed in Section 4.2.) We shall need the following two lemmas. We say that a graph G is *greedily* $(k, d)^*$-*choosable* if, for every k-list-assignment L to G and every ordering $v_1, v_2, \ldots, v_n$ of the vertices of G and every j $(1 \le j \le n - 1)$, it is true that every $(L, d)^*$-colouring of the induced subgraph $\langle v_1, \ldots, v_j \rangle$ can be extended to an $(L, d)^*$-colouring of the whole of G.

Lemma 3.1 *Let G be a $K_{1,s}$-minor-free graph.*
  (a) *G is $(k, d)^*$-choosable whenever $k(d + 1) \geq s$.*
  (b) *G is greedily $(k, d)^*$-choosable whenever $kd \geq s$.*
  (c) *G is greedily $(s, 0)^*$-choosable and greedily $(1, s - 1)^*$-choosable.*

Proof Let L be a k-list-assignment to G.

(a) Colour the vertices of G from their lists in such a way as to minimize the number of bad edges (that is, edges joining two vertices with the same colour). Suppose there is a vertex v with defect $\mathrm{def}(v) \geq d + 1$. Since v has degree at most $s - 1 < k(d + 1)$ by hypothesis, there must be a colour c in its list that is being used on at most d neighbours of v. Changing the colour of v to c would reduce the number of bad edges, and this contradiction completes the proof.

(b) Given an $(L, d)^*$-colouring of the induced subgraph $\langle v_1, \ldots, v_j \rangle$ and a colour $c \in L(v_{j+1})$, there are only two possible reasons why we may not be able to colour v_{j+1} with c: either v_{j+1} has at least $d + 1$ neighbours already coloured with c, or v_{j+1} has a neighbour v_i $(1 \leq i \leq j)$ such that v_i already has d neighbours coloured with c. In view of this, if we cannot colour v_{j+1} with any of the k colours in $L(v_{j+1})$, then it is easy to see that G has $K_{1,kd}$ as a minor, which is impossible if $kd \geq s$. It follows that if $kd \geq s$ then we can colour the vertices $v_{j+1}, v_{j+2}, \ldots, v_n$ in turn to obtain an $(L, d)^*$-colouring of G.

(c) Since a $K_{1,s}$-minor-free graph has maximum degree at most $s - 1$, it is clear that it is greedily $(s, 0)^*$-choosable (that is, properly s-choosable), since every uncoloured vertex can be given a colour from its own list that is different from the colours of all its neighbours. It is also greedily $(1, s - 1)^*$-choosable, since no vertex can have more than $s - 1$ like-coloured neighbours, regardless of what colour it is given.   □

To prove Conjecture $D(2, s, \mathrm{ch})$ we shall use the following lemma. Except for the $(s, 1)^*$-choosability of G, it seems just as easy to prove the result for $(K_1 + K_{1,s})$-minor-free graphs as for $K_{2,s}$-minor-free graphs; this is sufficient, since $K_1 + K_{1,s} = K_2 + \bar{K}_s$, and so every $K_{2,s}$-minor-free graph is $(K_1 + K_{1,s})$-minor-free.

Lemma 3.2 *Let G be a $(K_1 + K_{1,s})$-minor-free graph.*
  (a) *G is $(k, d)^*$-choosable whenever $(k - 1)d \geq s$.*
  (b) *G is $(s + 1, 0)^*$-choosable and $(2, s - 1)^*$-choosable.*
  (c) *G is $(s, 1)^*$-choosable if G is $K_{2,s}$-minor-free and $s \geq 2$.*

Proof (a) There is no loss of generality in supposing that G is connected. Let L be a k-list-assignment to G and let z be an arbitrary but fixed vertex in G. By Lemma 3.1(b), since $(k - 1)d \geq s$, every $K_{1,s}$-minor-free induced subgraph of G is greedily $(k - 1, d)^*$-choosable. We describe an algorithm for constructing an $(L, d)^*$-colouring of G, using what is referred to in Section

6.1.3 as the *boundary method*. We write $N_G(z)$ for the set of neighbours of z in G. Give z an arbitrary colour c_z from its list, and set $G_0 := G$ and

$$L_0(v) := \begin{cases} L(v) \setminus \{c_z\} & \text{if } v \in N_G(z), \\ L(v) & \text{if } v \in V(G - z) \setminus N_G(z). \end{cases}$$

Noting that the induced subgraph $\langle N_G(z) \rangle$ is $K_{1,s}$-minor-free and is therefore (greedily) $(k-1, d)^*$-choosable, form an $(L_0, d)^*$-colouring of $\langle N_G(z) \rangle$.

For $i = 0, 1, \ldots$, as long as there remains a vertex not adjacent to z in G_i, proceed as follows. Choose a colour c_i that has been used to colour at least one vertex of $N_{G_i}(z)$. Form G_{i+1} from G_i by contracting all edges zw such that w has been coloured with c_i. Form L_{i+1} from L_i by removing colour c_i from the list of every vertex in $N_{G_{i+1}}(z)$ that contains it; every such vertex either is already coloured with a colour different from c_i, or else still has at least $k-1$ colours in its list. Extend the existing partial colouring of $N_{G_{i+1}}(z)$ to an $(L_{i+1}, d)^*$-colouring of the $K_{1,s}$-minor-free induced subgraph $\langle N_{G_{i+1}}(z) \rangle$; this is possible by Lemma 3.1(b), since $(k-1)d \geq s$. Continue until all vertices of G are coloured. The result is an $(L, d)^*$-colouring of G, as required.

(b) This follows by the same argument as (a), but using Lemma 3.1(c) instead of Lemma 3.1(b). An alternative proof that G is $(2, s-1)^*$-choosable is given in Corollary 3.6.

(c) This follows by the same method, but more care is needed. At the initial stage, the induced subgraph $\langle N_G(z) \rangle$ is $K_{1,s}$-minor-free and is therefore $(s-1, 1)^*$-choosable (although not necessarily greedily so) by Lemma 3.1(a); thus we can form the required $(L_0, d)^*$-colouring of $\langle N_G(z) \rangle$. At each subsequent stage, after the formation of G_{i+1}, let C be a typical component of the subgraph of $\langle N_{G_{i+1}}(z) \rangle$ induced by the uncoloured vertices. If C has at least two vertices, let v, w be two adjacent vertices of C. Colour the vertices of C sequentially in such a way that the subgraph induced by the uncoloured vertices of C remains connected, and v, w are coloured last. Then each vertex can be coloured differently from all its neighbours, except that if v and w both have degree $s-1$ in $\langle N_{G_{i+1}}(z) \rangle$ then v may have to be given the same colour as w; but this is acceptable. Thus the only problem arises if some component C has just one vertex v, and v has degree $s-1$ in $\langle N_{G_{i+1}}(z) \rangle$. But then, in G_i, v has s neighbours in $N_{G_i}(z)$, which means that G has $K_{2,s}$ as a minor, contrary to hypothesis. Thus no problem can arise, and the algorithm can be followed until all vertices of G are coloured. □

It is not clear whether the conclusion of Lemma 3.2(c) remains true without the additional hypothesis that G is $K_{2,s}$-minor-free.

We shall now complete the proof of Conjecture $D(r, s, \text{ch})$, and hence of all the other conjectures mentioned in this Section, when $r \leq 2$.

Theorem 3.3 (a) *A $K_{1,s}$-minor-free graph is $(s-d, d)^*$-choosable whenever* $0 \leq d \leq s-1$.

(b)   *A* $(K_2 + \bar{K}_s)$*-minor-free graph is* $(s + 1 - d, d)^*$*-choosable whenever* $0 \le d \le s - 1$*, except possibly when* $d = 1$*.*

(c)   *A* $K_{2,s}$*-minor-free graph is* $(s + 1 - d, d)^*$*-choosable whenever* $0 \le d \le s - 1$*.*

Proof   The truth of (a) follows from Lemma 3.1(a), since $(s - d)(d + 1) \ge s$ whenever $0 \le d \le s - 1$. To prove (b), we use Lemma 3.2 together with the fact that $K_2 + \bar{K}_s = K_1 + K_{1,s}$. For $2 \le d \le s - 2$ the result follows from Lemma 3.2(a), since $(s - d)d \ge s$ in this range. Note that if $2 \le d \le s - 2$ then $s \ge 4$. For $d = 0$ or $s - 1$ the result follows from Lemma 3.2(b). Finally, the truth of (c) follows from Lemma 3.2(c) if $d = 1$, and for other values of d it follows from (b), since a $K_{2,s}$-minor-free graph is necessarily $(K_2 + \bar{K}_s)$-minor-free.

□

3.2   Some more general conjectures

Conjecture $A(r, H, \chi)$ below was made in [65], where it was proved for $r = 1$ and $r = 2$. In [65] it was stated in a weaker form, with "contractible to H" replaced by "isomorphic to H"; however, the proofs of this weaker result that were given for $r = 1$ and $r = 2$ actually prove the stronger result as well.

It is natural to try to extend this conjecture to choosability. In the following conjectures, + denotes "join", r is a positive integer, and H is a connected graph with at least one edge.

Conjecture $A(r, H, \chi)$ *Let G be a $(K_r + H)$-minor-free graph. Then G can be $(r + 1)$-coloured in such a way that no subgraph of G contractible to H has all its vertices the same colour.*

Conjecture $A(r, H, \mathrm{ch})$ *Let G be a $(K_r + H)$-minor-free graph and let L be an $(r + 1)$-list-assignment to G, where $r \le 2$. Then G has an L-colouring in which no subgraph of G contractible to H has all its vertices the same colour.*

Conjecture $B(r, H, \chi)$ *The same as Conjecture $A(r, H, \chi)$, but with K_r replaced by $K_{1,r-1}$.*

Conjecture $B(r, H, \mathrm{ch})$ *The same as Conjecture $A(r, H, \mathrm{ch})$, but with K_r replaced by $K_{1,r-1}$, and without the restriction to $r \le 2$.*

Conjecture $E(r, H, \chi)$ *The same as Conjecture $A(r, H, \chi)$, but with K_r replaced by $\bar{K}_r$.*

Conjecture $E(r, H, \mathrm{ch})$ *The same as Conjecture $A(r, H, \mathrm{ch})$, but with K_r replaced by $\bar{K}_r$, and without the restriction to $r \le 2$.*

Conjecture $A(r, K_2, \chi)$ is Hadwiger's conjecture for K_{r+2}-minor-free graphs, and so Conjecture $A(3, K_2, \chi)$ implies the Four-Colour Theorem. The choosability analogue of this, Conjecture $A(3, K_2, \mathrm{ch})$, would therefore be false since not every planar graph is 4-choosable: hence the restriction to $r \leq 2$ in Conjecture $A(r, H, \mathrm{ch})$. The B (for "bipartite") and E (for "edgeless") conjectures are weaker versions of the A conjectures in which the choosability version could conceivably be true for larger values of r. Of these conjectures, the A conjectures are the strongest and the E conjectures the weakest, with the B conjectures lying in between. The B conjectures seem natural since the graph $K_{1,r-1}$ is both complete-bipartite and minimal connected; but the analogous conjectures for any other family of r-vertex graphs would also lie between the A and the E conjectures.

Note that Conjectures $E(r, K_2, \chi)$ and $E(r, K_2, \mathrm{ch})$ are both true by Theorem 3.3(b); thus in the case $H = K_2$, in which Conjecture $A(r, H, \chi)$ reduces to Hadwiger's conjecture, the E conjectures are true (and quite easy to prove). This gives some indication of the difference in strength between the A and E conjectures.

This short section ends with a proof of Conjecture $A(1, H, \mathrm{ch})$, using the boundary method of Section 6.1.3. As before, we write $N_G(z)$ for the set of neighbours of z in G.

Lemma 3.4 *Let H be a connected graph with at least one edge, let G be a $(K_1 + H)$-minor-free graph, and let z be a vertex of G. Let L be a list-assignment to $G - z$ such that $|L(v)| \geq 1$ if $v \in N_G(z)$ and $|L(v)| \geq 2$ otherwise. Then $G - z$ has an L-colouring in which no subgraph of $G - z$ contractible to H has all its vertices the same colour.*

Proof We prove the result by induction on $|V(G)|$. Let c be a colour that is present in the list of at least one vertex in $N_G(z)$, and let $Z_c := \{v \in N_G(z) : c \in L(v)\}$. Form G' from G by contracting all edges zv such that $v \in Z_c$, and remove colour c from the list of every vertex in $N_{G'}(z)$ that contains c. By the induction hypothesis, the vertices of $G' - z$ can be coloured from their lists in such a way that no subgraph of $G' - z$ contractible to H has all its vertices the same colour. In $G - z$, give colour c to all vertices in Z_c, and give every other vertex the same colour that it has in $G' - z$. Since G is $(K_1 + H)$-minor-free, the subgraph induced by Z_c does not have H as a minor, and so no subgraph of $G - z$ contractible to H has all its vertices the same colour. □

Theorem 3.5 *Conjecture $A(1, H, \mathrm{ch})$ is true.*

Proof Let G be a $(K_1 + H)$-minor-free graph and let L be a 2-list-assignment to G. Choose a vertex $z \in V(G)$ and a colour $c_z \in L(z)$, and remove c_z from the list of every neighbour of z that contains c_z. By Lemma 3.4, $G - z$ can be coloured from these lists in such a way that no subgraph contractible to H has all its vertices the same colour. Colour z with colour c_z to obtain such an L-colouring of G. □

This gives an alternative proof of part of Lemma 3.2(b).

Corollary 3.6 *Every* $(K_1 + K_{1,s})$-*minor-free graph* G *is* $(2, s-1)^*$-*choosable.*

Proof By Theorem 3.5, if L is a 2-list-assignment to G then G has an L-colouring in which no subgraph isomorphic to $K_{1,s}$ has all its vertices the same colour; that is, G has an $(L, s-1)^*$-colouring.   □

4   Defective choosability of planar and related graphs

In this section we shall consider the defective colourability and defective choosability of six classes of graphs: planar, $K_{3,3}$-minor-free, K_5-minor-free, outerplanar, $K_{2,3}$-minor-free, and K_4-minor-free. The first class is the intersection of the second and third, and the fourth is the intersection of the last two. We summarize the results in Section 4.1, and prove the results for $K_{3,3}$-minor-free results in Section 4.2.

$K_{2,3}$-minor-free	Outerplanar	K_4-minor-free
$(4,0)$		
$(3,1)$	$(3,0)$	$(3,0)$
$(2,2)$	$(2,2)$	
Not $(3,0)$: K_4		
Not $(2,1)$: see right	Not $(2,1)$: $K_1 + 2K_{1,2}$	Not $(2,d)$:
Not $(1,d)$: see right	Not $(1,d)$: $K_{1,d+1}$	$K_1 + (d+1)K_{1,d+1}$

$K_{3,3}$-minor-free	Planar	K_5-minor-free
$(5,0)$ ([52])		
$(4,1)$	$(4,0)$ (4CT)	$(4,0)$ (4CT + WET)
$(3,2)$ ([65])	$(3,2)$ ([10], [65])	
Not $(4,0)$: K_5	Not $(3,1)$:	Not $(3,d)$: $K_1 + (d+1) \cdot$
Not $(3,1)$: see right	$K_1 + 2[K_1 + 2K_{1,2}]$	$[K_1 + (d+1)K_{1,d+1}]$
Not $(2,d)$: see right	Not $(2,d)$:	
	$K_1 + (d+1)K_{1,d+1}$	

Table 1: Values (k, d) for which every graph in the named class is $(k, d)^*$-colourable

4.1   Summary of results

If a graph G is $(k, d)^*$-colourable, then clearly G is $(k', d')^*$-colourable whenever $k' \geq k$ and $d' \geq d$; hence, in describing the pairs (k, d) for which G is

$(k, d)^*$-colourable, it suffices to list the pairs that are minimal in each coordinate. A similar remark applies when describing the pairs (k, d) for which G is $(k, d)^*$-choosable.

The pairs (k, d) such that every graph in one of our six classes is $(k, d)^*$-colourable are known and are tabulated (using the convention of the previous paragraph) in Table 1. Most of them were determined in [10] or [65], or are easy to see, or follow from deep results such as the Four-Colour Theorem (4CT) [3, 4, 51] with or without Wagner's Equivalence Theorem (WET) [64]. The $(4, 1)^*$-colourability of $K_{3,3}$-minor-free graphs is proved in Theorem 4.4 below. Examples are given in Table 1 to show that the figures there are best possible.

We can deduce from Table 1 that, for example, outerplanar graphs are $(2, 2)^*$-colourable *because they are $K_{2,3}$-minor-free*, but are $(3, 0)^*$-colourable (that is, properly 3-colourable) *because they are K_4-minor-free*. In a similar way, planar graphs are $(3, 2)^*$-colourable *because they are $K_{3,3}$-minor-free*, but are (properly) 4-colourable *because they are K_5-minor-free*.

$K_{2,3}$-minor-free	Outerplanar	K_4-minor-free
$(4, 0)$		
$(3, 1)$	$(3, 0)$	$(3, 0)$ (2-degenerate [11])
$(2, 2)$	$(2, 2)$ ([55], [12])	

$K_{3,3}$-minor-free	Planar	K_5-minor-free
$(5, 0)$	$(5, 0)$ ([58])	$(5, 0)$ ([54])
$(4, 1)$?	$(4, 1)$?	$(4, 1)$?
$(3, 2)$	$(3, 2)$ ([55], [12])	

Table 2: Values (k, d) for which every graph in the named class is $(k, d)^*$-choosable

The pairs (k, d) such that every graph in one of our six classes is $(k, d)^*$-choosable are now also known, except for uncertainty over $(4, 1)^*$-choosability in three cases. The results are tabulated in Table 2. For three of the classes, and possibly a fourth, they are the same as in Table 1; however, some differences from Table 1 are inevitable, since we have seen that not every planar graph is 4-choosable, and it follows that not every K_5-minor-free graph is 4-choosable.

Thomassen [58] proved that every planar graph is 5-choosable, and this result was extended to K_5-minor-free graphs by Škrekovski [54]. Škrekovski [55] proved that every planar graph is $(3, 2)^*$-choosable, and Eaton and Hull [12] gave a simpler version of this proof. We shall prove in Section 4.2 that both 5-choosability and $(3, 2)^*$-choosability extend to $K_{3,3}$-minor-free graphs.

It is not difficult to see that if $K_1 + k(d + 1)G$ is $(k, d)^*$-choosable (where $+$ denotes "join"), then G is $(k - 1, d)^*$-choosable. Also, if G is $K_{2,3}$-minor-

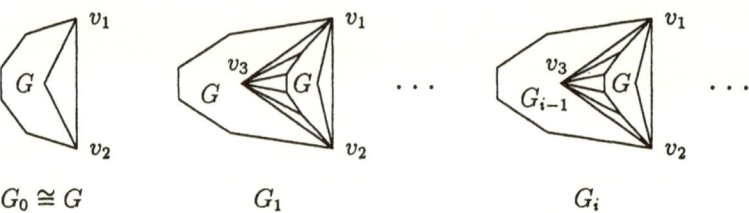

Figure 1: Construction of the graphs G_i

free, outerplanar or K_4-minor-free then $K_1 + rG$ is $K_{3,3}$-minor-free, planar or K_5-minor-free, respectively, for every positive integer r; thus some of the results for the first-mentioned three classes of graphs follow from those for the last-mentioned three classes. The results for $K_{2,3}$-minor-free graphs also follow from Theorem 3.3(c) and Corollary 3.6 above, as well as being proved by quite different methods in [67].

4.2 $K_{3,3}$-minor-free graphs

To extend the known results from planar to $K_{3,3}$-minor-free graphs we shall use a result of Wagner [64], which can be worded in the following form.

Lemma 4.1 ([64]) *If G is a 2-connected $K_{3,3}$-minor-free graph, then either G is planar, or $G \cong K_5$, or G has a cutset consisting of two adjacent vertices.*

For brevity, we say that a colouring λ is $\{v_1, v_2\}$-*proper* if, for $i \in \{1, 2\}$, v_i is not adjacent to any vertex $v \notin \{v_1, v_2\}$ such that $\lambda(v) = \lambda(v_i)$; note that v_1 and v_2 may nevertheless be adjacent to each other and have the same colour. A *near-triangulation* is a graph embedded in the plane in such a way that every face-boundary is a triangle except possibly for the outside face-boundary, which is a circuit of length at least 3.

Lemma 4.2 *Let G be a planar graph or K_5, let v_1, v_2 be two adjacent vertices of G, and let $(k, d) = (5, 0)$ or $(3, 2)$. Let L be a list-assignment to G, where $|L(v)| = 1$ if $v = v_1$ or v_2 and $|L(v)| \geq k$ otherwise, and assume that $L(v_1) \neq L(v_2)$ if $d = 0$. Then G has a $\{v_1, v_2\}$-proper $(L, d)^*$-colouring.*

Proof This is easy to see if $G \cong K_5$; so suppose G is planar. It suffices to prove the result when G is a near-triangulation and v_1 and v_2 are adjacent vertices in its bounding circuit. For $(k, d) = (5, 0)$ the result follows directly from the theorem of Thomassen [58]. For $(k, d) = (3, 2)$, Škrekovski [55] proved that for such G and L there is an $(L, d)^*$-colouring, but in order to deduce that there is a $\{v_1, v_2\}$-*proper* such colouring we need the following construction.

Let $G_0 := G$, and for $i \geq 1$ let G_i be formed by inserting a copy of G into the inside face $v_1 v_2 v_3$ of G_{i-1} that is incident with the edge $v_1 v_2$, and then retriangulating what is left of this face from v_3, as shown in Figure 1. Let L_i be the list-assignment to G_i in which each vertex is given the same list $L(v)$ as its corresponding vertex v in G. Then one can deduce from the theorem of Škrekovski [55] or Eaton and Hull [12] that every G_i has an $(L_i, 2)^*$-colouring. However, this implies that at least one of the five copies of G in G_4 has a $\{v_1, v_2\}$-proper $(L, 2)^*$-colouring. This completes the proof of Lemma 4.2. $\square$

Theorem 4.3 *Let G be a $K_{3,3}$-minor-free graph and let $(k, d) = (5, 0)$ or $(3, 2)$. Then G is $(k, d)^*$-choosable.*

Proof Using induction on $|V(G)|$, we prove the stronger result in which two adjacent vertices v_1, v_2 are precoloured, and we wish to find a $\{v_1, v_2\}$-proper $(L, d)^*$-colouring, exactly as in Lemma 4.2. Let L be a list-assignment to G that satisfies the hypotheses of Lemma 4.2. We may clearly suppose that G is connected. If G is planar or K_5 then the result follows from Lemma 4.2. So we may assume by Lemma 4.1 that G has a minimal cutset X comprising either one or two adjacent vertices, say $G = G_1 \cup G_2$ where $V(G_1 \cap G_2) = X$, $|V(G_1)| > |X|$, $|V(G_2)| > |X|$, and $v_1, v_2 \in G_1$. By the induction hypothesis, G_1 has a $\{v_1, v_2\}$-proper $(L, d)^*$-colouring λ. We must prove that we can extend λ to a $\{v_1, v_2\}$-proper $(L, d)^*$-colouring of G.

If $|X| = 1$, say $X = \{x\}$, choose a vertex $y \in N_{G_2}(x)$, choose a colour $\lambda(y) \in L(y) \setminus \{\lambda(x)\}$, and let $X' := \{x, y\}$; otherwise let $X' := X$. Let $L'(v) := \{\lambda(v)\}$ if $v \in X'$ and $L'(v) := L(v)$ if $v \in V(G_2) \setminus X'$. By the induction hypothesis again, G_2 has an X'-proper $(L', d)^*$-colouring λ'. If we now define $\lambda(v) := \lambda'(v)$ for each $v \in V(G_2) \setminus V(G_1)$, then λ is a $\{v_1, v_2\}$-proper $(L, d)^*$-colouring of G, as required. $\square$

Theorem 4.4 *Every $K_{3,3}$-minor-free graph is $(4, 1)^*$-colourable.*

Proof Let G be a $K_{3,3}$-minor-free graph in which two adjacent vertices v_1, v_2 are precoloured (not necessarily with different colours) and every other vertex v has the *same* list $L(v)$ of 4 colours, which includes the colour(s) of v_1 and v_2. If G is planar then it follows from the Four-Colour Theorem that G has a $\{v_1, v_2\}$-proper $(L, 1)^*$-colouring, and the same obviously holds if $G \cong K_5$. It now follows by induction on $|V(G)|$, using exactly the same argument as in Theorem 4.3, that the same conclusion holds for every $K_{3,3}$-minor-free graph G, so that every such graph is $(4, 1)^*$-colourable. $\square$

5 Additional constraints: planar and related graphs

A $(k, \cap \leq l)$-*list-assignment* is a k-list-assignment such that $|L(u) \cap L(v)| \leq l$ for each edge uv. A $(k, \cup \geq l)$-*list-assignment* is a k-list-assignment such that

$K_{2,3}$-minor-free	Outerplanar	K_4-minor-free
$((4,\cap\leq 4),0)$		
$((3,\cap\leq 3),1)$	$((3,\cap\leq 3),0)$	$((3,\cap\leq 3),0)$
$((3,\cap\leq 2),0)$		
$((2,\cap\leq 2),2)$	$((2,\cap\leq 2),2)$	
$((2,\cap\leq 1),1)$	$((2,\cap\leq 1),1)$	
$((2,\cap\leq 0),0)$	$((2,\cap\leq 0),0)$	$((2,\cap\leq 0),0)$
$((1,\cap\leq 0),0)$	$((1,\cap\leq 0),0)$	$((1,\cap\leq 0),0)$
$((4,\cup\geq 4),0)$		
$((3,\cup\geq 3),1)$	$((3,\cup\geq 3),0)$	$((3,\cup\geq 3),0)$
$((3,\cup\geq 4),0)$		
$((2,\cup\geq 2),2)$	$((2,\cup\geq 2),2)$	
$((2,\cup\geq 4),1)$	$((2,\cup\geq 4),1)$	
$((2,\cup\geq 5),0)$	$((2,\cup\geq 5),0)$	

Table 3: Values for which every graph in the named class is $((k,\cap\leq l),d)^*$-choosable or $((k,\cup\geq l),d)^*$-choosable

$|L(u)\cup L(v)|\geq l$ for each edge uv. We say that a graph G is $((k,\cap\leq l),d)^*$-*choosable* or $((k,\cup\geq l),d)^*$-*choosable* if G is $(L,d)^*$-choosable whenever L is a $(k,\cap\leq l)$-list-assignment or a $(k,\cup\geq l)$-list-assignment respectively. Both $((k,\cap\leq k),d)^*$-*choosable* and $((k,\cup\geq k),d)^*$-*choosable* have the same meaning as $(k,d)^*$-*choosable*, since the intersection and union constraints here have no effect; and every graph is $((k,\cap\leq 0),d)^*$-choosable, for every positive integer k and defect $d\geq 0$: just choose an arbitrary colour for each vertex from its own list.

Note that a $(k,\cap\leq k-t)$-list-assignment is necessarily a $(k,\cup\geq k+t)$-list-assignment, but not *vice versa* (unless every list has *exactly* k colours). Replacing the implicit condition $|L(v)|\geq k$ by $|L(v)|=k$ in the definition of $((k,\cup\geq l),d)^*$-choosability would convert it into an alternative definition of $((k,\cap\leq 2k-l),d)^*$-choosability, but the same change in the definition of $((k,\cap\leq l),d)^*$-choosability would make no difference to it.

It is easy to see that if a graph is $((k,\cup\geq l),d)^*$-choosable then it is $((k',\cup\geq l'),d')^*$-choosable whenever $k'\geq k$, $l'\geq l$ and $d'\geq d$. Hence in describing the triples (k,l,d) for which G is $((k,\cup\geq l),d)^*$-choosable, it suffices to list the triples that are minimal in each coordinate. A similar remark applies to $((k,\cap\leq l),d)^*$-choosability, except that here l should be maximized while k and d are minimized.

All possible choosability results of these types for the classes of outerplanar graphs, $K_{2,3}$-minor-free graphs and K_4-minor-free graphs are now known, and are tabulated (using the convention of the previous paragraph) in Table 3. The proofs and examples required to justify Table 3 are given in [67].

Much less is known about planar graphs (or $K_{3,3}$-minor-free or K_5-minor-free graphs; but once the problems are solved for planar graphs, it should not be too difficult to extend the solutions to these other two classes). For any given natural numbers l and d, the complete-bipartite graph $K_{2,r^2(2d+1)}$, where $r = \max\{2, l-1\}$, is a planar graph that is neither $((2, \cup \geq l), d)^*$-choosable nor $((2, \cap \leq 1), d)^*$-choosable, as explained in [67]. (This example is included in [67] because this graph is also K_4-minor-free.) Since planar graphs are 5-choosable, the only numbers of colours that it remains to consider are $k = 3$ and 4. Since all planar graphs are $(3, 2)^*$-choosable, the only types of colouring that it remains to consider are proper colourings and colourings with defect 1. There are thus eight problems remaining to be solved, which can be parametrized by three binary parameters: the minimum number of colours in each list (3 or 4), the type of constraint ($\cap$ or $\cup$), and the defect permitted (0 or 1). Here is what is known about these problems.

Theorem 5.1 (a) *The maximum number l such that every planar graph is:*

$((4, \cap \leq l), 0)^*$-*choosable*   *belongs to* $\{1, 2\}$,
$((4, \cap \leq l), 1)^*$-*choosable*   *belongs to* $\{2, 3, 4\}$,
$((3, \cap \leq l), 0)^*$-*choosable*   *belongs to* $\{0, 1\}$,
$((3, \cap \leq l), 1)^*$-*choosable*   *belongs to* $\{0, 1\}$.

(b) *The minimum number l such that every planar graph is:*

$((4, \cup \geq l), 0)^*$-*choosable*   *belongs to* $\{6, 7, 8, 9\}$,
$((4, \cup \geq l), 1)^*$-*choosable*   *belongs to* $\{4, 5, 6, 7, 8\}$,
$((3, \cup \geq l), 0)^*$-*choosable*   *belongs to* $\{6, 7, 8, 9, 10, 11\}$,
$((3, \cup \geq l), 1)^*$-*choosable*   *belongs to* $\{5, 6, 7, 8, 9, 10\}$.

We shall justify Theorem 5.1 in the rest of this Section. Both Gutner's [22] and Mirzakhani's [45] examples of non-4-choosable planar graphs are neither $((4, \cap \leq 3), 0)^*$-choosable nor $((4, \cup \geq 5), 0)^*$-choosable. No nontrivial upper bound for $((4, \cap \leq l), 1)^*$-choosability or lower bound for $((4, \cup \geq l), 1)^*$-choosability of planar graphs is known. Examples are given in [67] of planar graphs that are not $((3, \cap \leq 2), 1)^*$-choosable (or, therefore, $((3, \cap \leq 2), 0)^*$-choosable), $((3, \cup \geq 5), 0)^*$-choosable or $((3, \cup \geq 4), 1)^*$-choosable. The optimist's conjecture, assuming that there are no more examples to be found, would be the following.

Optimist's Conjecture *Every planar graph is* $((3, \cap \leq 1), 0)^*$-*choosable,* $((3, \cup \geq 6), 0)^*$-*choosable,* $((3, \cup \geq 5), 1)^*$-*choosable and* $(4, 1)^*$-*choosable.*

If true, this conjecture would solve all eight of the above problems, since

$((3, \cap \leq 1), 0)^*$-choosable $\Longrightarrow ((3, \cap \leq 1), 1)^*$-choosable,
$((3, \cup \geq 6), 0)^*$-choosable $\Longrightarrow ((4, \cup \geq 6), 0)^*$-choosable
$\Longrightarrow ((4, \cap \leq 2), 0)^*$-choosable,
and $(4, 1)^*$-choosable $\Longleftrightarrow ((4, \cap \leq 4), 1)^*$-choosable
$\Longleftrightarrow ((4, \cup \geq 4), 1)^*$-choosable.

The question of whether every planar graph is $((4, \cap \leq 2), 0)^*$-choosable has been asked by Kratochvíl, Tuza and Voigt [41], and the question of whether every planar graph is $((3, \cap \leq 1), 0)^*$-choosable has been asked by Škrekovski [56]. Incidentally, it seems that the idea of imposing constraints on the sizes of unions and intersections was introduced by Kratochvíl, Tuza and Voigt [41], in whose notation $(p, 1, r)$-choosability is the same as our $((p, \cap \leq p - r), 0)^*$-choosability $(r \leq p)$. Škrekovski [56] has introduced the variant idea of imposing a lower rather than an upper bound on the size of the intersections. He conjectures that (in an obvious variant of our terminology) every triangle-free planar graph is $((1, \cap \geq 3), 0)^*$-choosable and every planar graph is $((1, \cap \geq 4), 0)^*$-choosable. Here the initial 1 is needed only to ensure that the list of an isolated vertex is nonempty; for a graph with no isolated vertices, no lower bound on the list-sizes is needed other than that implied by the intersection condition. The second of these conjectures clearly implies the 4-colour theorem.

In the rest of this section we shall complete the proof of Theorem 5.1. No proof is needed that all planar graphs are $((3, \cap \leq 0), 0)^*$-choosable and $((3, \cap \leq 0), 1)^*$-choosable. Kratochvíl, Tuza and Voigt [41] proved that every planar graph is $((4, \cap \leq 1), 0)^*$-choosable, by using the fact that its edges can be oriented in such a way that every vertex has outdegree at most 3. Škrekovski [56] has given an alternative proof of the same result by a method analogous to Thomassen's proof [58] that every planar graph is 5-choosable. We shall complete the proof of Theorem 5.1(a) by using this same method to prove that every planar graph is $((4, \cap \leq 2), 1)^*$-choosable. To prove this for all planar graphs it suffices to prove it for near-triangulations, defined before Lemma 4.2. Recall from the same place that a colouring λ is $\{v_1, v_2\}$-*proper* if, for $i \in \{1, 2\}$, v_i is not adjacent to any vertex $v \notin \{v_1, v_2\}$ such that $\lambda(v) = \lambda(v_i)$. The $((4, \cap \leq 2), 1)^*$-choosability of planar graphs follows immediately from the following theorem, whose complicated statement is designed to facilitate its proof by induction.

Theorem 5.2 *Let G be a near-triangulation with $C : v_1v_2 \ldots v_n$ as its bounding circuit. Let L be a list-assignment to G such that $L(v_1) = \{a\}$, $L(v_n) = \{b\}$ (where possibly $b = a$), $|L(v_i)| \geq 3$ for each $i \in \{2, \ldots, n-1\}$, and $|L(v)| \geq 4$ for each vertex $v \notin C$. Suppose moreover that $|L(u) \cap L(v)| \leq 2$ for every edge uv. Then there exists a $\{v_1, v_n\}$-proper $(L, 1)^*$-colouring λ.*

Proof Suppose if possible that (G, L) is a counterexample to the theorem for which G has as few vertices as possible. If C has a chord v_iv_j, where $1 \leq i < j \leq n - 1$ or $2 \leq i < j \leq n$ and $j \neq i + 1$, then G is the union of two near-triangulations, say G_1 with bounding circuit $v_1 \ldots v_iv_j \ldots v_nv_1$ and G_2 with bounding circuit $v_iv_{i+1} \ldots v_jv_i$. By the minimality of G, there is a $\{v_1, v_n\}$-proper $(L, 1)^*$-colouring of G_1. And if we define $L'(v_i) := \{\lambda(v_i)\}$, $L'(v_j) := \{\lambda(v_j)\}$, and $L'(v) := L(v)$ for every vertex $v \in V(G_2) \setminus \{v_i, v_j\}$, then there is a $\{v_i, v_j\}$-proper $(L', 1)^*$-colouring of G_2. These two colourings together give the required $\{v_1, v_n\}$-proper $(L, 1)^*$-colouring of G.

So we may suppose that there is no such chord $v_i v_j$. We choose a colour c and a set X of one or two vertices, as follows. If there is a colour c belonging to $L(v_2) \setminus (L(v_3) \cup \{a\})$, choose such a c and take $X := \{v_2\}$. (This must happen if $n = 3$.) Otherwise, if there is a colour $c \in (L(v_2) \cap L(v_3)) \setminus (L(v_4) \cup \{a\})$, choose such a c and take $X := \{v_2, v_3\}$. If neither of these happens, then each of the colours in $L(v_2) \setminus \{a\}$, of which there are at least two, belongs to $L(v_3) \cap L(v_4)$; then there must be at least one further colour $c \in L(v_3) \setminus \{L(v_2) \cup L(v_4)\}$, so choose such a c and take $X := \{v_3\}$.

In each case, form a list-assignment L'' to $G - X$ by removing colour c from the list of every vertex that is adjacent in G to a vertex of X. If $G - X$ is 2-connected, then $G - X$ is a near-triangulation and $(G - X, L'')$ satisfies the hypotheses of the theorem, and so by the minimality of G there is a $\{v_1, v_n\}$-proper $(L'', 1)^*$-colouring of $G - X$. If $G - X$ is not 2-connected, then each block of $G - X$ is either K_2 or a near-triangulation smaller than G; so there is a $\{v_1, v_n\}$-proper $(L'', 1)^*$-colouring of the block of $G - X$ containing v_1 and v_n, and this can be extended to each remaining block of $G - X$ in turn. Finally, give colour c to the vertex or vertices in X to complete the required $\{v_1, v_n\}$-proper $(L, 1)^*$-colouring of G. $\qquad\square$

Finally, we prove the upper bounds in Theorem 5.1(b). Let the *weight* of an edge in a graph be the sum of the degrees of its endvertices. Kotzig [39] proved that every 3-connected planar graph G has an edge with weight at most 13, and if the minimum degree of G is at least 4 then G has an edge with weight at most 11. Borodin [5] extended this to normal plane multigraphs, where a plane multigraph is *normal* if no face-boundary consists of a circuit of two edges. To state his result, let $g(3) := 13$ and $g(4) := 11$.

Lemma 5.3 ([5]) *If $k \in \{3, 4\}$, then a normal plane multigraph with minimum degree at least k contains an edge with weight at most $g(k)$.*

We shall use this to prove the following result, which is almost certainly far from best possible, but which establishes the upper bounds in Theorem 5.1(b).

Theorem 5.4 *A planar graph is $((3, \cup \geq 11), 0)^*$-choosable, $((3, \cup \geq 10), 1)^*$-choosable, $((4, \cup \geq 9), 0)^*$-choosable and $((4, \cup \geq 8), 1)^*$-choosable.*

Proof In terms of the function $g(k)$, the theorem says that, for $(k, d) \in \{(3, 0), (3, 1), (4, 0), (4, 1)\}$, every planar graph is $((k, \cup \geq g(k) - 2 - d), d)^*$-choosable. For fixed k and d, let G be a minimal counterexample to this theorem, assuming that one exists, and let L be a $(k, \cup \geq g(k) - 2 - d)$-list-assignment to G for which no $(L, d)^*$-colouring exists. Then $|L(u) \cap L(v)| \geq 1$ for each edge $e = uv$, since otherwise $G - e$ would be a smaller counterexample. It follows that, for each edge $e = uv$,

$$|L(u)| + |L(v)| = |L(u) \cup L(v)| + |L(u) \cap L(v)| \geq g(k) - 1 - d. \qquad (5.1)$$

Also, each vertex v has degree $d_G(v) \geq |L(v)| \geq k$, since otherwise $G - v$ would be a smaller counterexample. (If $|L(v)| \geq d_G(v) + 1$, it must be possible to give v a colour from its list that is different from the colours of all its neighbours.) Thus, since $d \leq 1$ in (5.1), each edge uv has weight

$$d_G(u) + d_G(v) \geq |L(u)| + |L(v)| \geq g(k) - 2. \tag{5.2}$$

Suppose that a plane embedding of G has a nontriangular face. Let u, v, w be three consecutive vertices in the boundary of this face, chosen so that $d_G(v)$ is as small as possible. Then, by (5.2), $d_G(u) \geq \frac{1}{2}(g(k) - 2)$, so that $d_G(u) \geq \frac{1}{2}(g(k) - 1)$ since $d_G(u)$ is an integer and $g(k)$ is odd. Similarly $d_G(w) \geq \frac{1}{2}(g(k) - 1)$. It follows that if we add the edge uw to G then this new edge will have weight at least $g(k) + 1$ in the new graph.

Let us add edges to G in this way in order to convert it into a plane triangulation H, which is a normal plane multigraph, such that each edge in $E(H) \setminus E(G)$ has weight at least $g(k) + 1$ in H. By Lemma 5.3, H contains an edge $e = uv$ with weight at most $g(k)$; that is,

$$d_H(u) + d_H(v) \leq g(k). \tag{5.3}$$

By the construction of H, e must be an edge of G. By the minimality of G, there exists an $(L, d)^*$-colouring λ of $G - e$. We may assume that $\lambda(u) = \lambda(v)$, since otherwise λ is an $(L, d)^*$-colouring of G, which would be a contradiction.

In H, e is an edge of two triangular faces uvw_1 and uvw_2. Let ϵ be the number of different colours from $L(u) \cap L(v)$ that are of the form $\lambda(w_i)$ for a vertex w_i ($i \in \{1, 2\}$) such that uw_i and vw_i are both edges of G; so $\epsilon \in \{0, 1, 2\}$. Since $\lambda(u) = \lambda(v)$, a colour $\lambda(w_i)$ that contributes to ϵ must be different from $\lambda(u)$; thus

$$|L(u) \cap L(v)| \geq \epsilon + 1. \tag{5.4}$$

Let $d'(u)$ denote the number of different colours from $L(u)$ that are used by λ on vertices that are adjacent to u in $G - e$. Let $d'(v)$ be defined similarly for v. Note that if $\lambda(w_i)$ does not contribute to ϵ then either $\lambda(w_i) \notin L(u) \cap L(v)$ or $uw_i \notin G$ or $vw_i \notin G$ or $\lambda(w_1) = \lambda(w_2)$. Thus it is not difficult to see that

$$d'(u) + d'(v) \leq d_H(u) - 1 + d_H(v) - 1 + \epsilon - 2 \leq g(k) - 4 + \epsilon \tag{5.5}$$

by (5.3). However, $|L(u)| \leq d'(u) + 1$, since otherwise we could change the colour of u to be different from the colours of all its neighbours in G. A similar remark applies to v. Moreover, it is not possible that $d = 1$ and $|L(u)| = d'(u) + 1$ and $|L(v)| = d'(v) + 1$, since then we could change the colours of u and v if necessary so that they have no neighbours of the same colour in $G - e$, and this would give an $(L, d)^*$-colouring of G, contrary to the choice of G. It follows that

$$|L(u)| + |L(v)| \leq d'(u) + d'(v) + 2 - d \leq g(k) - 2 - d + \epsilon \tag{5.6}$$

by (5.5). However, $|L(u) \cup L(v)| \geq g(k) - 2 - d$ by hypothesis, and this and (5.4) show that

$$|L(u)| + |L(v)| = |L(u) \cup L(v)| + |L(u) \cap L(v)| \geq g(k) - 2 - d + \epsilon + 1,$$

contradicting (5.6). This completes the proof of Theorem 5.4. $\quad\square$

6 Appendix: methods of proof

We conclude this paper by reviewing briefly some of the main methods of proof used for choosability results, and some of the results that can be proved by using them. They can be divided roughly into methods that involve orienting the graph and those that do not. We start with the latter.

6.1 Methods that do not involve orientations

6.1.1 Degeneracy
A graph G is called k-*degenerate* if every subgraph H of G contains a vertex with degree (in H) at most k. It is not difficult to see that this is equivalent to saying that the vertices of G can be ordered in such a way that each vertex is adjacent to at most k vertices that come before it in the ordering.

It is easy to see that if G is k-degenerate then $\mathrm{ch}(G) \leq k + 1$: the argument is exactly as used in Section 1.3 to prove that all planar graphs are 6-choosable. In fact, as we have seen, all planar graphs are 5-choosable. It is not surprising that an argument as simple as the use of degeneracy does not always give sharp results. But it does sometimes. The proof that all planar graphs are 5-degenerate (and hence 6-choosable) can readily be adapted to prove that all triangle-free planar graphs are 3-degenerate and hence 4-choosable, a result first proved by Kratochvíl and Tuza [40]. Moreover this result is sharp: the existence of non-3-choosable triangle-free planar graphs was proved by Voigt [63]. On the other hand, since bipartite graphs are triangle-free, it follows immediately that every bipartite planar graph is 4-choosable, and this result is *not* sharp: Alon and Tarsi [2] proved that every bipartite planar graph is 3-choosable; see Section 6.2.1 below.

As apparently first pointed out by O. V. Borodin (see [32, p. 20]), the degeneracy argument can also be used to prove that $\mathrm{ch}(S) = \chi(S)$ for every compact surface S, orientable or nonorientable, except the sphere, where $\mathrm{ch}(S)$ and $\chi(S)$ denote the minimum values of $\mathrm{ch}(G)$ and $\chi(G)$, respectively, taken over all graphs G that can be embedded in S. Except when S is the Klein bottle, this follows from the results of Heawood [30] and Ringel and Youngs [50]; the former defined a number $H(S)$, now called the *Heawood number* of S, and showed that every graph embeddable in S is $(H(S) - 1)$-degenerate; and the latter proved that the complete graph $K_{H(S)}$ can be embedded in S. These results together establish that $\mathrm{ch}(S) = \chi(S) = H(S)$.

For the Klein bottle K, Heawood's argument shows that every graph embedded in K *either* contains a vertex with degree at most five *or* is 6-regular. Franklin [17] proved that K_7 cannot be embedded in K (but K_6 can be). Thus a 6-regular graph embeddable in K is not complete. However, both Vizing [61] and Erdős, Rubin and Taylor [15] proved the choosability analogue of Brooks's theorem [8], namely, that a graph with maximum degree Δ that is neither a complete graph nor an odd circuit is Δ-choosable. It follows from this that every graph embeddable in K is 6-choosable, so that $\mathrm{ch}(K) = \chi(K) = 6$.

6.1.2 The use of Hall's theorem

As mentioned in Section 2.3, Hall's theorem on systems of distinct representatives [28] was first used to prove results on choosability by Erdős, Rubin and Taylor [15]. It has subsequently been used by Gravier and Maffray [20, 21], Peterson and Woodall [47], Kostochka and Woodall [36], Ohba [46], and Enomoto, Ohba, Ota and Sakamoto [14].

For example, if v is a vertex of a graph G, let $\omega(G)$ denote the order of a largest clique in G, and let $\omega_G(v)$ denote the order of a largest clique containing v. In [36], following a similar proof in [47], Hall's theorem was used to prove the following simple lemma: *if every vertex v of G is given a list of at least $\omega_G(v)$ colours, in such a way that nonadjacent vertices always get disjoint lists, then G can be coloured from its lists.* (In this case, every vertex of G gets a different colour.) This was used to prove that $\mathrm{ch}(G) = \chi(G) = \omega(G)$ if G has neither $\bar{K}_3$ nor P_4 as an induced subgraph, and this in turn was used in the proof of two results mentioned in Section 2.3, that $\mathrm{ch}(G) = \chi(G)$ if G is an inflation of a graph with at most five vertices or is the inflation of the square of a graph with at most seven vertices; see Section 2.2 for the definition of *inflation*.

Although the applicability of Hall's theorem in this context is fairly limited, it is evidently used sufficiently frequently to justify its inclusion as a method of proof for choosability results.

6.1.3 The boundary method

A number of results can be proved by variants of the following method. Suppose we have coloured some vertices of a graph G, which induce a connected subgraph of G. Define the *boundary* to comprise those uncoloured vertices of G that are adjacent to coloured vertices. In the boundary method we colour only boundary vertices, which are allowed to have smaller lists than other vertices in the graph. In practice, when a vertex is coloured it is usually removed from the graph, either by deleting it or by contracting it into a residual vertex that represents the set of all the coloured vertices; but in either case, the boundary is defined (implicitly) as if the coloured vertices were still there.

We have seen several examples of this method in this survey. The first use of it was perhaps Thomassen's elegant proof [58] that all planar graphs are 5-choosable, mentioned in Section 1.3. Škrekovski's proof of the $((4, \cap \le 1), 0)^*$-choosability of planar graphs, mentioned in Section 5, and our proof of the

$((4, \cap \leq 2), 1)^*$-choosability of planar graphs in Theorem 5.2, both followed a very similar argument. In these cases the coloured vertices were deleted. Examples in which the coloured vertices were contracted into a residual vertex z were seen in Lemmas 3.2 and 3.4.

6.2 Methods involving orientations

There are two methods of proof that involve orienting the graph. They both start in the same way, but then diverge sharply. Suppose that one orients the edges of a graph G so as to form a digraph D, and then assigns lists of colours to the vertices in such a way that each vertex v has a list $L(v)$ containing more colours than its outdegree in D:

$$|L(v)| \geq d^+(v) + 1 \qquad \text{for each vertex } v. \tag{6.1}$$

If D has no directed cycles, then it is easy to see that G is L-colourable whenever (6.1) holds: just colour the vertices in order, always colouring a sink vertex of the subgraph induced by the uncoloured vertices; the degeneracy argument (Section 6.1.1) is a special case of this. Interval graphs form one of the few classes of graphs for which useful results can be obtained from acyclic orientations in this way: it is not difficult to see that if G is an interval graph then it has an acyclic orientation in which $d^+(v) \leq \omega(G) - 1$ for each vertex v, where $\omega(G)$ is the order of a largest clique in G, and it follows from this that $\mathrm{ch}(G) = \chi(G) = \omega(G)$ for every interval graph G.

However, usually one has to deal with digraphs containing directed cycles. The *Alon–Tarsi method* describes conditions on D that suffice to ensure that the vertices can be coloured from their lists whenever (6.1) holds. The *kernel method*, in its simplest form, does the same, although the conditions are very different. However, the kernel method seldom works in its simplest form, and in practice one is likely to have to look inside the lists and consider different cases according to what the lists contain. In the Alon–Tarsi method it is impossible to look inside the lists, and the trick is to find an orientation, of the original graph or of a supermultigraph of it, for which the Alon–Tarsi conditions hold.

One cannot help wondering whether there is some more general set of conditions that will unify these two methods into a single, more powerful, method. However, the conditions in the two methods are so different that it is difficult to imagine how they could be unified.

6.2.1 The method of Alon and Tarsi
This method can be summarized as follows; the proof that it works, using the graph polynomial, is fascinating, but it is explained in [1, 2] and elsewhere and so we do not include it here.

Let D_0 be an arbitrary orientation (the so-called *reference orientation*) of an undirected multigraph G. If D is any other orientation of G, let $a(D)$ be the number of edges that have opposite orientations in D and D_0, and define $\mathrm{sign}(D)$ to be 1 or -1 according as $a(D)$ is even or odd. Let $\mathcal{O}$ denote the set of all orientations of G in which every vertex v has the same outdegree $d^+(v)$ that it has in D_0, and let $\sigma(D_0) := \sum_{D \in \mathcal{O}} \text{sign}(D)$. Suppose that every vertex v of G is given a list $L(v)$ of at least $d^+(v) + 1$ colours. The main result of Alon and Tarsi is that, if $\sigma(D_0) \neq 0$, then G can be coloured from these lists.

There are two observations that may help to simplify the calculation of $\sigma(D_0)$. In the terminology of Alon and Tarsi, a subdigraph H of D_0 is called *Eulerian* if the indegree $d_H^-(v)$ of every vertex v is equal to its outdegree $d_H^+(v)$. Let $\text{ee}(D_0)$ and $\text{eo}(D_0)$ denote the numbers of spanning Eulerian subdigraphs of D_0 with an even and with an odd number of edges, respectively. It is proved in [1, 2] that $\sigma(D_0) = \text{ee}(D_0) - \text{eo}(D_0)$.

The second observation applies when G contains edge-disjoint complete subgraphs $C_1, \ldots, C_r$. Let $\mathcal{O}'$ denote the subset of all orientations in $\mathcal{O}$ that are acyclic on every clique C_i. It is shown in [1] that $\sigma(D_0) = \sum_{D \in \mathcal{O}'} \text{sign}(D)$; that is, in calculating $\sigma(D_0)$, we need only consider orientations that are acyclic on every C_i.

These two observations are both useful, although it seems unlikely that they can be used simultaneously. The second observation seems to be more useful when proving results about edge-choosability, since line graphs naturally decompose into edge-disjoint unions of complete graphs.

An easy application of the Alon–Tarsi method, contained in their original paper [2], is the result (mentioned in Section 6.1.1) that every planar bipartite graph is 3-choosable. Several further results proved by this method are mentioned in Section 2.3.

6.2.2 The kernel method

A *kernel* in a digraph D is a set K of nonadjacent vertices such that every vertex in $V(D) \setminus K$ is joined by an arc to at least one vertex in K. The kernel method, in its simplest form, can be stated as the following theorem, which is a special case of a generalization by Galvin [18, Lemma 2.1] of an observation attributed by Alon and Tarsi [2, Remark 2.4] to Bondy, Boppana and Siegel.

Theorem 6.1 *If D is an orientation of a graph G, and every induced subdigraph of D has a kernel, and (6.1) holds, then G can be coloured from its lists.*

Proof Choose a colour c that is present in the list of at least one vertex. Let D_c denote the subdigraph of D that is induced by the vertices that have c in their lists. By hypothesis, D_c has a kernel K. Give colour c to the vertices in K and delete these vertices from G and from D. Remove colour c from the list of every remaining vertex in D_c. Since each such vertex was joined by an arc to at least one vertex in K, its outdegree has now decreased by at least 1, and so (6.1) still holds for $D - K$ with these reduced lists. Repeat this process until all vertices are coloured. $\square$

In practice, it is rather rare that every induced subdigraph of D has a kernel. Digraphs with this property are called *kernel-perfect*. Not much seems to be known about kernel-perfect digraphs in general, although some progress has been made in the case of orientations of line graphs of multigraphs [7, 44].

The most significant application to date of the kernel method is Galvin's proof [18] of the list-edge-colouring conjecture for bipartite multigraphs (Section 2.3), achieved by showing that the line graph of a bipartite multigraph has a kernel-perfect orientation. This uses the kernel method in its simplest form, that is, Theorem 6.1. It is much more common to deal with graphs that are not kernel-perfect. In this case the conclusion of Theorem 6.1 will still hold if every induced subgraph of every subdigraph D_c has a kernel, but it is necessary to use other methods to deal with list-assignments for which this is not true. The name *kernel method* is used in this paper to describe extended methods of this type, which are loosely based on Theorem 6.1. Several results proved by the kernel method are mentioned in Section 2.3.

References

[1] N. Alon, Restricted colorings of graphs, *Surveys in Combinatorics, 1993* (ed. K. Walker), *London Math. Soc. Lecture Note Series* **187**, Cambridge University Press, Cambridge, 1993, 1–33.

[2] N. Alon and M. Tarsi, Colorings and orientations of graphs, *Combinatorica* **12** (1992), 125–134.

[3] K. Appel and W. Haken, Every planar map is four colorable: Part I, Discharging, *Illinois J. Math.* **21** (1977), 429–490.

[4] K. Appel, W. Haken and J. Koch, Every planar map is four colorable: Part 2, Reducibility, *Illinois J. Math.* **21** (1977), 491–567.

[5] O.V. Borodin, On the total coloring of planar graphs, *J. Reine Angew. Math.* **394** (1989), 180–185.

[6] O.V. Borodin, A.V. Kostochka and D.R. Woodall, List edge and list total colourings of multigraphs, *J. Combin. Theory Ser. B* **71** (1997), 184–204.

[7] O.V. Borodin, A.V. Kostochka and D.R. Woodall, On kernel-perfect orientations of line graphs, *Discrete Math.* **191** (1998), 45–49.

[8] R.L. Brooks, On colouring the nodes of a network, *Proc. Cambridge Phil. Soc.* **37** (1941), 194–197.

[9] A. Chetwynd and R. Häggkvist, A note on list-colorings, *J. Graph Theory* **12** (1989), 87–95.

[10] L.J. Cowen, R.H. Cowen and D.R. Woodall, Defective colorings of graphs in surfaces: partitions into subgraphs of bounded valency, *J. Graph Theory* **10** (1986), 187–195.

[11] G.A. Dirac, A property of 4-chromatic graphs and some remarks on critical graphs, *J. London Math. Soc.* **27** (1952), 85–92.

[12] N. Eaton and T. Hull, Defective list colorings of planar graphs, *Bull. Inst. Combin. Appl.* **25** (1999), 79–87.

[13] M.N. Ellingham and L. Goddyn, List edge colourings of some 1-factorizable multigraphs, *Combinatorica* **16** (1996), 343–352.

[14] H. Enomoto, K. Ohba, K. Ota and J. Sakamoto, Choice number of some complete multi-partite graphs, *Combinatorica*, to appear.

[15] P. Erdős, A.L. Rubin and H. Taylor, Choosability in graphs, *Proc. West Coast Conference on Combinatorics, Graph Theory and Computing, Arcata, 1979, Congr. Numer.* **26** (1980), 125–157.

[16] H. Fleischner and M. Stiebitz, A solution to a colouring problem of P. Erdős, *Discrete Math.* **101** (1992), 39–48.

[17] P. Franklin, A six-color problem, *J. Math. Phys.* **13** (1934), 363–369.

[18] F. Galvin, The list chromatic index of a bipartite multigraph, *J. Combin. Theory Ser. B* **63** (1995), 153–158.

[19] S. Gravier, A Hajós-like theorem for list coloring, *Discrete Math.* **152** (1996), 299–302.

[20] S. Gravier and F. Maffray, Choice number of 3-colorable elementary graphs, *Discrete Math.* **165/166** (1997), 353–358.

[21] S. Gravier and F. Maffray, Graphs whose choice number is equal to their chromatic number, *J. Graph Theory* **27** (1998), 87–97.

[22] S. Gutner, The complexity of planar graph choosability, *Discrete Math.* **159** (1996), 119–130.

[23] H. Hadwiger, Über eine Klassifikation der Streckenkomplexe, *Vierteljahresschr. Naturforsch. Ges. Zürich* **88** (1943), 133–142.

[24] H. Hadwiger, Ungelöste Probleme Nr. 26, *Elemente Math.* **13** (1958), 128.

[25] R. Häggkvist and A. Chetwynd, Some upper bounds on the total and list chromatic numbers of multigraphs, *J. Graph Theory* **16** (1992), 503–516.

[26] R. Häggkvist and J. Janssen, New bounds on the list-chromatic index of the complete graph and other simple graphs, *Combin. Prob. Comput.* **6** (1997), 295–313.

[27] G. Hajós, Über eine Konstruktion nicht *n*-färbbarer Graphen, *Wiss. Z. Martin-Luther-Univ. Halle-Wittenberg Math.-Natur. Reihe* **10** (1961), 116–117.

[28] P. Hall, On representatives of subsets, *J. London Math. Soc.* **10** (1935), 26–30.

[29] D. Hanson, G. MacGillivray and B. Toft, Choosability of bipartite graphs, *Ars Combin.* **44** (1996), 183–192.

[30] P.J. Heawood, Map colour theorem, *Quart. J. Pure Appl. Math.* **24** (1890), 332–338.

[31] A.J.W. Hilton and P.D. Johnson, The Hall number, the Hall index, and the total Hall number of a graph, *Discrete Applied Math.* **94** (1999), 227–245.

[32] T.R. Jensen and B. Toft, *Graph Coloring Problems*, Wiley-Interscience, New York, 1995.

[33] M. Juvan and B. Mohar, List colorings of outerplanar graphs, unpublished manuscript, 1996.

[34] M. Juvan, B. Mohar and R. Thomas, List edge-colorings of series-parallel graphs, *Electr. J. Combin.* **6** (1) (1999), R42.

[35] M. Juvan, B. Mohar and R. Škrekovski, List-total colorings of graphs, *Combin. Prob. Comput.* **7** (1998), 181–188.

[36] A.V. Kostochka and D.R. Woodall, Choosability conjectures and multi-circuits, *Discrete Math.*, to appear.

[37] A.V. Kostochka and D.R. Woodall, Total choosability of multicircuits I, manuscript, 1999.

[38] A.V. Kostochka and D.R. Woodall, Total choosability of multicircuits II, manuscript, 2000.

[39] A. Kotzig, Contribution to the theory of Eulerian polyhedra (in Slovak), *Mat.-Fyz. Čas. (Math. Slovaca)* **5** (1955), 101–113.

[40] J. Kratochvíl and Z. Tuza, Algorithmic complexity of list colorings, *Discrete Appl. Math.* **50** (1994), 297–302.

[41] J. Kratochvíl, Z. Tuza and M. Voigt, Brooks-type theorems for choosability with separation, *J. Graph Theory* **27** (1998), 43–49.

[42] J. Kratochvíl, Z. Tuza and M. Voigt, New trends in the theory of graph colorings: choosability and list coloring, *Contemporary Trends in Discrete Mathematics, From DIMACS and DIMATIA to the Future, Proceedings of the DIMATIA-DIMACS Conference, 1997* (ed. R.L. Graham), American Math. Soc., Providence, 1999, 183–197.

[43] R.B. Levow, Coloring restrictions, *Theory and Applications of Graphs. Proceedings, Michigan 1976* (ed. Y. Alavi and D.R. Lick), *Lecture Notes in Math.* **642**, Springer, Berlin, 1978, 347–352.

[44] F. Maffray, Kernels in perfect line-graphs, *J. Combin. Theory Ser. B* **55** (1992), 1–8.

[45] M. Mirzakhani, A small non-4-choosable planar graph, *Bull. Inst. Combin. Appl.* **17** (1996), 15–18.

[46] K. Ohba, On chromatic-choosable graphs, manuscript, 1999.

[47] D. Peterson and D.R. Woodall, Edge-choosability in line-perfect multigraphs, *Discrete Math.* **202** (1999), 191–199.

[48] M.J. Plantholt and S.K. Tipnis, On the list chromatic index of nearly bipartite multigraphs, *Australas. J. Combin.* **19** (1999), 157–170.

[49] A. Prowse and D.R. Woodall, Choosability of powers of circuits, manuscript, 2000.

[50] G. Ringel and J.W.T. Youngs, Solution of the Heawood map-coloring problem, *Proc. Natl. Acad. Sci. USA* **60** (1968), 438–445.

[51] N. Robertson, D. Sanders, P.D. Seymour and R. Thomas, The four-colour theorem, *J. Combin. Theory Ser. B* **70** (1997), 2–44.

[52] N. Robertson, P.D. Seymour and R. Thomas, Hadwiger's conjecture for K_6-free graphs, *Combinatorica* **13** (1993), 279–361.

[53] C.E. Shannon, A theorem on coloring the lines of a network, *J. Math. Phys.* **28** (1949), 148–151.

[54] R. Škrekovski, Choosability of K_5-minor-free graphs, *Discrete Math.* **190** (1998), 223–226.

[55] R. Škrekovski, List improper colourings of planar graphs, *Combin. Prob. Comput.* **8** (1999), 293–299.

[56] R. Škrekovski, A note on choosability with separation for planar graphs, *Ars Combin.*, to appear.

[57] T. Slivnik, Short proof of Galvin's theorem on the list-chromatic index of a bipartite multigraph, *Combin. Prob. Comput.* **5** (1996), 91–94.

[58] C. Thomassen, Every planar graph is 5-choosable, *J. Combin. Theory Ser. B* **62** (1994), 180–181.

[59] Z. Tuza, Graph colorings with local constraints—a survey, *Discuss. Math. Graph Theory* **17** (1997), 161–228.

[60] Z. Tuza and M. Voigt, Every 2-choosable graph is $(2m, m)$-choosable, *J. Graph Theory* **22** (1996), 245–252.

[61] V.G. Vizing, Vertex colorings with given colors (in Russian), *Metody Diskret. Analiz* **29** (1976), 3–10.

[62] M. Voigt, List colourings of planar graphs, *Discrete Math.* **120** (1993), 215–219.

[63] M. Voigt, A not-3-choosable planar graph without 3-cycles, *Discrete Math.* **146** (1995), 325–328.

[64] K. Wagner, Über eine Eigenschaft der ebenen Komplexe, *Math. Ann.* **114** (1937), 570–590.

[65] D.R. Woodall, Improper colourings of graphs, *Graph Colourings* (eds. R. Nelson and R.J. Wilson), *Research Notes in Math.* **218**, Longman, Harlow, 1990, 45–63.

[66] D.R. Woodall, Edge-choosability of multicircuits, *Discrete Math.* **202** (1999), 271–277.

[67] D.R. Woodall, Defective choosability results for outerplanar and related graphs, manuscript, 2000.

School of Mathematical Sciences
University of Nottingham
Nottingham NG7 2RD
United Kingdom
douglas.woodall@nottingham.ac.uk

For EU product safety concerns, contact us at Calle de José Abascal, 56–1°,
28003 Madrid, Spain or eugpsr@cambridge.org.

www.ingramcontent.com/pod-product-compliance
Ingram Content Group UK Ltd.
Pitfield, Milton Keynes, MK11 3LW, UK
UKHW010853090126
466816UK00011B/221